EVAPORITE SEDIMENTOLOGY

Prentice Hall Advanced Reference Series

Physical and Life Sciences

BINKLEY *The Pineal: Endocrine and Nonendocrine Function*
CAROZZI *Carbonate Rock Depositional Models: A Microfacies Approach*
EISEN *Mathematical Methods and Models in the Biological Sciences: Linear and One-Dimensional Theory*
EISEN *Mathematical Methods and Models in the Biological Sciences: Nonlinear and Multidimensional Theory*
FRASER *Stratigraphic Evolution of Clastic Depositional Sequences*
JEGER, ED. *Spatial Components of Plant Disease Epidemics*
MCLENNAN *Introduction to Nonequilibrium Statistical Mechanics*
PLISCHKE AND BERGERSEN *Equilibrium Statistical Physics*
SORBJAN *Structure of the Atmospheric Boundary Layer*
VALENZUELA AND MYERS *Adsorption Equilibrium Data Handbook*
WARREN *Evaporite Sedimentology: Importance in Hydrocarbon Accumulation*

JOHN K. WARREN
University of Texas at Austin

EVAPORITE SEDIMENTOLOGY

Importance in Hydrocarbon Accumulation

PRENTICE HALL, *Englewood Cliffs, New Jersey 07632*

Library of Congress Cataloging-in-Publication Data

Warren, John K. (date)
Evaporite sedimentology : importance in hydrocarbon accumulation /
John K. Warren.
p. cm.—(Prentice Hall advanced reference series)
Bibliography: p. 251
Includes index.
ISBN 0-13-292335-1
1. Evaporites. 2. Organic geochemistry. I. Title.
QE471.15.E8W37 1989 88-5914
553.2′8—dc19 CIP

Editorial/production supervision
and interior design: Mary Rottino
Cover design: Wanda Lubelska Design
Manufacturing buyer: Mary Ann Gloriande

Top cover drawing based in part on work by Eugster and Hardie, 1978.

Prentice Hall Advanced Reference Series

Printed in the United States of America

10 9 8 7 6 5 4 3 2 1

ISBN 0-13-292335-1

PRENTICE-HALL INTERNATIONAL (UK) LIMITED, *London*
PRENTICE-HALL OF AUSTRALIA PTY. LIMITED, *Sydney*
PRENTICE-HALL CANADA INC., *Toronto*
PRENTICE-HALL HISPANOAMERICANA, S.A., *Mexico*
PRENTICE-HALL OF INDIA PRIVATE LIMITED, *New Delhi*
PRENTICE-HALL OF JAPAN, INC., *Tokyo*
SIMON & SCHUSTER ASIA PTE. LTD., *Singapore*
EDITORA PRENTICE-HALL DO BRASIL, LTDA., *Rio de Janeiro*

Dedication

To my children Kieren, Matthew, and Tristan

Lake Frome, South Australia

Air arched, enormous, lining the ribs of salt,
With a drench of white light,
So huge, from such infinities of height,
You walk on the sky's beach.

J.K.W.

Contents

Preface

I have written this book as a practical textbook for geologists engaged in petroleum exploration, as well as for any geologist working with evaporitic sediments. Its purpose is to introduce the reader to the sedimentology of evaporites and demonstrate how sedimentological principles can be applied to ancient evaporites even when the evaporite salts have flowed into diapirs and other structures.

The first chapter presents current ideas on evaporites and brine chemistry and a comprehensive summary of the diversity found in ancient depositional settings. Most evaporite salts are important to hydrocarbon accumulations for two reasons: their ability to form a widespread seal to underlying reservoirs (Chapters 2, 3, and 4); and their ability to generate hydrocarbon trapping structures as the salt flows into diapirs and other related structures (Chapter 6). To many geologists the term evaporite means only evaporitic salts, e.g., gypsum, anhydrite, and halite; it should include any mineral precipitated from a brine. This means evaporitic limestones and dolomites as well as the salts should be included in a discussion of evaporite sedimentology. For this reason, I have included, in Chapter 5, a comprehensive discussion of dolomites and dolomitization models as well as economically important dolomite formations.

Like most sedimentological texts the approach is to study modern evaporite occurrences and summarize the principal observations into a depositional model, then use the model to interpret ancient counterparts. The heart of the book is a little different from most other sedimentological texts in that it also outlines how models based solely on modern counterparts can falsely limit our interpretation of ancient evaporites. There are no modern counterparts to the variety of depositional settings found in most ancient evaporite basins—there are no modern examples of evaporite platforms, slopes or deep-water deposits. For such settings the rocks themselves contain the evidence. In terms of interpretation, the lack of a one-for-one Holocene analog is not as hopeless as it first appears; most depositional textures found in ancient evaporites can also be found in modern sediments, although on scales that are orders of magnitude smaller than their ancient counterpart.

John K. Warren
University of Texas at Austin

Acknowledgments

This book would not have been possible without the support of the Geology Foundation of the University of Texas. I would like to thank all the members of the Foundation; their moral and financial support allowed me to complete this book. Thanks also to fellow "soft-rock" faculty members and students of the University, as well as Charlotte Schreiber of Lamont Doherty and others whose discussions and disagreements helped shape many of the ideas and concepts outlined in the pages that follow. Special thanks to my wife, Jennifer, who typed and compiled the early drafts of the manuscript.

CHAPTER 1

General Concepts

INTRODUCTION

Economic Significance of Evaporitic Settings

Facies analysis in evaporitic sediments combines aspects of general sedimentology, petroleum geology, hydrology, hydrogeochemistry, and biology as it strives for an understanding of the processes controlling sediment deposition and diagenesis in various saline settings. This type of analysis is more than an academic pursuit, as evaporites overlie carbonates that contain an estimated one-half of the world's known petroleum reserves (Kirkland and Evans, 1981). About 70% of the world's giant oil fields in carbonate rocks bear a relationship to evaporites (Zhang Yi-gang, 1981). The association among evaporites, carbonates, and hydrocarbons is more than fortuitous as evaporites constitute less than 2% of the world's platform sediments (Fig. 1.1).

Bedded ancient evaporites overlying or up-dip from porous sediments form excellent seals to the underlying potential reservoirs. Buried halite units are often remobilized into salt domes, which in turn create potential reservoirs in the surrounding sediments. The subsurface movement of pore waters can leach evaporites in carbonate and siliciclastic matrices and so create areas of secondary porosity. Evaporite diagenesis can release large quantities of magnesium-rich

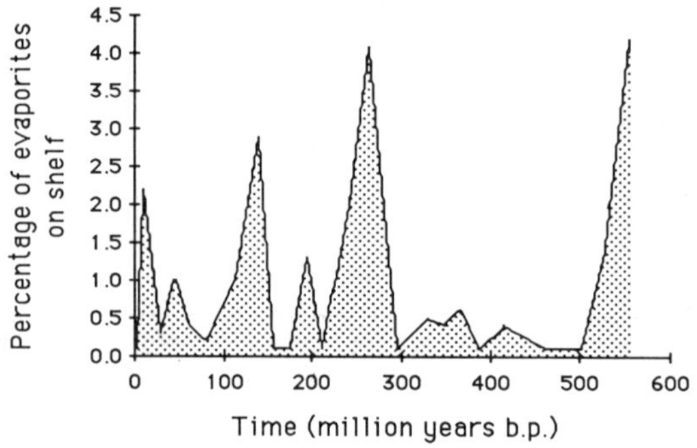

Figure 1.1 Distribution of evaporites as a percentage of the sediments on the world's platforms from the Cambrian to the present (based on data in Ronov et al., 1980).

brine and so form sucrosic dolomite in adjacent limestones, an excellent potential reservoir. An association may exist between sedimentation and source rock formation, as most modern evaporites are deposited adjacent to, or within, areas of high organic productivity and accumulation (Warren, 1986a). Evaporites are by definition deposited in arid environments lacking fresh water, hence sediments do not contain a high proportion of organic matter derived from higher plant debris. Thus evaporitic source rocks are more likely to be oil-prone than gas-prone. In the following chapters I describe the fine detail of evaporitic sedimentation and explore some of the reasons behind the association of evaporites with oil and gas occurrence.

Brines from evaporite dissolution can be important base-metal carriers and have been vital in the formation of many ore deposits. Salts such as sylvite, glauberite, gypsum, halite, and trona are the source of many industrial and agricultural minerals. Evaporitic salts have been mined and quarried for the last 6000 years. Trona was used by the ancient Egyptians for mummification some 2000 to 4000 years b.p. Halite is an excellent preservative for meat and vegetables; before refrigeration it was the sole means of long-term food storage. Throughout history salt was important in human affairs; wars have been fought to possess salt lakes in America, Africa and the Middle East.

History of Evaporite Research

Until the early 1960s ancient evaporites were often considered as enigmatic deposits more akin to metamorphic rocks than to sediments. Prior to this time most geological studies concentrated on the large-scale distribution of the evaporites and their probable geochemical paragenesis. Many geologists assumed diagenesis had so altered rock fabric that original depositional textures were completely obliterated. More recent work has shown that this is not the case, and many ancient evaporites contain textural indicators of the original depositional setting. In the last two decades a quantum leap has occurred in our understanding of the primary and diagenetic processes active in modern and ancient evaporites. It began with detailed studies of sabkhas in the Arabian (Persian) Gulf (Illing et al.,

1965) and bedded salts in the Dead Sea (Neev and Emery, 1967). Then came the Deep Sea Drilling Project discovery of widespread bedded evaporites in Late Miocene sediments beneath the Mediterranean seafloor (Hsu et al., 1977). This was followed by the documentation of evaporite pseudomorphs in rocks 3.45 billion years old in Western Australia (Dunlop, 1978). Since then modern subaqueous gypsum has been described from West and South Australia (Warren and Kendall, 1985).

The history of evaporite sedimentology is just beginning. Much of the progress of the last two decades is based on the selective application of depositional models to ancient evaporitic sequences. Current depositional models are derived by combining aspects of modern evaporitic sedimentation in areas as widespread as Africa, Antarctica, Australia, the Middle East, and the southwestern United States. The result is a composite suite of vertical sequences and lateral facies associations which can be used to interpret similar features in ancient depositional successions. Selective application of depositional models to ancient evaporitic sequences is occurring in all fields of sedimentology. In the search for the subtle stratigraphic trap it is changing hydrocarbon exploration from a retrodictive to a predictive science. Depositional modeling is an important part of this book and will be used throughout each chapter. First, modern depositional settings are described, the important observations are then gathered into depositional models with characteristic vertical sequences, and finally ancient sequences are interpreted using the models.

EVAPORITE MINERALOGY

Phanerozoic evaporitic successions often contain salt units with distinct mineralogies which can be used to decipher hydrogeochemical origins and in some cases the original depositional setting of the basin (Table 1.1; Eugster and Hardie, 1978; Hardie, 1982, 1984). Using this method, the mineralogy of the deposit is a much more important piece of information than the percentage of potassium versus sodium in the unit. Mineralogy indicates the hydrogeochemical conditions at the time of precipitation. Even when the mineralogical suite is known, the method is not faultless. Evaporite minerals are subject to intense and often early diagenesis involving partial or complete mineral replacement. Yet it may still be possible to use petrographic evidence to determine the original mineralogy as Lowenstein (1982) has done in the Permian Salado Formation of west Texas and New Mexico. The following discussion of evaporite mineralogy draws heavily on papers by Dean (1978a), Eugster and Hardie (1978), Eugster and Jones (1978), Holser (1979), and Hardie (1984) and references therein.

Evaporite units are usually defined by the dominant mineral in the bed; for example a halite unit usually contains more than 50% halite, the mineral. Holser (1979) argued that it is better to define an evaporite unit by the appearance in any proportion of a critical new mineral or its pseudomorphs. A rock does not then

TABLE 1.1 (A) MAJOR EVAPORITE MINERALS AND THEIR CHEMICAL COMPOSITIONS (after Hardie, 1984)

anhydrite	$CaSO_4$	leonhardtite	$MgSO_4{\cdot}4H_2O$
aphthitalite (glaserite)	$K_2SO_4{\cdot}(Na,K)SO_4$	leonite	$MgSO_4{\cdot}K_2SO_4{\cdot}4H_2O$
antarcticite	$CaCl_2{\cdot}6H_2O$	loewite	$2MgSO_4{\cdot}2Na_2SO_4{\cdot}5H_2O$
aragonite	$CaCO_3$	magnesian calcite	$(Mg_xCa_{1-x})CO_3$
bassanite	$CaSO_4{\cdot}\frac{1}{2}H_2O$	mirabilite	$Na_2SO_4{\cdot}10H_2O$
bischofite	$MgCl_2{\cdot}6H_2O$	nahcolite	$NaHCO_3$
bloedite (astrakanite)	$Na_2SO_4{\cdot}MgSO_4{\cdot}2H_2O$	natron	$Na_2CO_3{\cdot}10H_2O$
burkeite	$Na_2CO_3{\cdot}2Na_2SO_4$	pentahydrite	$MgSO_4{\cdot}5H_2O$
calcite	$CaCO_3$	pirssonite	$CaCO_3{\cdot}Na_2CO_3{\cdot}2H_2O$
carnallite	$MgCl_2{\cdot}KCl{\cdot}6H_2O$	polyhalite	$2CaSO_4{\cdot}MgSO_4{\cdot}K_2SO_4{\cdot}2H_2O$
dolomite	$CaCO_3{\cdot}MgCO_3$	rinneite	$FeCl_2{\cdot}NaCl{\cdot}3KCl$
epsomite	$MgSO_4{\cdot}7H_2O$	sanderite	$MgSO_4{\cdot}2H_2O$
gaylussite	$CaCO_3{\cdot}Na_2CO_3{\cdot}5H_2O$	schoenite (picromerite)	$MgSO_4{\cdot}K_2SO_4{\cdot}6H_2O$
glauberite	$CaSO_4{\cdot}Na_2SO_4$	shortite	$2CaCO_3{\cdot}Na_2CO_3$
gypsum	$CaSO_4{\cdot}2H_2O$	sylvite	KCl
halite	$NaCl$	syngenite	$CaSO_4{\cdot}K_2SO_4{\cdot}H_2O$
hanksite	$9Na_2SO_4{\cdot}2Na_2CO_3{\cdot}KCl$	tachyhydrite	$CaCl_2{\cdot}2MgCl_2{\cdot}12H_2O$
hexahydrite	$MgSO_4{\cdot}6H_2O$	thernadite	Na_2SO_4
kainite	$MgSO_4{\cdot}KCl{\cdot}\frac{11}{4}H_2O$	thermonatrite	$NaCO_3{\cdot}H_2O$
kieserite	$MgSO_4{\cdot}H_2O$	trona	$NaHCO_3{\cdot}Na_2CO_3{\cdot}2H_2O$
langbeinite	$2MgSO_4{\cdot}K_2SO_4$	van'thoffite	$MgSO_4{\cdot}3Na_2SO_4$

TABLE 1.1(B) EVAPORITE MINERALS AND THEIR PARENT BRINES (AFTER HARDIE, 1984)

Brine composition and its source	Major saline minerals	Key minerals
i. Na-K-CO_3-Cl-SO_4 NON-MARINE WATERS (particularly meteoric)	alkaline earth carbonates, gaylussite, pirssonite, mirabilite, thenardite, trona, nahcolite, natron, thermonatrite, shortite(?), burkeite, aphthitalite, halite, sylvite, hanksite.	Na_2CO_3 minerals
ii. A)Na-K-Mg-SO_4-Cl SEAWATER AND NON-MARINE WATERS (meteoric/volcanogenic) OR MIXED MARINE AND NON-MARINE WATERS	alkaline earth carbonates, gypsum, anhydrite, mirabilite, thenardite, glauberite, polyhalite, epsomite, hexahydrite, kieserite, bloedite and other Na-Mg sulfates, leonite and other K-Mg sulfates, kainite, carnallite, halite, sylvite, bischofite.	$MgSO_4$ minerals or $NaSO_4$ minerals
B)Na-K-Mg-Ca-Cl NON-MARINE WATERS (particularly hydrothermal)	alkaline earth carbonates, gypsum, anhydrite, sylvite, carnallite, bischofite, tachyhydrite, antarcticite.	$KCl \pm CaCl_2$ minerals in the absence of Na_2SO_4 and $MgSO_4$ minerals

need to contain 50% halite to be in the halite facies; it need only contain halite in any amount. The very fact that halite could precipitate at all is the critical factor in the appearance of a new facies. The advantage of this usage is that it can define salinity and pore fluid changes much better than can defining units on proportions of the dominant minerals. The disadvantage of Holser's method is that it is of less use in the subsurface, where massive evaporite beds are seldom cored. Often, all that is available are chips or cross-plots from the wire-line logs. Mineralogy determined from wire-line data and chips reflects the dominant mineralogies or the least soluble salts in a unit. To date, no consistent classification exists for evaporites; most workers (including the author) describe the unit based on the dominant mineral. For sulfates the best nongenetic classification is that of Maiklem et al. (1969). For evaporite beds dominated by other minerals a nongenetic classification with widespread acceptance just does not exist.

Marine Mineral Assemblages

Most ancient evaporites are composed of salts originally formed by the concentration of seawater-derived brines (Table 1.2). Modern seawater is a well-mixed aqueous solution with salinities averaging 35 ± 2 per mille. The salinity of seawater is given throughout the book as parts by weight of dissolved salts per mille (i.e. parts per thousand (‰)). A salinity of 35 per mille corresponds to a specific gravity of 1.026 at 16°C. Over most of the Pacific, Atlantic, and Indian Oceans, the salinity lies between 34 and 36 per mille, but a progressive increase occurs toward the centers of the permanent anticyclonic area in tropical latitudes. Salinities in these regions are the highest found in the open ocean and hover around 38 per mille in such areas as the Sargasso Sea. Higher salinities occur in enclosed seas such as the Mediterranean, where salinity rises from 36 per mille near the Straits of Gibraltar to more than 39 per mille near Cyprus. A similar situation exists in the Red Sea, where salinities are above 40 per mille. The salinities of

TABLE 1.2 MINERAL SEQUENCE FROM THE EVAPORATION OF A SIMPLE SEAWATER SOLUTION AT 25° C. CONCENTRATION FACTOR IS RELATIVE TO AVERAGE SEAWATER.

Mineral	Concentration factor	% Water loss	Brine density
K-Mg Salts	63 x	98.7%	1.29
Halite	11 x	90%	1.214
Gypsum (anhydrite)	5 x	80%	1.126
$CaCO_3$	2–3x	50%	1.10
Seawater	1 x	0%	1.04

ancient epeiric seaways were far more variable, especially in arid settings, where salinities ranged into, and stayed at, halite- and gypsum-precipitating levels for extended periods. This led to the deposition of widespread shallow subaqueous evaporite units extending over hundreds of square kilometers (e.g., San Andres Formation on the Northern Shelf of the Permian Basin, Texas; Chapters 2 and 3)

Modern seawater chemistry is dominated by Na^+ and Cl^-, with lesser amounts of SO_4^{2-}, Mg^{2+}, Ca^{2+}, K^+, CO_3^{2-}, and HCO_3^-. Using the classification of Eugster and Hardie (1978) seawater is a Na—(Mg)—Cl—(SO_4) water. The chemical makeup and the proportions of the major ions of seawater are constant in all the world's oceans. When seawater evaporates it deposits a predictable suite of minerals (Tables 1.1b and 1.2). The first mineral to precipitate is $CaCO_3$ usually as aragonite; this begins when the solution reaches twice the concentration of seawater (40 to 60 per mille). Next to precipitate is gypsum ($CaSO_4 \cdot 2H_2O$), when the brine is about five times the concentration of seawater (130 to 160 per mille). At 11 to 12 times the concentration of seawater (340 to 360 per mille), salt or halite (NaCl) drops out of the brine. After halite the bittern salts (potassium and magnesium salts) precipitate at concentrations more than 60 times that of the original seawater. The types of bittern salts and the salinities at which they precipitate from concentrated seawater vary according to temperature or levels of organic matter in the remaining brine.

Using a detailed study of mineral paragenesis in conjunction with computer modeling, Harvie et al. (1980) have shown that even the complex salts of the bittern stage can be geochemically modeled. They explained much of the variability present in the bittern salts of the Zechstein deposits in Germany using a combination of equilibrium batch evaporation interspersed with times of fractionation. A full discussion of the actual geochemical model can be found in Harvie et al. 1984. The important difference between their model and earlier geochemical models is that they recognized syndepositional backreactions of mineral precipitates with concentrated brines. As the system is concentrated to a level slightly before halite saturation, anhydrite replaces gypsum. As evaporation continues anhydrite is partially replaced by glauberite which is in turn replaced by polyhalite and levels of anhydrite are further reduced. These replacements change the brine composition so that kainite does not supersaturate, a problem which plagued earlier models of potash precipitation. Their model not only explains the absence of primary kainite in the Zechstein deposits but also explains the pseudomorphous replacement of gypsum and anhydrite by glauberite and polyhalite and the replacement of glauberite by polyhalite. Previous authors attributed these replacements to thermal and solution metamorphism.

Nonmarine Mineral Assemblages

In a continental setting, rivers and groundwaters are the source of ions deposited in the evaporite basin, seawater takes little or no part in supplying the ions. The chemistry of continental waters is far more variable than seawater, consequently a more diverse suite of evaporite minerals is found in continental evaporitic basins (Harvie et al., 1982). As continental waters flow into a basin, they pass over and through many different lithologies. Their dissolved load is derived by leaching the more soluble minerals they contact on their way into the evaporative basin. Flow through limestones produces waters rich in Ca^{2+} and HCO_3^-, dolomites also generate Mg^{2+}, igneous and metamorphic rocks yield silica-rich Ca—Na—HCO_3 waters. Sulfides will contribute sulfate ions whereas basic and ultrabasic rocks often produce alkaline Mg—HCO_3 waters. Eugster and Hardie (1978) distinguished five major water types in continental evaporite basins:

1. Na—CO_3—Cl
2. Na—CO_3—SO_4—Cl
3. Na—SO_4—Cl
4. Na—Mg—SO_4—Cl
5. Ca—Mg—Na—Cl waters

As any one of these waters concentrates within a particular evaporite basin, it deposits a characteristic suite of evaporite minerals (Fig. 1.2). Regardless of water type, the first minerals to precipitate are the alkaline earth carbonates: low-magnesian calcite, high-magnesian calcite, aragonite, and dolomite. Gypsum is usually next. Subsequent precipitates depend on the brine inflow chemistry; for example, trona forms in waters dominated by sodium and bicarbonate ions. Minerals commonly formed at this stage include trona, glauberite, polyhalite, thermonatrite, and halite (Table 1.1b). Chemical makeup of the brackish inflow water is the most important control on the subsequent mineralogy of the lake evaporites (Fig. 1.2). The evaporation pathway of a closed basin brine is determined by the proportions of calcium, magnesium, and bicarbonate ions. Associated with many continental evaporites are suites of authigenic minerals formed by the diagenetic action of near-surface pore waters. The best studied are the sodium silicates, such as magadiite; the borosilicates, such as searlesite; and the zeolites of the alkaline saline lakes, such as clinoptilolite, erionite, and analcime (Eugster, 1980).

Chemical modelling of brine–mineral interactions in continental settings is not as reliable as for marine settings due to a far greater number of variables in the inflow waters. In highly concentrated continental brines, the chemical behavior of many important ions such as aluminium and boron is not as well understood. However Felmy and Weare (1986) applied the model of Harvie et al (1984) to the natural borate deposits of Searles Lake, California, with a fair degree of success. They investigated the possibility that straightforward evaporation of present day Owen's River water could explain the mineralogy of the upper salt layer and its parting mud layer. Their initial model allowed complete equilibration between the brine and previously precipitated minerals including the back reaction of Na_2CO_3 in the brine with previously precipitated calcite to form pirssonite.This predicted a

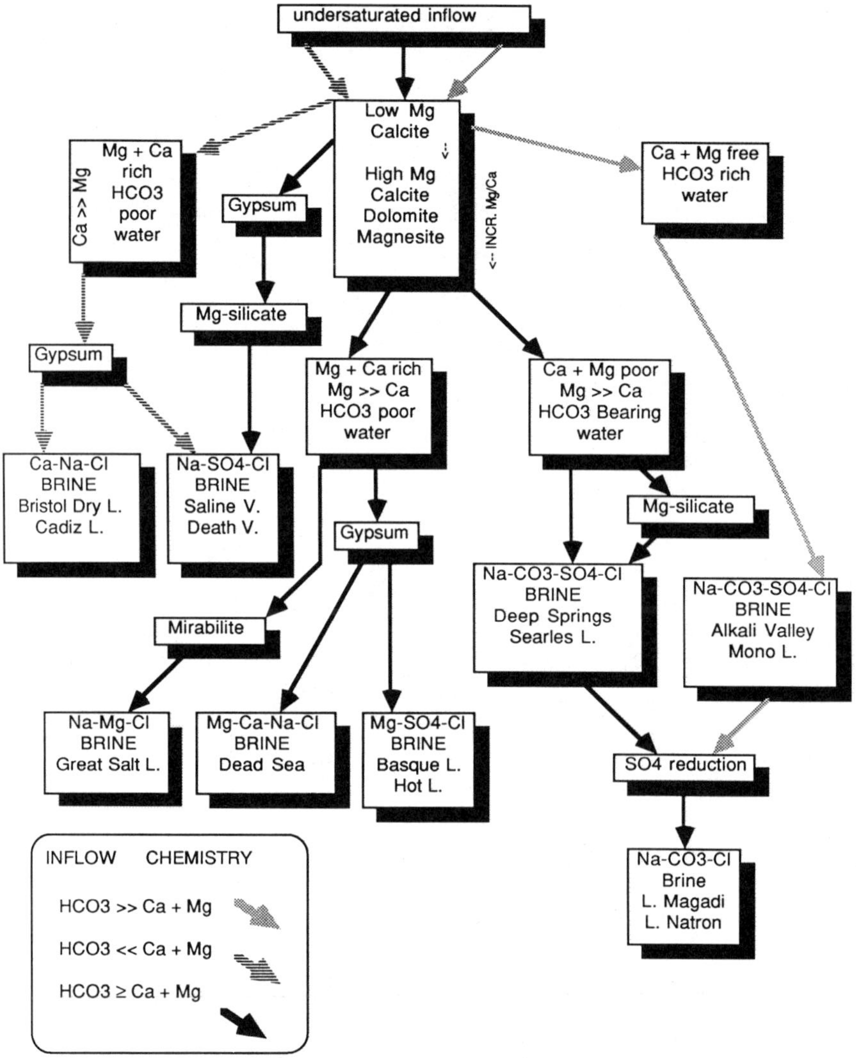

Figure 1.2 Brine evolution in a closed basin and associated evaporite mineral sequence. Note how the inflow chemistry controls the subsequent brine evolution in all lakes (after Eugster and Hardie, 1978).

large amount of pirssonite ($Na_2Ca(CO_3)_2 \cdot 2H_2O$) and no trona in the upper salt, whereas the actual salt is characterized by large amounts of trona and little pirssonite. In the actual deposit most of the calcite resides in the relatively impermeable clay layer below the upper salt. When they used a geologically more realistic model preventing a backreaction with calcite, and suppressed the formation of dolomite and magnesite, but otherwise allowed the system to equilibrate, they obtained good agreement with the mineralogical sequence in Searles Lake.

Marine versus Nonmarine Minerals

Mineralogy alone will not always distinguish between marine and nonmarine evaporites; halite and gypsum often dominate both settings. Similarly, minerals such as glauberite, polyhalite, epsomite, bloedite, sylvite, and tachyhydrite are not diagnostic of either setting (Table 1.1b). On the other hand, seawater is dominated by Na^+, Cl^-, and SO_4^{2-} ions, so saline carbonates never form in a marine evaporite. All the carbonate ions are used up early in the seawater precipitation series. Trona, gaylussite, pirssonite, burkeite, northupite, hanksite, and similar phases are restricted to saline continental settings where ample bicarbonate remains in solution after most of the calcium has been precipitated as alkaline earth carbonates. The lack of saline carbonate minerals in most ancient evaporites argues for little change in the chemical makeup of the world's oceans for the last billion years (Holser, 1979). The occurrence of calcite pseudomorphs after saline carbonates in Cambrian limestones in South Australia is due to the continental setting of this ancient saline playa (White and Youngs, 1980). Sodium carbonate mineral pseudomorphs and Magadi-type cherts also occur in organic-rich laminites in lacustrine rocks of Middle Devonian age in the Orcadian Basin of Scotland (Parnell, 1986).

Hardie (1984) disagreed with the idea that most ancient evaporites are marine derived. He argued that evaporite basins are extremely restricted depositional settings and that in many basins, even in the modern Arabian Gulf sabkhas, inflowing continental brines are far more important during early deposition and diagenesis than most geologists realize.

DIAGENESIS

Evaporites are difficult sediments to use in interpreting depositional setting, as the primary textures are altered by recrystallization and replacement during late, early, and even syndepositional diagenesis (Table 1.3; Hardie 1984). Pseudomorphs and reaction rims are common, as are relict mineral cores encased in the mineral that replaced it. Documented examples include pseudomorphs of anhydrite after subaqueous gypsum, reaction rims of polyhalite around earlier gypsum or anhydrite, relicts of carnallite in secondary sylvite, and pseudomorphs of sylvite plus kieserite after earlier langbeinite. Gypsum deposited at the surface is converted to anhydrite with burial, and back to gypsum with reemergence. During the dewatering associated with burial, other salts, such as halite and the bittern salts, can recrystallize or transform to other mineral phases. Textures are also altered during the deformation and recrystallization associated with rheotropic flow. Buried salt can flow and fold into intricate salt structures on a scale ranging from millimeters to kilometers. Buried gypsum can develop enterolithic folds as it converts to anhydrite. Some salts flow more readily than others; in bedded evap-

TABLE 1.3 DIAGENETIC SALINE MINERALS THAT CAN GROW SYNDEPOSITIONALLY AT TEMPERATURES BELOW 30°C (AFTER HARDIE, 1984)

Mineral	Syndepositional origin
anhydrite ($CaSO_4$)	dehydration of early formed gypsum as aH_2O of evaporating brine decreases.
glauberite ($CaSO_4 \cdot Na_2SO_4$)	back-reaction of evaporating Na_2SO_4-brine with early formed gypsum or halite.
polyhalite ($2CaSO_4 \cdot MgSO_4 \cdot K_2SO_4 \cdot 2H_2O$)	back-reaction of evaporating K-Mg-SO_4 brine with early formed gypsum, anhydrite or glauberite.
kieserite ($MgSO_4 \cdot H_2O$)	dehydration of early formed hexahydrite as aH_2O of evaporating brine decreases.
sylvite (KCl)	incongruent dissolution of carnallite in undersaturated waters.
gaylussite ($CaCO_3 \cdot Na_2CO_3 \cdot 5H_2O$)	back-reaction of evaporating alkaline brine with early formed aragonite or calcite.

orites a layer composed of carnallite and halite can be isoclinally folded, yet sit between two relatively undeformed anhydrite layers (Realmonte Mine in Miocene potash salts, Sicily).

Gypsum–Anhydrite and Dewatering

The calcium sulfate minerals gypsum and anhydrite are important, often puzzling phases that change according to prevailing temperatures, pressures, and salinities (Kinsman, 1974; Holser, 1979). Gypsum formation is favored by lower temperatures, lower pressures, and relatively lower salinities; anhydrite is favored by higher temperatures, pressures, and salinities (Fig. 1.3). At earth surface conditions gypsum is by far the most common calcium sulfate phase. Anhydrite can precipitate and grow by capillary evaporation in the high salinities and temperatures of the Arabian Gulf sabkhas, or it can replace gypsum already present in the sabkha (Warren and Kendall, 1985). At times after a heavy rainstorm, the partially hydrated form of gypsum (bassanite—$CaSO_4 \cdot \frac{1}{2}H_2O$) is as common as anhydrite on the sabkha surface. Anhydrite is less likely in many other modern coastal evaporitic areas, as temperatures are modified by the coastal setting; in such settings gypsum is the more likely sulfate. Penecontemporaneous anhydrite is more common under the extreme temperature regime found in continental deserts. Even there, in many near-surface settings, gypsum not only grows but persists at temperatures and brine concentrations where anhydrite should be the

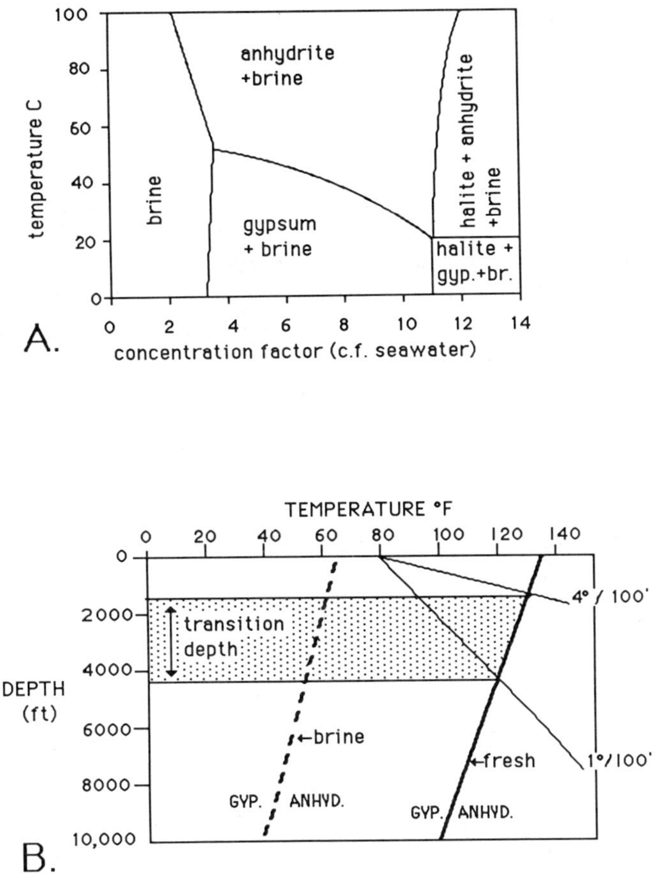

Figure 1.3 Gypsum—anhydrite transition: (a) stability of the calcium sulfate minerals in terms of temperature and salinity (after Hardie, 1967); (b) stability of the calcium sulfate minerals in terms of temperature and depth showing the importance of the salinity of the pore fluids (after Holser, 1979).

stable phase. In this respect anhydrite is an exceptional mineral; it is easier to make it in the laboratory under conditions equivalent to natural surface settings than it is to find it in nature. Some types of organic matter and the high relative humidity of many evaporitic settings probably discourage the surface formation or preservation of penecontemporaneous anhydrite.

When bedded gypsum is buried and the temperature rises above 60°C it is transformed to nodular anhydrite (Fig. 1.3b). In a hydrologically-open system the process is complete by a depth of 1000 m (3000 to 4000 ft) corresponding to a pressure of about 300 atm. Transformation can occur at much shallower depths if the pore fluids are saline and not until much greater depths if the unit is encased in shale and so overpressured. The reverse process, the conversion of anhydrite back to gypsum, occurs during erosion and exposure as the sulfate reenters the meteoric realm and is bathed in cooler and fresher pore fluids. The conversion reaction is

$$CaSO_4 \cdot 2H_2O \rightarrow CaSO_4 + 2H_2O$$

$CaSO_4$-saturated water is released upon burial as the gypsum is transformed to anhydrite, adding a water-filled volume of 38% to the porosity of the sulfate unit. This water of dehydration must alter its surrounds. The most likely effect is a decrease in strength of the anhydrite unit due to increased lubrication and perhaps overpressuring. If the water cannot drain freely from the dewatering gypsum, the bed converts to a quicksandlike consistency. This explains some of the intense deformation and enterolithic texture in many ancient bedded anhydrites. Enterolithic anhydrite can form in several different ways:

1. In the supratidal zone of an Arabian Gulf sabkha
2. During burial as gypsum dehydrates to anhydrite
3. As a near surface, soft sediment slumping into the basin
4. Crumpling as diagenetic anhydrite hydrates back to gypsum, a 64% volume increase

Enterolithic anhydrite formed tectonically, or as slumps in bedded dewatering gypsum, tends to show constant vergence of the fold axes. Thus constant vergence can be used to distinguish some forms of burial anhydrite from sabkha anhydrite. In the latter case the orientation of the enterolith crests is almost always random. During burial if the water of dehydration drains a little slower than when enterolithic structures form, the sediment can retain enough rigidity to form aligned, nodular anhydrite pseudomorphs or ghosts after gypsum (Warren and Kendall, 1985).

Water of crystallization also affects units subjacent to the dewatering anhydrite. As it drains out of the sulfate unit it lubricates nearby growth faults and other relatively shallow slippage features, including nappes. The escaping water is saturated with respect to $CaSO_4$ but undersaturated with respect to other evaporite minerals, including any nearby carbonates (Sass and Ben-Yaakov, 1977) as well as halite and the bittern salts. As it flows through the surrounding rocks it can leach minerals to create secondary porosity or bring about mineralogical transformations. During brine reflux in the shallow subsurface of the Palo Duro Basin, the flow of such waters dissolved marine carbonate to form anhydrite-filled molds (Bein and Land, 1982). With dewatering a gypsum bed 10 m thick will convert to a 6.2-m anhydrite bed. This releases 4.9 m^3 of water per square meter of cross-sectional area. At 30°C this volume of $CaSO_4$-saturated water can dissolve 0.8 m^3 of halite or 5.4 m^3 of carnallite or convert 8.1 m^3 of carnallite to sylvite (Blatt, 1982, p. 354).

$$\underset{\text{carnallite}}{KMgCl_3 \cdot 6H_2O} + 4H_2O \rightarrow \underset{\text{sylvite}}{KCl} + Mg^{2+} + 2Cl^- + 10H_2O$$

If these waters are also relatively hot, they can carry and transform immature, shallow hydrocarbons (Warren, 1986a).

Dewatering a thick gypsum unit causes a decrease in sediment volume and thickness as the unit compacts with a loss of 38% water (and associated porosity). Usually, this compaction will occur evenly across the whole evaporite unit. If the unit is widespread and of equal thickness the end result is not all that noticeable. But in certain depositional settings dewatering can enhance hydrocarbon migration pathways by differential compaction. Consider the example of a carbonate buildup capped and surrounded by a $CaSO_4$-rich evaporite unit (Fig. 1.4). Evaporites off the structure would most likely be thick sequences of sub-

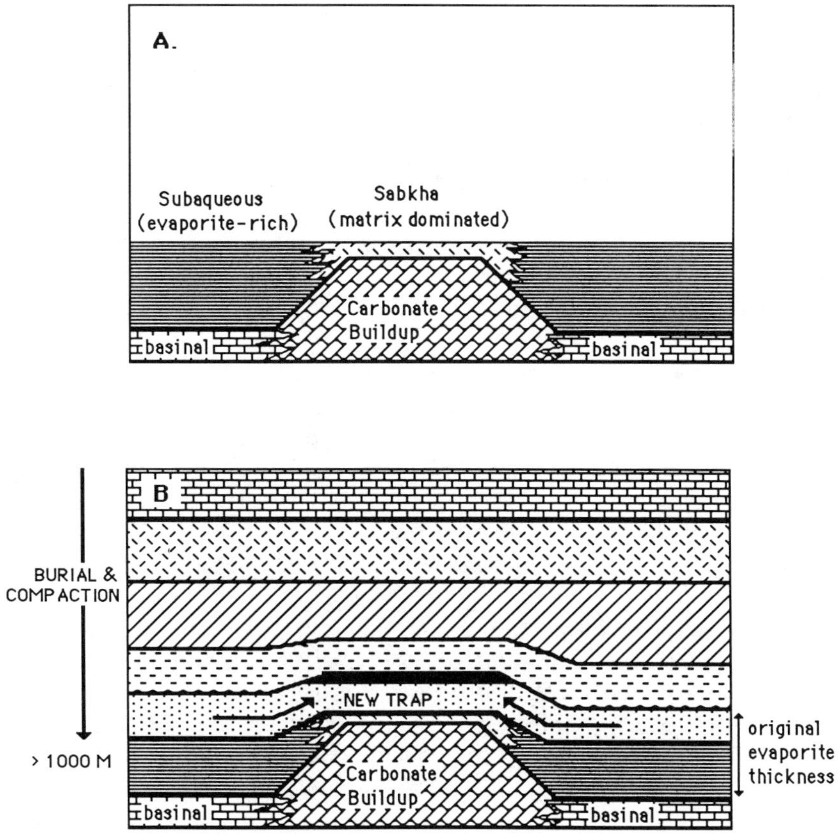

Figure 1.4 Schematic of the effects of gypsum dewatering on reservoir formation by differential compaction: (a) depositional setting; (b) burial compaction has created a new reservoir due to the greater volume loss in the subaqueous pure sulfate regions.

aqueous, relatively pure gypsum and halite; the onstructure cap would most likely be an anhydritic, matrix-dominated sabkha succession (Fig. 1.4a; Warren and Kendall, 1985). When evaporite deposition ceased in the basin the upper surface would have been a near-horizontal plane. As the succession was buried, the gypsum-rich, off-structure evaporite sequence would dewater to around 60% of its original thickness. The on-structure matrix-rich sabkha sequence would also compact, but not to the extent of the off-structure succession (Fig. 1.4b). The result would be a differentially compacted sequence draping over the buildup with excellent migration pathways into the region above the buildup. No matter if the underlying carbonates were tight, the differential compaction of an evaporite unit above the buildup could create potential reservoirs above the reef, a bonus if the reef were to be drilled.

EVAPORITE DISTRIBUTION AND CLIMATE

Evaporites form in areas on the world's surface where a potential exists for more water to leave the basin by evaporation than enters the basin by a combination of rainfall, surface, and subsurface inflow. Thus evaporite deposits are most often found in the arid and semiarid deserts of the world. Today such areas form two belts between 15 and 45° north and south of the equator (Fig. 1.5a and c). Modern evaporite deposits are also forming in the cold polar deserts of Antarctica, but the amount of salt in these regions pales to insignificance when compared to the volumes forming closer to the equator. The two main belts of evaporite deposition correspond to circumtropical areas of cold dry descending air in a Hadley Cell, a result of large scale atmospheric circulation (Fig. 1.5a and b). Solar radiation is most intense directly above the equator and lessens toward the poles. The equatorial belt is heated more than the adjacent latitudes, so the air above the equator warms and rises (Fig. 1.5a). While it rises it cools, so losing much of its water vapor as rain to the tropical jungles. This dry air cools and compresses as it moves away from the equator; finally, it sinks back to the earth's surface at around 30° latitude. Cool descending dry air absorbs water, and so creates the present desert belts of the world (Fig 1.5c).

Other local climatic conditions can create deserts with their associated evaporites in areas outside the major desert belts. The playas and midlatitude deserts of central Asia and central Australia are in part a product of their isolation from the nearest ocean. Continentality, the geographic isolation of an area by large distances from the oceans creates deserts. Deserts also form in the rain shadows of high mountain ranges; the deserts of Patagonia are on the lee side of the Andes, whereas the deserts of Nevada and Utah lie in the rain shadow of the Sierra Nevada. Cold upwelling ocean currents account for the deserts of Baja California, southwest and northwest Africa, and the Peruvian and north Chilean deserts. Areas of cold upwelling ocean currents create cool onshore winds with a low moisture bearing capacity. As such winds move overland, they warm and so have

A.
B.
C.

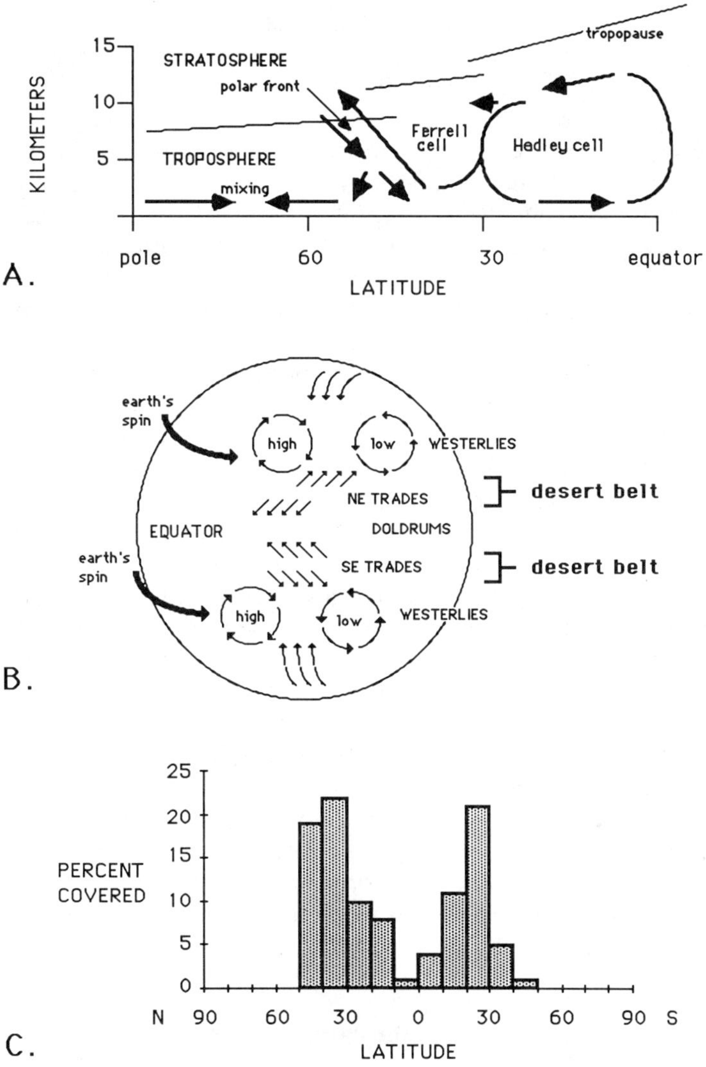

Figure 1.5 Evaporites and climate. (a) Longitudinal cross section through the earth's atmosphere showing the major circulation cells. Two belts of cold dry descending air at 30° N and S of the Equator create the arid zones of the world. (b) Effects of this circulation on the wind belts of the world. Deserts and their associated saline minerals form in the trade wind belts. (c) Latitudinal distribution of the world's modern evaporitic sediments as a percentage of total sediment.

their moisture-bearing capacity increased. This takes up moisture from the land surface to create desert conditions in these nearshore areas.

Local climate along with the associated evaporation rates causes important effects within evaporitic basins, as it controls mineralogy and preservation potential. Aerodynamically, evaporation rate is a product of wind velocity and the difference in water vapor pressures between the main air mass and air above the surface of the evaporating brine (Myers and Bonython, 1958). Water vapor pressure over the brine decreases with increasing salinity. At very high salinities the pressure difference between the brine and the air immediately above the brine can become so small that evaporation effectively stops. In nature this happens rarely, as the air above an evaporating brine body is in continual motion, driven by winds blowing across the basin. In reality the vapor pressure and air humidity over a brine body are determined by the humidity of air blown in from the surrounding region. In coastal areas beneath humid marine air, such as the sabkhas of the Arabian (Persian) Gulf, a tendency exists for the lower salinity evaporites, gypsum and halite, to be the sole precipitates and to preserve mainly gypsum/anhydrite as the sabkha evaporites (Kinsman, 1976). Most modern marine evaporite basins rarely contain salts more saline than halite; to produce primary salts such as polyhalite, epsomite, carnallite, and kieserite requires a degree of aridity so extreme that it is difficult to envision many modern marine marginal settings ever reaching the bittern stage. The only modern area of potash salt deposition occurs in Qarhan Salt Lake in Qaidam Basin, a basin situated in western China some 1500 km from the nearest ocean (Kezao and Bowler, 1985). Even there, the potash comes from the dissolution an earlier Plio-Miocene potash lake, formed in a continental transform setting. In ancient extremely arid marine settings, such as when marine waters first enter incipient rifts in desert belts, the deposited evaporite suite may have reached the potash—magnesia facies. Hence the occurrence of primary tachyhydrite ($CaMg_2C_{15} \cdot 12H_2O$) in the Cretaceous rift evaporites of Brazil and the Congo (Wardlaw, 1972).

Temperature is another important control on evaporite mineralogy. Glauberite ($Na_2SO_4 \cdot 10H_2O$) is a winter precipitate in Great Salt Lake, Zuni Salt Lake, and Death Valley. It forms by the chilling of seawater-like brines. Its presence or its pseudomorphs in ancient evaporitic settings probably indicate high latitude or continental deposition.

ANCIENT EVAPORITES

Evaporite pseudomorphs occur in rocks as old as early Archean but are far more common, as are the actual salts, in Phanerozoic strata. The earliest evaporite pseudomorphs are found in the Pilbara of Western Australia in greenstone belts some 3.45 billion years old (Fig. 1.6; Barley et al., 1979; Lowe, 1983). The sequences are now extensively silicified and baritized but were deposited as shallow-water carbonates and bottom-nucleated evaporites containing desiccation breccias, edgewise conglomerates, climbing ripples, scour-and-fill structures, and stromatolites. Shallow water barite-chert sequences occur in rocks of equivalent age in the Barberton Mountains of South Africa. The occurrence of Precambrian evaporites has caused a reevaluation of models of a reducing Archean atmo-

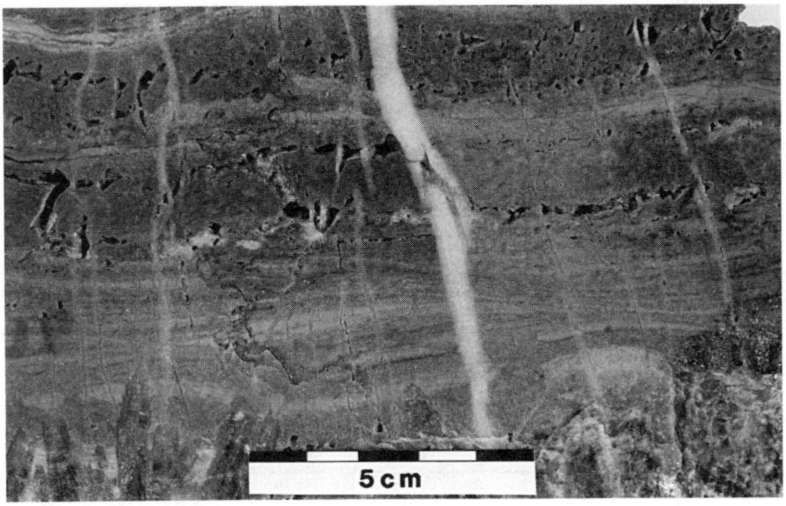

Figure 1.6 Evaporite [gypsum(?)] pseudomorphs from the 3.45-billion-year-old Warrawoona Group, North Pole, Western Australia. The pseudomorphs seen as upright crystals at the base of the plate are now composed of barite.

sphere. The presence of free sulfate, not sulfide ions, in some surface waters of the ancient world implies at least locally oxygenated conditions during the Archean (Dimroth and Kimberly, 1976; Clemmey and Badman, 1982).

Proterozoic rocks contain far more abundant evidence of evaporites than Archean sequences. Walker et al. (1977) presented evidence of widespread shallow-water sulfates (gypsum) in the 1400 to 1600 million-year-old carbonates of northern Australia. Evaporite pseudomorphs and presumed solution breccias also occur in the Lower Proterozoic (1700 to 1900 million-year-old) rocks of Great Slave Lake in northern Canada. These rocks were laid down in shallow marine, coastal, and alluvial environments under an oxygenated atmosphere.

Most Precambrian evaporites are preserved as pseudomorphs, not as the original salt. In the Phanerozoic rock record, extensive accumulations of evaporite minerals are far more common, but the amount of evaporite salts laid down in a set time interval varies a great deal depending on the geological period when deposition occurred (Fig. 1.1; Ronov, 1968, Ronov et al., 1980). Widespread episodes of evaporite deposition took place in the Upper Cambrian, the Permian, the Jurassic, and the Miocene, with lesser but significant episodes in the Silurian, the Devonian, the Triassic, and the Eocene. Even the lesser accumulations are on a vast scale not found in Quaternary sequences, where evaporites constitute less than 0.1% of shelf sediments.

Ancient evaporites were exceptional among platform sediments in their very high depositional rates. When conditions were right, a 100-m thick unit could be deposited in something like 1000 years (Table 1.4; Schreiber and Hsu, 1980). This rate is one to three orders of magnitude greater than most shelf depositional rates

TABLE 1.4 COMPARISON OF THE VARIOUS DEPOSITIONAL RATES IN SOME SEDIMENTARY SETTINGS

Sediment type	Depositional rate (m/m.y.)
Deep sea clay	1
Deep sea carbonate ooze	1
Shelf carbonate and mud (non-reef)	10–30
High-productivity clastic slope	40
Deep sea turbidite	100–1000
High-productivity marine diatomite	400–1000
Shallow-water reefal limestone	1000–3000
Sabkha (supratidal) evaporite	1m thick section progrades 1 km/1000 years
Subaqueous gypsum	10000–40000
Subaqueous halite	10000–100,000
Diapir growth (Zagros Mtns., Iran)	150,000–200,000

and indicates extremely rapid basin infill during evaporitic episodes. The Zechstein Basin was filled in 2 to 10 percent of the time that it existed (Schreiber and Hsu, 1980). The remainder of the time was taken up by episodes of erosion and nondeposition. The Late Miocene Mediterranean Messinian evaporite sequence, at least one kilometer thick, was deposited in less than 200,000 years. Both the Zechstein and Mediterranean Basins were filled with brines which shallowed and deepened throughout their depositional history. Rates of water-level change in these and other evaporite basins were at times far more rapid than any open marine sea level changes, even the glacial-induced changes of the last 2 million years. This means that most ancient evaporite deposits show the effects of depositional and diagenetic overprints related to rapid water level changes; deep-water evaporites can be sabkharized, and shallow-water deposits can be eroded and resedimented into deep water settings. The end result is usually a stacked shallowing-upward sequence separated by hiati and unconformities.

The rest of this book argues for the use of selective uniformitarianism, that is, the present is the key to the past when the model is wisely applied. However, the farther back in time one goes, the more difficult it becomes to apply one-for-one Quaternary counterparts, especially in marine-margin settings. Some geologists argue that many of the basin scale processes active in the Quaternary were of less or no importance in ancient sequences. The Quaternary is a time characterized by rapid changes in sea level brought about by the waxing and waning of the polar ice sheets. For example, in the last 18,000 years, sea level has risen 120 to 140 m and for some 100,000 years prior, it was consistently below its present level. The rapid to-and-fro of the Pleistocene sea across carbonate shelves and platforms with amplitudes of 100 to 140 m has prevented the deposition of extensive shallow water evaporites. The expanding and shrinking of the ice sheets first began to influence sea level some 12 to 14 million years ago; prior to this, the last major ice age spanned the Permo-Carboniferous. In the intervening period of some 200 million years, the oceanic water level was not subjected to such large amplitude and high-frequency changes as have occurred in the mid to late Tertiary. However, some authors argue that ice volume fluctuations have left an impression as far back as the Cretaceous, especially in the mid to late Tertiary (Matthews and Poore, 1980; Miller et al., 1985).

The three-dimensional distribution of evaporites and carbonates deposited in times of more stable sea level (smaller amplitude or slower fluctuations) may have been controlled by factors other than glacioeustacy. The stability of the continental margin and the creation of intrashelf basins, the formation of huge restricted basins by continental collision or rifting, changes in oceanic circulation patterns, and fluctuating rates of seafloor spreading may have been extremely important in the creation of widespread evaporite deposition. Yet today these causes are swamped by the overriding influence of rapid, high-amplitude, icesheet-induced sealevel changes. This may explain why modern evaporite depositional settings never rival the extensive and varied deposits of the rock record (Warren, 1986b).

Hydrogeochemical interpretations of ancient evaporite mineralogies, based on the constancy of seawater chemistry, also become less reliable as one goes farther back in time. Most workers would agree that the chemistry of the ocean has not changed much over the last 500 million years, although some suspect that the Mg/Ca ratio may have evolved through the Phanerozoic (Holser, 1979; Zenger, 1972). Tucker (1982) argued that dolomite may have been a primary or at least penecontemporaneous mineral phase in late Precambrian oceans. During the Archean, oceanic chemistry may have differed greatly from the Na—Cl-dominated waters of today. Maisonneuve (1982) argues that early Archean ocean waters were chlorobicarbonated solutions (Fig. 1.7). If this is true, early Archean evaporites would contain different mineral suites to evaporites deposited from Phanerozoic oceans. To model Archean evaporites (or more probably their

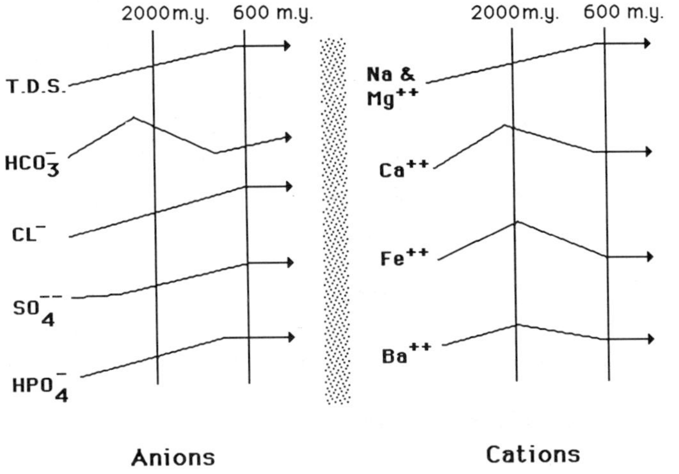

Figure 1.7 Variation in the proportions of the major cations and anions in the world's oceans from the Archean to the present (after Maisonneuve, 1982).

pseudomorphs) may require a study of the textures of salts forming in Na—HCO_3^- dominated saline lakes of the modern African rift valleys and then scaling up the model to oceanic evaporite proportions.

MODELS OF EVAPORITE BASINS

Throughout geological history evaporites, like carbonates and siliciclastics, formed in a wide range of depositional settings from pedogenic to lacustrine on the continents, and from the supratidal to the deep subaqueous in the marine realm (Fig. 1.8). Geologists have only begun to realize the significance of evaporite

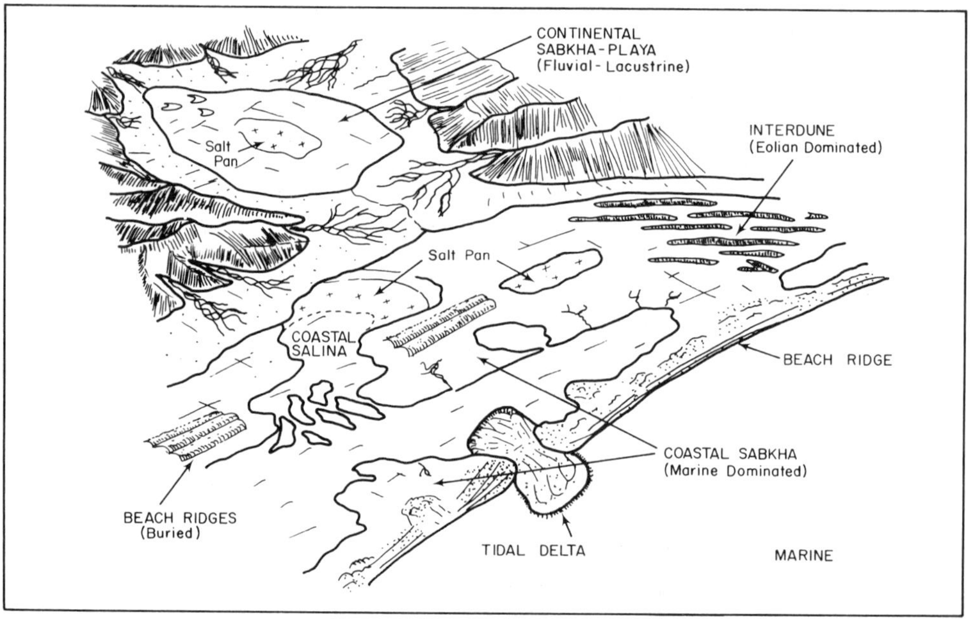

Figure 1.8 Spectrum of modern evaporite settings available for geological study. Deepwater and platform evaporites are absent in modern depositional settings (in part after Kendall, 1984).

depositional textures in the last two decades; only since the advent of studies in modern depositional environments such as the Arabian Gulf sabkhas, and the subaqueous salinas of Australia (Warren and Kendall, 1985). Studies in recent evaporite basins have shown that certain crystal textures give an indication of water depth (Fig. 1.9). Displacive evaporites (calcium sulfate and halite) tend to grow in the capillary zones of mud flats surrounding a brine body. Aligned giant crystals (1 to 100 cm long) and mechanically reworked evaporites are most often subaqueous deposits (Babel, 1987). In large, ancient subaqueous basins the evaporites were transported and reworked in the same way as carbonates and silici-

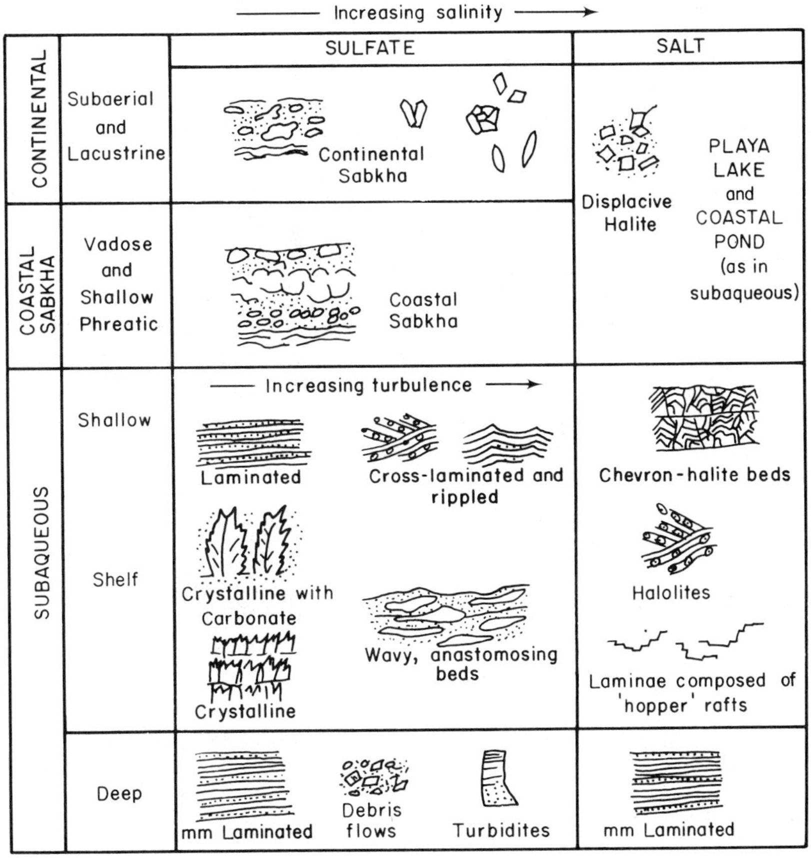

Figure 1.9 Summary of the various evaporite textures indicative of particular physical environments (after Kendall, 1984; Schreiber, 1986; Warren and Kendall, 1985).

clastics are today. Shoreward transport of evaporite particles formed sabkha sequences, basinward reworking into deep-water deposited debris flows and turbidites.

A study of ancient evaporite deposits underscores the limitations of a strict application of the Law of Uniformitarianism. The processes and textures of evaporite deposition and diagenesis can be studied in modern settings, but models so derived should not limit concepts of the scale or setting of ancient sequences. Widespread evaporite deposition requires particular climatic and tectonic conditions that do not coexist today, hence, no modern counterparts exist on the size and scale of ancient evaporites. When conditions suitable for deposition have occurred in a sedimentary basin, thick laterally extensive sequences were deposited very rapidly; in the late Miocene in the Mediterranean, a 1-km- thick sequence of halite was deposited in less than 200,000 years (Schreiber, 1986).

Ancient evaporite deposits possess thicknesses and horizontal extents two to three orders of magnitude greater than those of modern evaporites. Thus depositional models based solely on Quaternary evaporitic settings are not always applicable directly to ancient sequences.

In the Ordovician, the Late Permian, and the Cretaceous there were epeiric seaways across North America. Then it was possible to cross vast expanses of continental platform and never see exposed land. At other times in the Ordovician and the Permian the same areas were covered by a huge complex of evaporitic lagoons and mud flats. Many ancient continental platforms were much wider than today (thousands of km versus tens of km) and shallow seas extended far into continental interiors. Evaporites laid down on such platforms were deposited across vast areas of more than one million square kilometers. The term tidal-flat evaporite is probably inappropriate for mud flats in epeiric settings, as tidal exchange was minimal or non-existent over these wide evaporitic areas (Hallam, 1981). Most mechanically transported epeiric sediment was carried across the shallow-water areas of the platform and associated evaporitic mud flats by storm induced floods of marine water, not by tidal currents. Major ocean currents were nonexistant in the evaporitic areas of epeiric seas due to the damping effect of shallow-water shoals (Hallam, 1981; Irwin, 1965). Thus, the term strandline, rather than shoreline or tidal zone, is appropriate in describing the water or brine-sheet edge along the more landward parts of these epeiric seas and platforms.

Ancient marine-associated evaporites were deposited in two major settings:

Platform Evaporites. *Stratiform units (<50 m thick, often <5 to 10 m thick) of mixed subaerial(ephemeral)/shallow-water evaporites. Deposited on a continental platform and often intercalated with other shoal-water marine-influenced platform sediments. Platform evaporites not deposited as part of a basin-wide setting, passed seaward into time-equivalent open-marine deep-water sediments.*

Basin-wide Evaporites. *Thick, basin-wide units (often >50 m thick) of deep-water/shallow water evaporite deposits containing textural evidence of many different depositional settings including platform, slope, and basin. When evaporite deposition took place the whole basin was evaporitic.*

Both settings have characteristic associations of sediments and vertical sequences, yet they may overlie one another, or pass laterally into each other. Platform evaporites deposited about the edge of an evaporite basin can pass laterally into time-equivalent deep-water evaporites of the basin center (e.g., the Zechstein evaporites of the northern Europe). There is often more than one time of evaporite deposition in a single sedimentary basin. Older evaporites deposited in a basin-wide setting may underlie younger evaporites deposited in a platform setting (e.g., in the Gulf of Mexico the Jurassic Louann Salt was deposited as a basin-wide evaporite and overlain by the Cretaceous Ferry Lake Anhydrite, a platform evaporite; in the Arabian Gulf the Precambrian Hormuz Salt was deposited as a basin-wide evaporite, and overlain by the Jurassic Hith Anhydrite, a platform evaporite).

The wider extent and greater thickness of ancient evaporites compared to those of today was due to three main factors (Warren, 1986b):

1. *Warmer climate worldwide,* creating wider latitudinal belts suitable for evaporite deposition and preservation.
2. *Huge shallow epeiric seas,* often forming large inland seaways in arid areas, so creating areas of extensive platform and intracratonic evaporites. This type of evaporite deposition was most common during times of long-term tectonic quiescence. Relative sea level curves at such times were overall transgressive but broken up by short intervals of stillstand or slight fall. It was during these shorter intervals that the platform evaporites were laid down. Sediments were deposited as stacked platform cycles of thin evaporite units (usually less than 30 to 60 ft (10 to 20-m) thick) interlayered with open marine shelf carbonates or siliciclastics. Permian strata of west Texas and New Mexico were deposited as platform evaporites/carbonates/siliciclastics. Shoal-water evaporites of the Arab cycles and the overlying Hith Anhydrite filled and sealed the intrashelf basins of the Arabian platform during the Jurassic.
3. *Set up of the correct tectonic conditions* for the formation of saline giants; extensive basin-wide evaporite deposits with thicknesses exceeding 300 ft (100-m) and extending across whole basins. These areas formed during tectonically active times when tectonism generated rifts or during times of continental collision and/or transtensional faulting. The last saline giants formed on the earth's surface in the mid to late Miocene, when the restriction associated with the collision of Africa and Eurasia converted the Mediterranean Basin into a complex of shallow saline lakes and salt flats (Hsu et al., 1977), and the Zagros basin into an evaporite-filled lacustrine(?) depression.

The problem of scale always confuses a strictly uniformitarian approach: today, the large hypersaline shallow-water expanses of ancient epeiric seas do not exist (Irwin, 1965); nor do equivalents of the huge evaporite-depositing basins which covered the floor of the Mediterranean some 5.5 million years ago. This does not mean that models based on Quaternary studies are useless. Depositional processes active today were also active in these larger ancient settings; thus sedimentary structures forming today formed in an identical fashion in ancient evaporites. Diagenetic processes observed today also share their equivalents in the ancient. To explain ancient sequences, most Quaternary models require scaling up, or composite models have to be constructed from two or more modern settings (e.g., Handford, 1981; Muir et al., 1980; Ericksson and Warren, 1983).

PLATFORM EVAPORITES

Platform evaporites were deposited either on mud flats or subaqueously as saltern deposits. *Mud flat* sediments formed shoaling upward cycles 1 to 5 m thick, but they are not easily correlated across the platform due to the limited continuity of individual facies in the depositional setting. They were deposited as a facies mosaic of both saline and dry mud flats that separated local brine pans (Chapter 2). Evaporite mineralogy varied according to the degree of aridity at the time of deposition, and the chemistry of inflow waters. Surface brines were ephemeral, or extremely shallow, and so sediments often contained evidence of subaerial exposure such as mudcracks, erosional, or karstic surfaces. Displacive evaporites—nodular, enterolithic, and diskoidal gypsum or anhydrite, and skeletal halite—were growing beneath the mud flats. Evaporitic mud-flat beds were often interlayered with eolian or fluvial sediments. Examples discussed in the book include the Lower Clear Fork Formation of the Texas Panhandle, the Minnelusa Formation in Wyoming, and the northern regions of the Palo Duro Basin during deposition of the Upper San Andres Formation (Chapter 2). Modern counterparts, that is, counterparts in terms of size and scale of individual facies within the mosaic but not the overall scale of ancient mud flats, can be found in the sabkhas of the Arabian Gulf and the subaqueous coastal-lakes of South and West Australia. (Chapters 2 and 3)

Salterns were areas of widespread subaqueous evaporite deposition that occurred on platforms and in basin-wide settings. Saltern is a new term to describe extensive shallow-subaqueous evaporites; there is no known modern counterpart. Saltern sediments were often cyclic and made up of shoaling units 2 to 50 m thick, sometimes capped by thin sabkha sequences. Individual units were dominated by subaqueous, often laminated and layered textures; laminae were often crosscut or interrupted by coarse-grained to giant, vertically aligned crystals of calcium sulfate or halite. Cycles tended to be widespread and correlatable over distances of tens to hundreds of kilometers.

Saltern evaporites in platform settings were deposited in lagoonal or intrashelf basin areas; either on rimmed carbonate shelves or on platform ramps (carbonate terminology of Read, 1985). Platform cycles often had open-marine or restricted-marine shoal-water carbonates at their base, passing up into the saltern evaporites. Ancient examples include the Cretaceous Ferry Lake Anhydrite in the northern Gulf of Mexico, and the P1 and P2 evaporites of the San Andres Formation on the northwest shelf of the Permian Basin, and the Lower San Andres Formation of the Texas Panhandle (Chapter 3). In the subsurface and in wire-line logs an ability to correlate clean evaporites over large areas and to define the mineralogical purity as more than 80% evaporite (on cross-plots) tends to separate saltern evaporites from mudflat evaporites.

Mud flats and salterns were often part of the same formation. Mud flats tended to dominate about the edges of the depositional basin or subaerial areas within salterns. Slight changes in climate, subsidence rate or relative sea level could rapidly change the platform from a saltern-dominated to a mud flat-dominated area of deposition. There are no modern equivalents for the scale of saltern deposition although shallow-water evaporite textures can be observed in many small coastal lakes along the coasts of Australia and the Mediterranean.

Evaporites on ancient platforms were laid down in one of two settings, either on a *ramp* or on a *rimmed-shelf* (Fig. 1.10; Chapters 2 and 3). Such platforms were sometimes covered by evaporites, deposited as a series of stacked shoaling cycles, usually associated with open-marine carbonates or continental siliciclastics below and above the evaporite stage. The only exception to this association was a platform that was drowned soon after evaporite deposition; there the platform evaporites were covered by dark low-energy wackestones and mudstones deposited in deep-water in a sediment-starved basin.

Evaporitic Ramps

Extensive shallow epicontinental or epeiric seas usually formed on platform ramps (Ahr, 1973). Evaporitic ramps were characterized by very low slopes into open-marine waters with no substantial break in slope on the platform. They were first modelled by Shaw (1964) and Irwin (1965). Modern counterparts are the carbonate ramps of the southern Arabian Gulf and the Yucatan platform (Read, 1985), but there are no modern counterparts to ancient evaporite-covered ramps. Irwin's (1965) epicontinental-sea model proposes there were three zones of varying hydraulic energy on ancient platform ramps (Fig. 1.11):

1. *Zone X*. A wide low-energy zone occurring on the open-marine side of the platform and situated below wave-base. Sediments were mainly bioturbated mudstones and wackestones.
2. *Zone Y*. A narrow, intermediate, high-energy belt beginning where waves first began to bottom on the seafloor, it extended landward to the limit of any tidal action. Sediments in this zone included sand and grainstone shoals, reef mounds and patch reefs.
3. *Zone Z*. An extremely shallow low-energy zone on the landward side of zone Y. It was often hundreds of kilometers wide and characterized by limited water circulation with occasional stagnation; sometimes this led to the deposition of organic-rich laminated sediments. In arid settings, zone Z was at times covered by hypersaline water and extensive platform evaporites were deposited.

Slopes down the ramp were as low as 1 in 50,000 compared to modern shelf gradients between 1 in 500 and 1 in 2500. Behind the zone Y shoal, the zone Z slopes were even less and had bottom angles of only a few seconds of arc, thus huge areas of zone Z were covered by extremely shallow water, often only centimeters deep. Tidal effects were negligible. Water levels varied only during

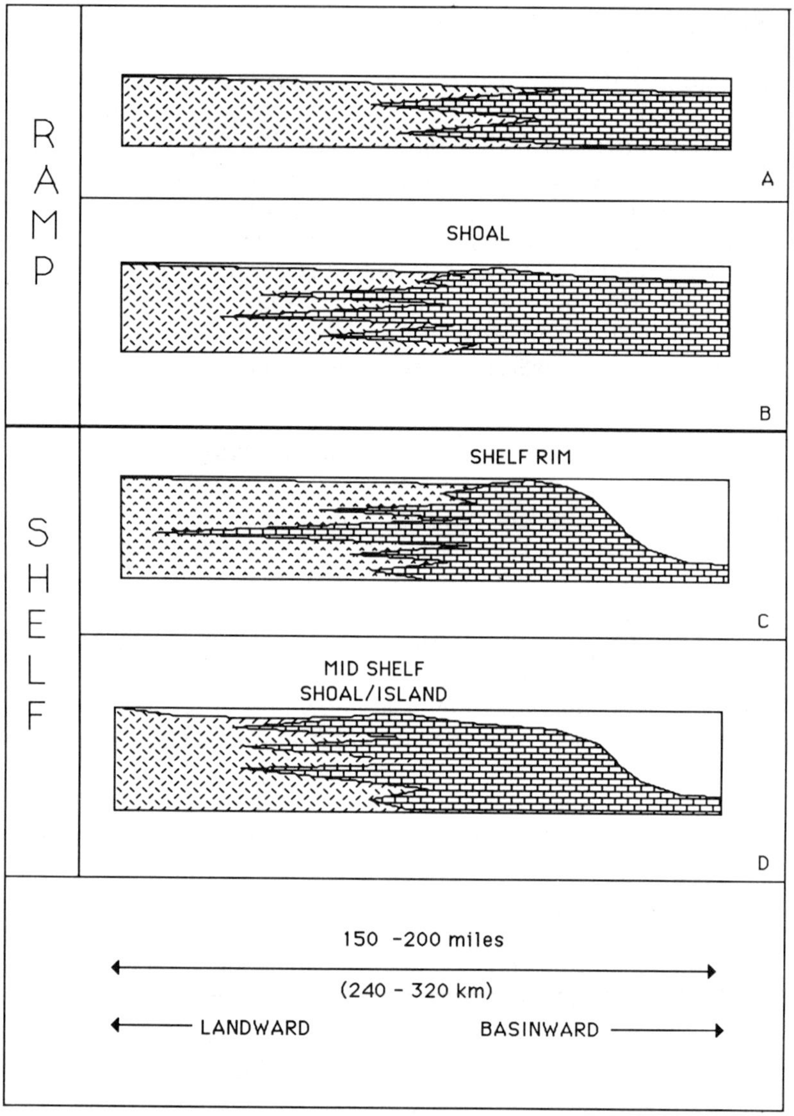

Figure 1.10 Depositional styles—ramp and shelf. (a) Gently deepening ramp. (b) Ramp with mid-platform shoal (zone Y). (c) Rimmed-shelf with topographically highest part located at the shelf break. (d) Shelf with deep rim, topographically highest part is a shoal or island-crest facies located on the mid shelf.

storms or hurricanes; as sheets of seawater were washed into zone Z, the strandline migrated tens of kilometers inland. The influx freshened the subaqueous brine sheets while supplying a new influx of marine clastic sediment to zone Z. The evaporite-forming area in zone Z was often hundreds of kilometers wide,

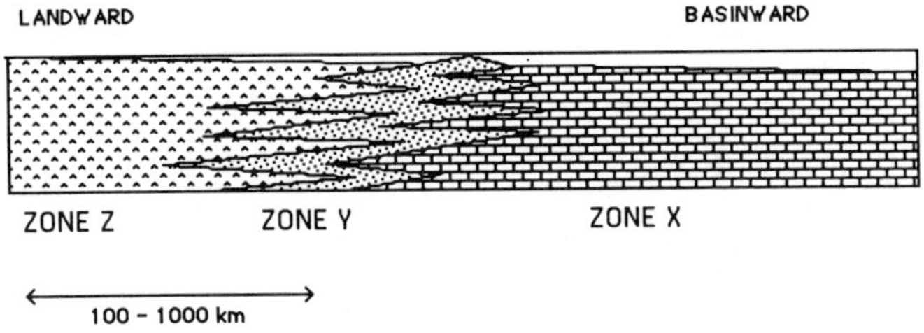

Figure 1.11 Ramp Profile: X–Y–Z zones shown in basinward to landward transect.

far wider than any evaporitic area on the modern land surface. Today the evaporite-depositing zone Z area in the Arabian Gulf (a modern ramp) is only 10 to 15 km wide. During the Permian, in times of widespread evaporite deposition on the Palo Duro ramp in West Texas, one could have traveled 50 to 150 km south and seaward from the strandline of the brine sheet and never crossed through water which was much more than a meter or two deep, nor less saline than gypsum-saturated (Chapter 3).

Much of the wave energy of the ramp platform was dissipated some distance basinward from the zone Z strandline; it spent itself against high energy shoals, bars, and reefs of zone Y. Shallow wave-agitated facies of the zone Y passed down the slope into deeper-water open-marine facies without a rapid change in slope angle. However, even on the ramp there was some degree of topographic relief about zone Y, and some restriction of water flow across it, otherwise surface waters in zone Z could have rarely concentrated to levels suitable for evaporite deposition.

Vertical sequences of evaporites deposited in zone Z were very much dependent on the relative physiography of zone Y. If zone Y formed a well-developed shoreline with only extensive washover flats behind, then zone Z was a mosaic of mud flats and small brine pans. Sediments were predominantly sabkha with occasional lenses of subaqueous deposition (Fig. 1.12a). If on the other hand, zone Y was a shoal backed by an evaporite lagoon, then the succession in zone Z would consist of saltern evaporites with an increasing proportion of subaerial indicators (dessication cracks, karst, erosion features, displacement textures) toward the top of the unit and toward the strandline (Fig. 1.12b). Units P1 and P2 of the San Andes Formation in the Northwestern platform of the Permian Basin and the Palo Duro Basin illustrate the saltern-ramp end member (Chapter 3). Extensive areas of saltern evaporites were sometimes backed by mud flats or sand-sheets of arid-zone siliciclastics. Little if any open-marine carbonate was ever transported this far back onto the ramp, and water conditions were too saline for most carbonate secreting organisms with the possible exception of cyanobacterial mats. The Lower Clear Fork Formation of the Palo Duro Basin illustrates the mud flat end member (Chapter 2); it was an evaporitic mud flat passing landward into a desert complex of wadis and dunes. With time and a relative rise in sea level it evolved from a mud-flat ramp to a ramp dominated by saltern evaporites (Chapter 2).

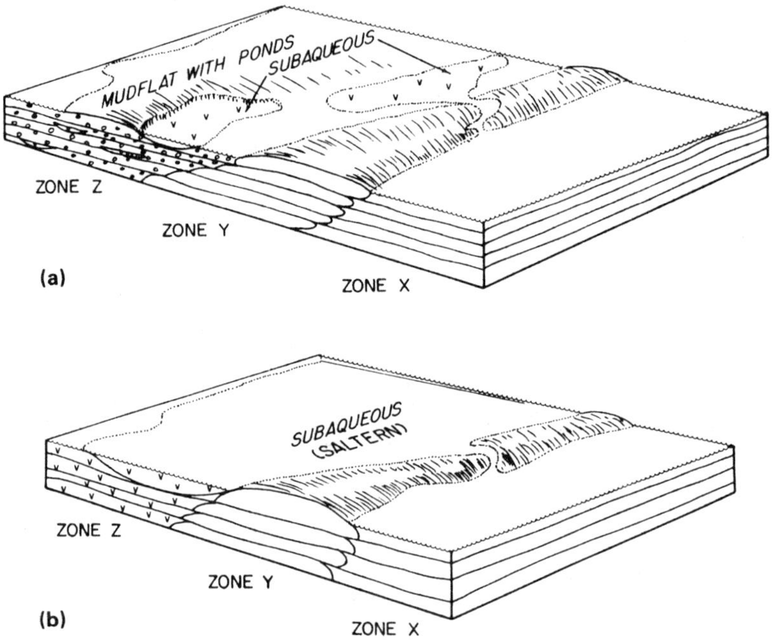

Figure 1.12 Ramp models. (a) Ramp–with sabkha/salina mud flat complex in zone Z. (b) Ramp–with subaqueous saltern units in zone Z.

In Yates Field, west Texas, the Permian sequence of mud-flat evaporites and carbonates was complicated by periodic influxes of siliciclastic sand. Locally these quartz sands, particularly in the Queen Formation, show evidence of subaqueous channeling. These strata have been interpreted as tidal-channel fill but have also been interpreted as an eolian-fill blowing into a restricted lagoon (Mazzullo et al., 1984). Recent work by Spencer and Warren (1986) in the Queen Formation in Yates Field on the Central Basin Platform concluded that the siliciclastic sands were deposited on a coastal mud flat fed by eolian sands in a fashion very similar to the deposition now taking place on the landward side of Al Zibaya sabkha in coastal Abu Dhabi. These same authors also showed that later in the same area, during the deposition of the Seven Rivers Formation, a number of siliciclastic beach ridges, fed by eolian sands, formed a series of laterally-accreting beach ridges and spits. The ridges grew to where they enclosed a number of marine embayments that became restricted enough to deposit the pure salina sulfates of the basal Seven Rivers Formation.

Evaporitic Rimmed-shelves

Rimmed-shelves were shallow-water carbonate platforms with an outer wave-agitated edge marked by a pronounced increase in slope (commonly a few degrees to more than 60°) into deep water. The carbonate rim was continuous to semicontinuous, characterized by high-energy reefs, reef mounds, and grainstone shoals (Fig. 1.10). Such shelves required organic binding or early cementation to maintain the rim and did not form on siliciclastic-dominated platforms. During evaporite deposition the platform area behind the rim was characterized by lagoonal evaporites that passed landward into mud flat evaporites or arid zone siliciclastics.

The carbonate rim was responsible for the restriction of the backreef or lagoon; often the degree of restriction was controlled by the continuity of the rim. To form an evaporitic unit in the lagoonal area behind the rim required the crest of the rim to be at or just above sea level, with any channels cutting across the rim to be sufficiently restricted to prevent the free exchange of marine and lagoonal waters. Otherwise, the conditions behind the shelf rim or shoal would have never become saline enough to deposit evaporites. As in ramp platforms, two end members were deposited behind the rim, either saltern sequences or mud-flat (sabkha)-dominated successions. The dominance of one or the other reflected the response of lagoonal hydrology to changes in relative sea level (Fig. 1.13). If rim growth kept pace with a slow rise in relative sea level, the lagoonal evaporite fill also kept pace with sea level, and stacked mud-flat sequences or intercalated

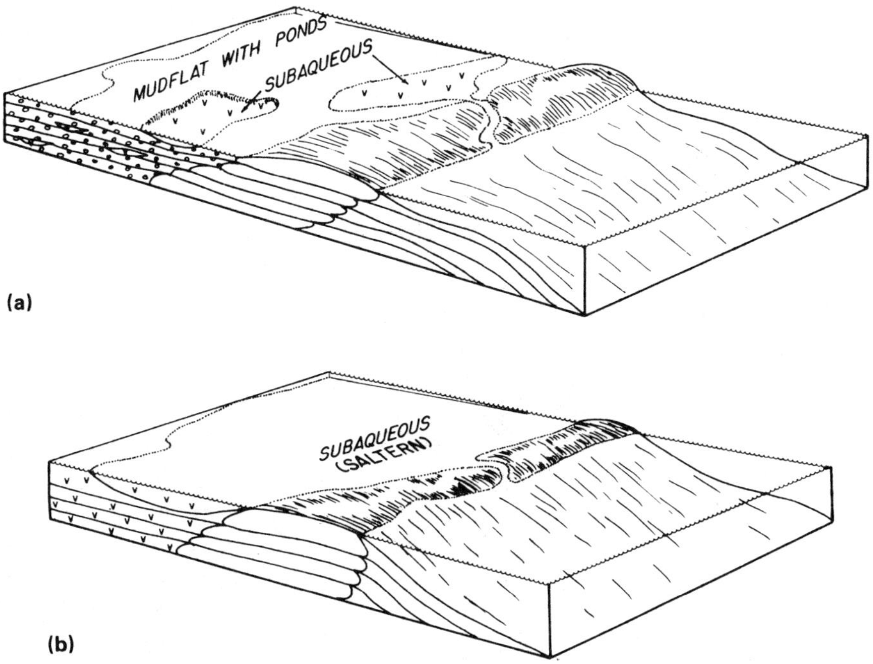

Figure 1.13 (a) Rimmed shelf where, during and after a rise in sea level, the lagoon keeps pace with upward growth of the shelf rim to deposit a sabkha/salina mud-flat complex behind the rim. (b) Rimmed shelf where lagoon catches up to rise in sea level after the rim has caught up and restricted inflow to the lagoon. This deposits a subaqueous (saltern) evaporitic sequence behind the reef flat.

sabkha/brine pan sequences were laid down as a mud-flat mosaic (Fig. 1.13a). The mud-flat complexes would pass basinward into the shoal-water carbonate shelf-rim with no intervening saltern evaporite area. In a prograding setting the open marine boundstones of the rim pass directly upward into thin intercalated sabkha/salina units or stacked sabkha cycles with no intervening saltern sequences.

If the rim did not initially keep pace with a rapid rise in sea level, the shoal-water carbonates deposited on the lagoonal side of the rim were not all that different from those on the seaward side of the rim; especially when the rim was incipiently drowned, and then later caught up to the new higher sea level (Fig. 1.13b). While the rim was growing and catching up to a new, higher sea level the waters in the lagoon could freely exchange with the open ocean and so a diverse open-marine biota flourished within the lagoon. As the rim caught up to the new sea level, the conditions in the lagoon became more restricted as the exchange with open ocean waters was reduced; sediments were now restricted—marine carbonates. With continued restriction across the rim or shoal, the lagoonal waters became sufficiently saline to deposit evaporites and fill the lagoon with a shoaling upward saltern succession perhaps capped by a thin sabkha sequence. The lagoonal evaporite unit was as thick as the lagoon was deep at the inception of the evaporite deposition. Such subaqueous saltern evaporites were continuous across the lagoon and so are easily correlated from well to well. Often, the upper part of the carbonate portion of the cycle is extensively dolomitized by refluxing brine that sank through the platform sediments as the saltern was deposited (Chapter 5). The ten or so subaqueous cycles of the Ferry Lake Anhydrite in the Northern Gulf of Mexico region and the evaporites in the Queen and Seven River Formations of the Guadalupian backreef in the Permian Basin are both examples of stacked shelf units showing this response (Pittman, 1985; Sarg, 1981).

The western rim of the Delaware basin now exposes the Capitan Formation in the Guadalupe Mountains. In Permian time the Capitan was laterally equivalent to the Queen, Seven Rivers, and Tansill Formations. In the 1950s and 60s it was considered by many to be the type example for a rimmed-barrier shelf, with the reef as the topographically highest point on the shelf (Newell et al., 1972). Subsequent work has shown the Permian platform was indeed a shelf with a reef-rim, but the reef was a subtidal mud-mound and not the shallowest part of the shelf (Fig. 1.14; Read, 1985). The shelf during Capitan time is an example of a deep-rimmed shelf, with a mid-shelf shoal. On the back side of the mud-mound reef there were extensive reef-flats, shallowing landward and passing into an island crest facies composed of pisolites and tepees. The islands were backed by a huge expanse of evaporitic mud flats and salterns (Ward et al., 1986; Handford et al., 1984). On the seaward side of the deep-rim was a steep reef-slope passing out into the deep basin. The reef-rim never grew to sea level but was a laterally extensive subtidal mud mound. There is no modern equivalent to this type of shelf. It may well be that the reef biota never grew to sea level nor formed closer to the shoreline because of the debilitating effect of increased salinities closer to

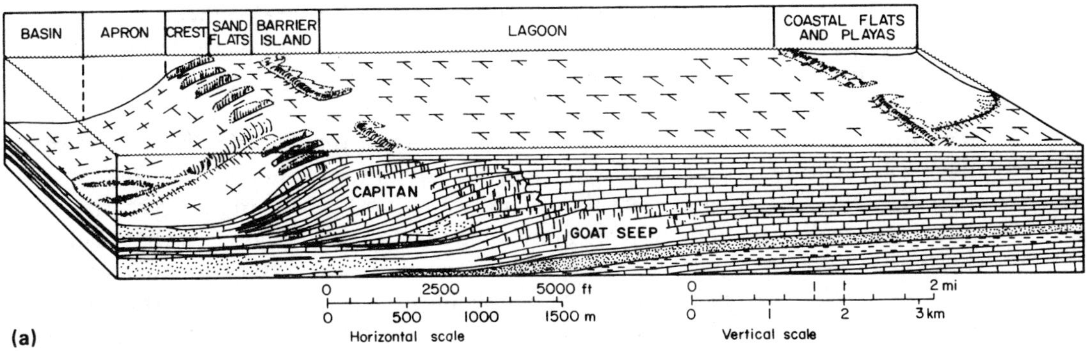

(a)

FACIES BELTS and OUTCROP NOS.	BASIN (1, 2, and 3)	SHELF EDGE: APRON (4)	SHELF EDGE: CREST or REEF (5)	SHELF (BACK REEF): SAND FLATS (6)	SHELF (BACK REEF): BARRIER ISLANDS (7)	LAGOON (8)	COASTAL MUDFLATS AND PLAYAS (9)
SEDIMENTS	Silt and sandstone, Lime mudstone and carbonate debris	Breccias, lime mud, grain and boundstone	Lime mud, grain and boundstone	Lime to dolomitic wacke and grainstone	Dolomitic, pisolitic, intraclastic grain and wackestone	Dolomitic mudstones and wackestones	Gypsum, and /or Anhydrite, Quartz sands and silts. Dolomitic mudstones and wackestones
BEDDING and SEDIMENTARY STRUCTURES	Sandstone channel fills, siltstone drapes, low-energy bedforms and some grading Limestone-thin laminated and graded beds Limestone debris,-chaotic and fining upward	Irregular thick beds of graded, cross bedded, chaotic and dense limestone	Massive and biohermal limestones with fractures	Widespread heterogenous medium beds, cross bedded to dense	Widespread medium beds with fenestrae and tepees	Shoaling cycles of widespread dolomitic mudstone and wackestone beds capped by layered and laminated mudstones and evaporites (mainly subaqueous saltern)	Shoaling cycles of widespread evaporites (mainly sabkha), quartz sands and silts, and laminated vuggy dolomites
POROSITY	Interparticle, in silt and sandstone, 18-20% productive, 17% nonproductive	Intraskeletal, vuggy and fractured Average 4%		Interparticle, fenestral and fractured. Average 8%		Intercrystalline, fenestral, and moldic in dolomite. Interparticle in sands. 9-16% productive. 9% non-productive	None
RESERVOIR POTENTIAL and SEALS	Sandstone reservoirs, silts and limestones	Water prone				Stacked reservoirs in sands and carbonates. Seal, updip evaporite	Evaporite prone, with patchy reservoirs. Evaporite seal
WIDTH OF FACIES BELTS	70 miles (112 km)	1 mile (1.6km	0.5mi (0.8km)	0.5mi (0.8 km)	2 miles (3.2 km)	2-8 miles (3-13km)	25 miles (40km)

(b)

Figure 1.14 (a) Block diagram of Permian Basin (west Texas) rimmed shelf sedimentation during Guadalupian time. (b) Characteristics of various settings (after Ward et al., 1986)

the evaporitic platform. These same saline waters probably favored the growth of pisolites in the island crest facies (Warren, 1983). The lack of extensive evaporite-platforms on the back-reef side of modern carbonate-shelves may be why this type of mid-shelf pisolitic shoal has no modern counterpart.

During Guadalupian time in the Permian Basin of west Texas these shelf evaporites often formed up-dip and overlying seals to extensive shelf carbonate and siliciclastic reservoirs (Fig. 1.14; Ward et al., 1986); in other words evaporites were intercalated with carbonates and siliciclastics throughout the shelf sequence. Evaporites in the Seven Rivers Formation are probably the best understood of the Guadalupian sequences (Sarg, 1981; Spencer and Warren, 1986). The evaporite facies consists predominantly of laterally extensive bedded-mosaic gypsum/anhydrite units (1 to 7 m thick) deposited in a saltern environment, and interbedded with dolomite and quartz siltstone beds. Some of the Seven Rivers evaporites, especially about the edge of evaporite lagoons, were deposited in a matrix-rich mud-flat setting. Together these evaporites form the major up-dip seal and trap for the hydrocarbons of the Permian Basin (Ward et al., 1986). The evaporites were separated from the mid-shelf rim of the Capitan Formation by a porous grainstone facies (potential reservoir) which forms an abrupt facies change both into the "reef" and into the evaporite saltern. Toward the reef rim this grainstone facies thickens to form a low angle roll-over (island-crest) between the Capitan and the lagoonal evaporites. Porosities in the reservoirs are characteristically due to the presence of intercrystalline and moldic pores in bioturbated dolowackestones and dolopackstones (Chapter 5). Moldic porosity formed by the dissolution of fusilinids and bivalve fragments. Porosities in carbonate reservoirs generally range between 7 and 15%, and permeabilities, although sometimes higher, are characteristically less than 10 md. In single reservoir zone, the permeability may vary significantly whereas the porosity remains relatively constant; this is often the result of evaporite plugging. Sandstones are also productive in Guadalupian shelf sediments with average porosity of 18%. Like the carbonates these sands are transitional into evaporite plugged zones, with the latter forming the trap and seal.

On some ancient platforms, such as the Arabian Gulf during the Jurassic, there were widespread depressions on the platform known as *intrashelf basins*. These were areas of local deepening with water depths attaining 10–30 meters. Bottom waters were anoxic with occasional oxygenation, thus they were sites of organic enrichment in the bottom sediments. The Hanifa and Jubaila Formations of Saudi Arabia (Ayers et al., 1982) and the Diyab Formation in the Abu Dhabi area (Loutfi and El-Bishlawy, 1986) are both source rocks deposited in Oxfordian/mid-Kimmeridgian intrashelf basins. They are two of the major suppliers of hydrocarbons to the extensive grainstone reservoirs of the Arab cycles. In both Saudi Arabia and offshore Abu Dhabi the source-reservoir association was sealed by an overlying complex of sabkhas and then salterns on a ramp profile (Murris, 1980). Personal observation of cores from producing intervals in the Arab Formation in some of the giant fields of the eastern Arabian Gulf has shown that sabkha sequences dominate the intercalated evaporites deposited over the crests of the structure, and saltern evaporites are more common in time-equivalent evaporite units located off the structural crest. In some fields this syndepositional shoaling of the structure was generated by Jurassic flow of the underlying basinal Hormuz Formation evaporites (Chapter 6). In other fields the syndepositional shoaling was localized by deep-seated basement faults. Similar, more localized separation of sabkhas on topographic highs and subaqueous evaporites on topographic lows can be seen in the Grayburg/Seven Rivers Formations of Yates Field, west Texas (Spencer and Warren, 1986).

Siliciclastic ramps or rimmed-shelves associated with platform evaporites are relatively rare in the rock record. In most shelf settings the unit deposited atop a shelf sand-sheet was an open-marine carbonate and lies beyond the scope of this paper (see Read, 1985). Siliciclastic-associated platform evaporites usually formed when the sea transgressed a desert sand-sea (erg) and are best interpreted using the distally steepened ramp model of Read (1985). When evaporites or evaporitic carbonates were deposited atop a sand sea it was usually in a rifted basin; e.g., when the basin-wide Zechstein evaporites were deposited atop the Rotliegende sands of the North Sea region (Glennie and Buller, 1983).

BASIN-WIDE EVAPORITES

Basin-wide evaporites are thick and contain textural evidence of a variety of depositional settings ranging from deep-water to platform to continental. They can be found in many ancient sedimentary basins such as the Delaware Basin of west Texas, the Zechstein Basin of Europe, the Louann Salt Basins in the Gulf of Mexico, the Hormuz Salt Basins in the Arabian Gulf and the Late Miocene Messinian Sub-basins in the Mediterranean. Total evaporite fill in a basin was often hundreds to thousands of meters thick and composed of shoaling evaporite cycles tens to hundreds of meters thick. Saltern evaporites were common in the "platform" areas of the basin, or across the whole basin in the latter stages of infill, or in times of basin-wide evaporative drawdown. Such saltern successions were characterized by depositional signatures identical to those found in the evaporitic platform sequences just described; although saltern units in evaporite basins tend to stack into much thicker units (>50 m). They are not interbedded as often with open-marine sediments, and are not restricted to the platform areas of the basin.

Deeper-water evaporites are characteristic of basin-wide evaporite deposits and were commonly deposited during the early stages of each episode of basin-wide infill. Evaporitic basins did not fill with salts in a single depositional episode, but filled as a number of stacked depositional cycles. Deeper-water deposits, near the platform edge in the basin, were often characterized by slope deposits; a wedge of reworked evaporites, slumps and turbidites (Chapter 4). In the basin center, there were laterally continuous, fine-grained, laminated deep-water evaporites, often intercalated between saltern and mud-flat units laid down during times of evaporative drawdown (Chapter 4).

There are no modern equivalents to the basin-wide style of evaporite deposition just as there are no modern deep-water or saltern evaporite deposits. In the subsurface, basin-wide evaporites are easily distinguished from platform evaporites by their purity, thickness, and basin-wide occurrence; features easily studied in wire-log suites and seismic sections.

Ancient basins containing thick sequences of bedded evaporites are of three subtypes (Fig. 1.15; Kendall, 1984):

1. Shallow water–Deep Basin
2. Deep water–Deep Basin
3. Shallow water–Shallow Basin

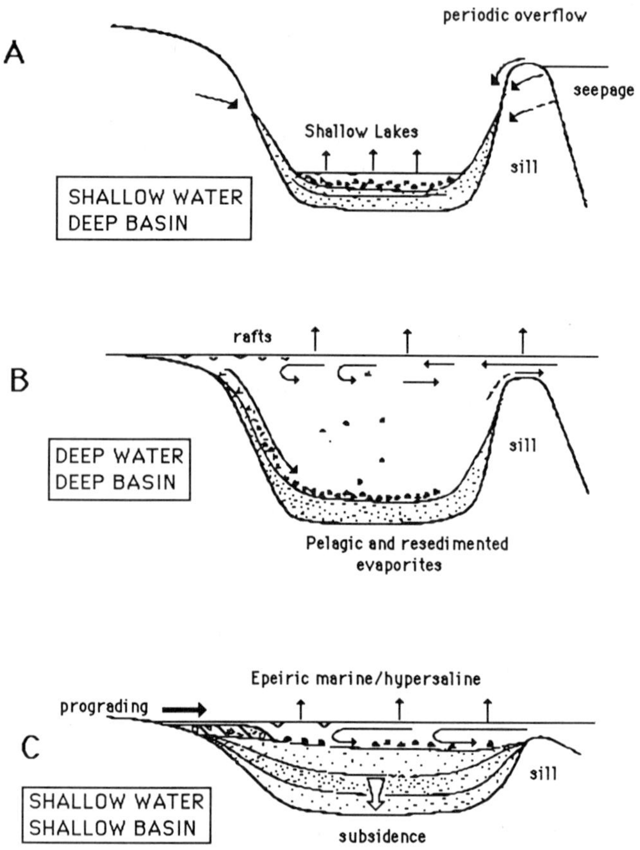

Figure 1.15 Basinal settings: (a) shallow-water–deep basin; (b) deep-water-deep basin; (c) shallow-water–shallow basin (in part after Kendall, 1984).

Each basin is dominated by a characteristic suite of evaporite sequences described in the appropriate chapters. The *shallow water–deep basin* sequence is dominated by shallow water evaporites deposited as a series of shoaling saltern and mud-flat cycles in a basin with a base level hundreds to thousands of meters below sea level. Often this type of basin is supplied by evaporative drawdown, pulling seawater into the basin as a series of marginward seeps and the occasional waterfall (Fig. 1.15a). The evaporite unit was often underlain by deep-water marine sediments deposited before the basin was isolated from a continuous supply of seawater. Because evaporative drawdown maintains the brine level well below the level of the nearby ocean, the evaporite succession is often overlain by deep-water sediments deposited once open-marine waters gain entrance to the basin. The Messinian evaporites of the Mediterranean are probably the best documented examples of this type of basin-wide evaporite fill dominated by shoal water sequences (Chapter 3).

Deep water–deep basin evaporites have basin centers dominated by "deep-water'' evaporites (Fig. 1.15b). The central basin areas are dominated by mm-laminated fine-grained salts while the basin slopes are dominated by reworked platform evaporites deposited as slumps and turbidites (Chapter 4). The basin is often surrounded by a platform covered with saltern evaporites. The basin center usually starts off deep and so the basinal evaporites are underlain by laminated deep-water carbonates and siliciclastics. Seawater flows into the basin as a perennial surface inflow; the evaporite unit will fill the basin to a level equivalent to the level of the supplying ocean. Consequently, the evaporite unit *in toto* is a shallowing upward succession often composed of smaller shallowing-upward cycles. The evaporite fill starts off as deep-water laminated and resedimented salts; they pass up section into shallow-water salterns, mud flats, or continental playas. The Castile Formation and the transition into the overlying Salado Formation in the Delaware Basin in west Texas is a good example of this setting. The basin started off deep with the deposition of the varved sediments of the Castile Formation and with time filled and shallowed into the shoal-water/playa deposits of the Salado Formation (Chapter 4). Any sediments which immediately postdate the evaporites of this overall shallowing sequence will also be shoal water or mud-flat deposits. This is the case with the shoal-water/mud-flat deposits of Rustler Dolomite where it overlies the Salado Formation in the Permian Basin, west Texas. Another good example are the basin-center deposits of the north and south Zechstein basins in Europe (Chapter 4).

The *shallow water–shallow basin* is the third variety of evaporitic basin (Fig. 1.15c). The basin starts off shallow and remains shallow throughout the deposition of the evaporite sequence. Basin subsidence is slow enough to maintain restriction and yet deposit a thick sequence of salts. Sedimentary signatures are dominated by interfingering saltern and mud-flat successions often cross-cut by karst and erosional surfaces. The whole sedimentation package behind the shoal that restricts seawater inflow to the basin is composed of a number of shoaling-upward evaporite cycles. Shallow water–shallow basin evaporites are often found as a rift valley or aulocogen fill where the basal succession in the rift is continental and passes up section into marine evaporites. This is the general sequence beneath Atlantic continental margins (Evans, 1978) and the sequence that is currently forming in the Ethiopian sector of the African Rift. It was also characteristic of most intracratonic evaporite basins such as the Michigan and Williston Basins (Chapters 2 and 3). Unlike the other basin-wide evaporite deposits these evaporites are commonly intercalated with shallow-water open-marine carbonates and siliciclastics. In some tectonic settings, such as an intracratonic basin lying within epeiric seaway, it can be very difficult to distinguish this type of basin-wide evaporite from a platform evaporite. The distinction requires a thorough knowledge of the basin's tectonic history.

In reality, the filling of most ancient evaporitic basins is never as simple or as predictable as the three-end member classification suggests; it is given here as a guide, not as a gospel. Changes in relative sea level, or climate, can play havoc with simple models of evaporite basin type; ancient evaporite basins were never filled with simple sequences. Relative water level in an evaporite basin is never constant; it rises and falls in response to intensity of tectonism, climate, fluctuations in polar ice volumes (sea level), and changes in the degree of restriction in the inflow zones. If we simplify the model by ignoring fluctuations in the volume of siliciclastic sediment input related to water level changes in the basin, then a slight rise in relative sea level can introduce a pulse of marine deep-water carbonates into the deep water–deep basin evaporites. A fall in basin water level, related to increasing restriction in the inflow areas of the basin, perhaps due to a slight fall in relative sea level or tectonic uplift, can convert a deep water–deep basin setting into a shallow water–deep basin setting. The effect of a sea level rise on the now shallow water–deep basin setting is just as unpredictable. It may introduce another pulse of deep-water evaporites. Or, if the rise was sufficient to fill the basin to the spill point with marine-derived water, it would deposit deep-water carbonates in the basin center. Or, if the basin only partially filled with marine water, a shallow-water restricted-marine carbonate sequence could be deposited.

Large evaporite provinces such as the Mediterranean Basin during the Late Miocene, or the Permian Basins of Texas and northern Europe in the Late Permian, are actually composed of a series of subbasins. Each subbasin in an evaporitic province responds in its own fashion, depending on the topography, tectonics, and climate. During the Messinian in the Mediterranean some evaporite basins were filling with predominantly shallow-water evaporites while at the same time subjacent basins were filling with deep-water successions. This complexity of evaporites, changing with time and intrabasin location, does not lend itself to such a simple three-part classification. Nor can we come to modern sediments for counterparts as basin-wide evaporites are not found in the present epoch. The evaporite sedimentologist must look first to the rocks for the basic evidence of his interpretation.

During burial some basinal evaporites, especially thick halite units, were remobilized into extensive salt structures (Chapter 6). Sometimes remobilization was early, as it was during the Jurassic in the Gulf of Mexico. Depositional styles and facies patterns of many Jurassic Smackover Formation reservoirs and the later Cretaceous and Tertiary strata were controlled by the ongoing syndepositional movement of the underlying Jurassic Louann Salt (Seni and Jackson, 1983a, b, and c). In other evaporite basins the timing of salt movement was later as in the Arabian Gulf and the North Sea. In the Arabian Gulf, the Precambrian Hormuz Formation did not move until the Jurassic and Cretaceous when it controlled the depositional style of many of the Mesozoic reservoirs (personal observation). In the North Sea most of the movement of the Permian Zechstein salt was Mesozoic or Tertiary. Thus the recognition of salt structures, either in well correlation or in seismic lines, corresponds to a basin-wide interpretation of the evaporite unit. However, not all basin-wide evaporite units have migrated into salt structures with burial (Chapter 6).

SUMMARY

The scale of evaporite deposition has varied across geological time. In the Phanerozoic, times of widespread evaporite deposition occurred during the Upper Cambrian, the Permian, the Jurassic, and the Miocene. At such times evaporites covered areas on the earth's surface more than two orders of magnitude larger than areas of evaporite deposition of the present day. Yet the processes depositing all those evaporites can be observed today in the semiarid and desert areas of the world, wherever a potential exists for more water to leave a sedimentary basin than enter it. Former times of widespread evaporite deposition were due to warmer worldwide climates, epeiric seas, and tectonically induced basin restriction, all conditions not active on a large scale anywhere on the present earth surface. Intense study in evaporitic settings during the last two decades has moved in emphasis from the geochemical to the sedimentological.

Ancient evaporites can form in a variety of depositional settings ranging from deep water to subaerial to lacustrine. Each setting possesses its own vertical signature and association of lithofacies. Evaporites are susceptible to diagenesis from the time they are first laid down, and this often destroys much of the early depositional textures. However, with a knowledge of the variety of ancient evaporite settings, as well as the controls on depositional and diagenetic fabrics, it is possible to look through this overprint and interpret the original depositional setting.

CHAPTER 2

Sabkhas Marine and Continental

INTRODUCTION

Sabkha is an Arabic word for salt flat. It is an equilibrium geomorphic surface whose level is dictated by the local water table. Sabkhas (*sabkhat* is the correct plural) are occasionally covered by ephemeral shallow water, but for most of the time they are subaerial mud flats. They were documented first in the Arabian (Persian) Gulf by Curtis et al. (1963) but were subsequently described along the coast of Baja California, Mexico (Phleger, 1969; Shearman, 1970), the coast of Sinai (Gavish, 1974; Levy, 1977), and many other areas of the world.

If the word *sabkha* is used objectively to imply "salt flat," there are both coastal and continental sabkhas. Some coastal sabkhas, such as those along the Trucial Coast, pass laterally into continental sabkhas without any noticeable change in surface morphology (Kinsman, 1969). The marine portion of the sabkha is characterized by marine sediments with largely marine-derived groundwaters and the continental portion by nonmarine sediments and groundwaters. Kendall (1979) used *continental playa* as a term equivalent to *continental sabkha;* other workers use *sabkha* in a narrower sense to describe marine sabkhas exemplified by the Arabian Gulf sabkhas. I define sabkhas as marine and continental mudflats where displacive and replacive evaporite minerals are forming in the capillary zone above a saline water table. The advantage of this definition is its utility in the subsurface, where it can be applied using the vertical sequence in a single core.

The surrounding depositional sequence does not need to be interpreted to use the term *sabkha.*

As more modern evaporitic salt flats are described, it becomes difficult to retain narrower usage of the word *sabkha,* exemplified by the Arabian Gulf coastal sabkha. Just as major siliciclastic depositional systems exhibit a continuous spectrum of morphologic types (river versus tide versus wave-dominated deltas), so there are significant variations among sabkhas (Handford, 1981a). In sabkhas the physical processes thought to be most important (besides evaporation) include those operative under the following conditions

1. Marine coastal
2. Fluviolacustrine
3. Eolian dominated

Dominance of one of these processes gives rise to marine coastal sabkhas, or continental sabkhas, or interdune sabkhas, respectively (Fig. 2.1). An eolian

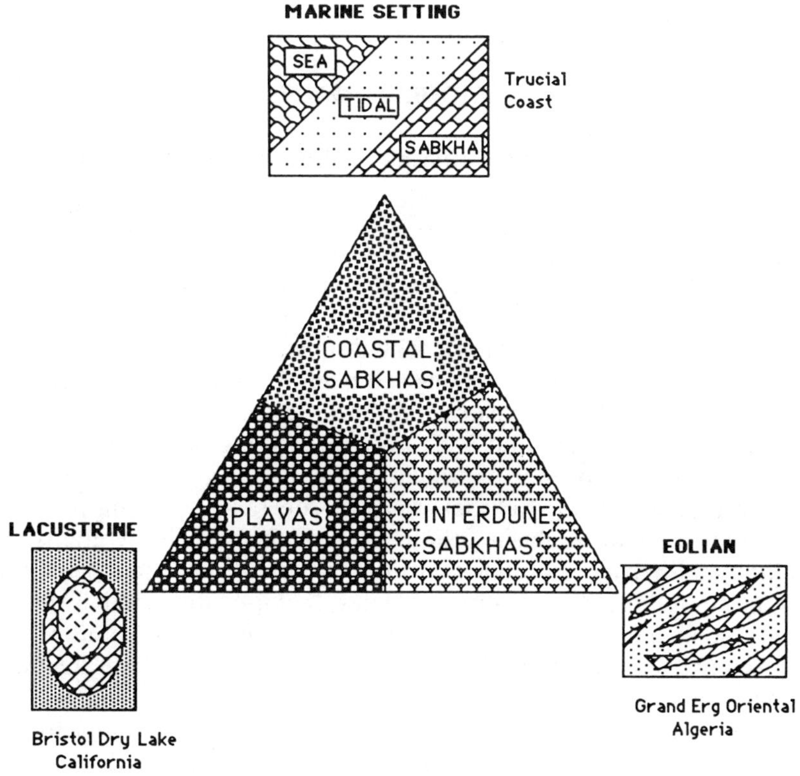

Figure 2.1 Tripartite classification of sabkhas showing that the sabkha depositional setting covers a series of depositional settings ranging from the coastal to the continental (after Handford, 1981a).

sabkha can be a subfacies of both marine coastal and continental sabkhas. Sedimentary responses to dominant physical processes create sabkhas composed of combinations of

1. Terrigenous clastics
2. Carbonate–sulfate (anhydrite–gypsum minerals)
3. Soluble salts (halite, sylvite, polyhalite, etc.)

Any sabkha classification must consider the sabkha system as an environmental and lithological continuum. The only single criterion to describe ancient sabkhas is their variability. Use of the Arabian Gulf sabkha to generate an all encompassing model in ancient sequences should be avoided. Many ancient sabkhas evolved from marine coastal to eolian to fluviolacustrine-dominated systems during a single progradational episode in the coastal plain (Warren and Kendall, 1985). The remainder of this chapter deals with documented examples of this continuum, first in the Holocene and then in economically important ancient environments.

MODERN MARINE SABKHAS

Marine Sabkhas of the Arabian (Persian) Gulf

Modern marine sabkhas are forming along the coasts of many stable cratons in arid to semiarid settings. Coastal sabkhas stand along the western and southern coasts of the Arabian Gulf, the coasts of Australia, Northern Africa, and Israel (Table 2.1). The best studied marine sabkhas in the world today occur in the Arabian Gulf (Persian Gulf). The Arabian Gulf is a modern peripheral foreland basin which has been receiving sediment from the Tertiary to the present. It is an asymmetric collision-basin with a well-developed ramp on the cratonic southern side of the Arabian Gulf and a steeply dipping platform passing into the Zagros Mountains on the north. Much of the southern area of the modern basin is a classic example of a carbonate-evaporite ramp. Starting in the micrites of the deeper parts of the basin, one passes south through the grainstones of the mainland and barrier shorelines into the famous coastal sabkha of the southern Gulf (the shorelines of the Emirates, Qatar, Saudi Arabia and Kuwait). Inflow of siliciclastic sediment into the Arabian Gulf is primarily longitudinal entering the Gulf from the Shatt Al Arab delta and the northeastern side of the Gulf. The Shatt Al Arab is the confluence of the Tigress, the Euphrates and the Karoon rivers. It forms a huge wedge of fluvio-deltaic sediments at the northwestern end of the basin. The northeastern side of the Gulf is supplied with small amounts of siliciclastics from alluvial fans sourced in the ephemeral streams of the Zagros Mountains.

The Arabian Gulf sabkha is approximately 320 km (200 miles) long and up to

TABLE 2.1 DOCUMENTED MODERN SEA-MARGIN EVAPORITES

Sabkha	Reference
Abu Dhabi, Arabian Gulf	Butler, 1970; Purser, 1973b
Sabkha Matti, Arabian Gulf	Purser, 1973b
Bardawil Lagoon, North Sinai coast	Levy, 1977
Western Nile Delta	West et al., 1979
Gulf of Suez coast	Gavish, 1980
Tunisian coast	Perthuisot et al., 1972
Northwest Australian coast	Warren and Kendall, 1985
Spencer Gulf, Australia	Ferguson et al., 1982
Baja California, Mexico	Castens-Siedell, 1984
Laguna Madre, Texas	Elliott, 1986
Salina	**Reference**
South Coast, Australia	Warren, 1982a,b
Hutt & Leeman Lagoons, Western Australia	Arakel, 1980
Lake MacLeod, Western Australia	Logan, 1981, 1982
Solar Lake, Gulf of Elat	Aharon et al., 1977
Ras Muhammad, South Sinai coast	Kushnir, 1981
Tunisian coast	Perthuisot et al., 1972
Pleistocene coast, Sicily	Schreiber and Kinsman, 1975
Pekelmeer, Netherland Antilles	Lucia, 1968

24 km (15 miles) wide (Fig. 2.2a). It extends along the coast of the United Arab Emirates between the Qatar Peninsula and Ras Ghanada; on average it has prograded 1.5 km (1 mile) every 1000 years. The evaporitic mud flat (sabkha) occurs landward of coralgal reefs, sponge banks, oolite shoals, skeletal grainstone banks, and shallow lagoons, all deposited atop drowned Pleistocene dune sheet-sands. The sabkhas southeast of Abu Dhabi City are especially well studied (Fig. 2.2a) and are the best modern counterparts for ancient arid-zone mud-flat sequences deposited along shallow-water stable platforms with a relatively low eolian input.

The Arabian Gulf sabkha is unique in the large volumes of nodular anhydrite and Holocene dolomite currently forming in the supratidal zone. This uniqueness first attracted the attention of the geological community during the 1960s. The initial works by Curtis et al. (1963), Illing et al. (1965), Evans et al. (1969), Kinsman (1963, 1965, 1969), Shearman (1963), Butler (1966, 1969, 1970), and Kendall and Skipwith (1969) provided the observational data on which all our present sabkha models are based. Subsequent work in the Arabian Gulf sabkha usually considered only a specific problem within the sabkha setting, such as the hydrological setting (Patterson, 1972; Hsu and Schneider, 1973; McKenzie et al. 1980; Patterson and Kinsman, 1981), or dolomitization (McKenzie, 1981; Patterson and Kinsman, 1982), or it has reinforced the model of these earlier workers (Bush, 1973; Butler et al. 1982).

To understand sabkha deposition requires an understanding of the climatic and hydrological factors that create the conditions suitable for sabkha progradation. The Arabian Gulf is probably the warmest sea in the world; in the open Gulf the surface temperatures range from 20°C in February to 34°C in August. On the open platform in the vicinity of Abu Dhabi Island the water temperatures range from 23 to 34°C and in the inner lagoon areas from 15 to 40°C, with an annual average temperature of 29°C. The shallow inner water masses can show daily changes of 10°C. Air temperatures on the sabkha mud flat can be as low as 5°C in the winter and as high as 50°C in summer, with average temperatures across the sabkha ranging from 23 to 33°C. Sediment surface temperatures on the sabkha are even more variable, with reported values up to 60 to 80°C on the surface of the algal mats (Kinsman, 1965). Shallow subsurface temperatures range from 22 to 40°C. In summer, the sabkha is extremely humid, especially at night, when the humidity often reaches 100%. Salinities in the open Gulf are 3 to 4 per mille above normal open marine salinities although in the vicinity of Abu Dhabi, where the coastal islands restrict exchange with the open Gulf, the lagoon salinities range between 54 and 67 per mille. To date, no freestanding gypsum-precipitating pools have been found in the inner lagoon, although outcropping gypsum mush formed as a capillary zone precipitate is common in the upper intertidal zone (pers. obs.).

Evaporation on the sabkha is aided by prevailing onshore winds. In winter the intensity of the onshore winds increases as the Shamal wind blows seawater and sediment from the lagoon onto the sabkha. Thus the Shamal provides lagoonal carbonates to the sabkha as onshore winds drive sediment-laden lagoonal water onto the sabkha surface. At other times offshore winds blow continental dune sand into the back side of the sabkha (see Fig. 6.11 in Chapter 6). Thus the sabkha is furnished with sediment from two source areas, carbonates from the lagoon and siliciclastics from the hinterland. Many studies and models have tended to ignore the siliciclastic component although areas exist on the modern Abu Dhabi sabkha (Al Zibaya sabkha) where the siliciclastic component is the major matrix material. Rain is infrequent but torrential, especially near the Oman Mountains; flooded wadis can carry continental sediments out onto the sabkha. As in any desert the annual rainfall is extremely variable, in the 6-year period from 1958 to 1964 the highest annual reading on the sabkha was 6.73 cm, the lowest 0.33 cm. Rain floodwaters can sometimes cover portions of the sabkha for 2 to 3 months and will dissolve any surface and near-surface halite crusts.

Physiographically, the sabkha is characterized by its billiard table flatness (Fig. 2.2b). One can stand on the sabkha and see nothing but the perfectly smooth spread of the sabkha surface out to the horizon in all directions (Fig. 2.3a). The sabkha extends some 15 to 20 km from the extremely shallow waters of the lagoon, across the algal mats and mangroves of the intertidal zone into low beach ridges and up onto the extensive flat surface of the sabkha. The present sabkha surface reflects the last 7000 years of depositional history (Evans et al., 1969; Kinsman, 1969). Deposition began some 7000 years ago when a rapid transgression infilled interdunal Pleistocene lagoons and depressions with marine-reworked, predominantly eolian sediments. About 4000 years ago the Arabian Gulf

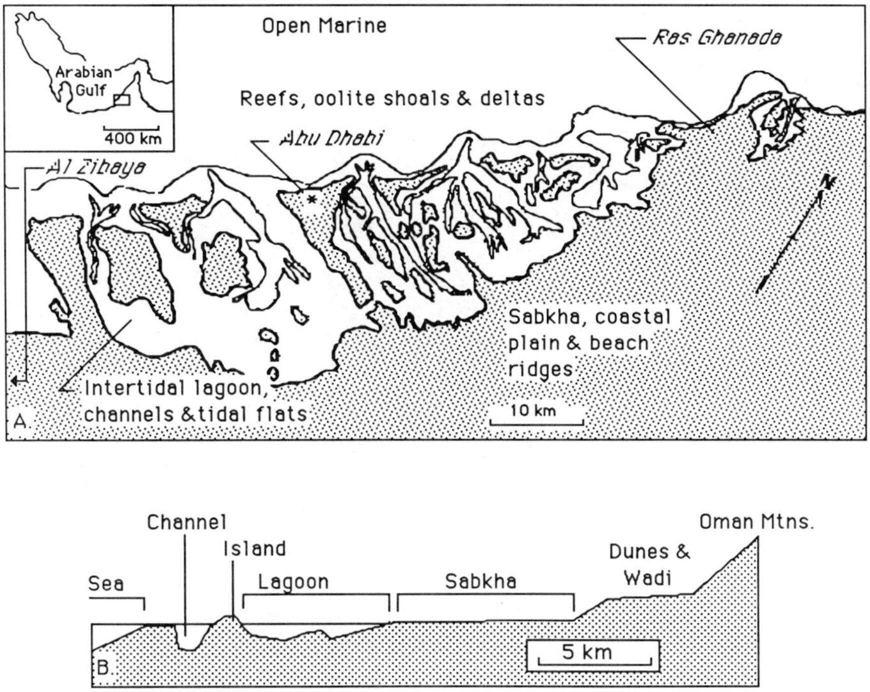

Figure 2.2 (a) General sediment distribution in the Abu Dhabi sabkha of the Arabian (Persian) Gulf (after Warren and Kendall, 1985); (b) generalized cross section of the sabkha physiography showing the flat nature of the sabkha surface.

shoreline reached its maximum Holocene strandline. This was followed by progradation of the sabkha sequence to its present configuration. Wind deflation and local groundwater levels account for the flatness of the sabkha and have played an important role in creating the now well-known sabkha depositional signature.

Geological Setting and the Sabkha Model

In the 20 years since its first documentation the sabkha model has come to dominate the thinking of many geologists working with ancient sequences containing nodular anhydrite. As with any model based on a modern depositional system, modifications and simplifications must be made for the model to show utility in interpreting ancient sequences. The sabkha depositional system is divided into three areas: the subtidal, the intertidal, and the supratidal (Fig. 2.4). The sediment successions deposited in these areas are assigned to subtidal, intertidal, and supratidal facies. In a prograding sequence the three facies are stacked one atop the other to form the well-known shallowing-upward sabkha cycle termed the "Trilogy" by Warren and Kendall (1985). Numerous subfacies can be identified within this tripartite division.

(a)

(b)

(c)

(d)

(e)

(f)

Figure 2.3 Arabian Gulf sabkha, Abu Dhabi. (a) sabkha surface covered in an ephemeral halite crust; (b) lagoonal sediments of Khor Al Bazam showing peloidal grainstones and packstones with growing clumps of brown algae covered with calcareous epibionts (photo courtesy of Chris Kendall); (c) algal mat with smooth laminated mat in the tidal channels and pustular mat in the intervening areas, southern Khor Al Bazam; (d) cuspate, laterally accreting beach ridge on seaward side of Al Zibaya sabkha, Abu Dhabi; (e) core with displacive gypsum, sabkha southeast of Abu Dhabi City; (f) nodular and enterolithic anhydrite, Al Zibaya sabkha.

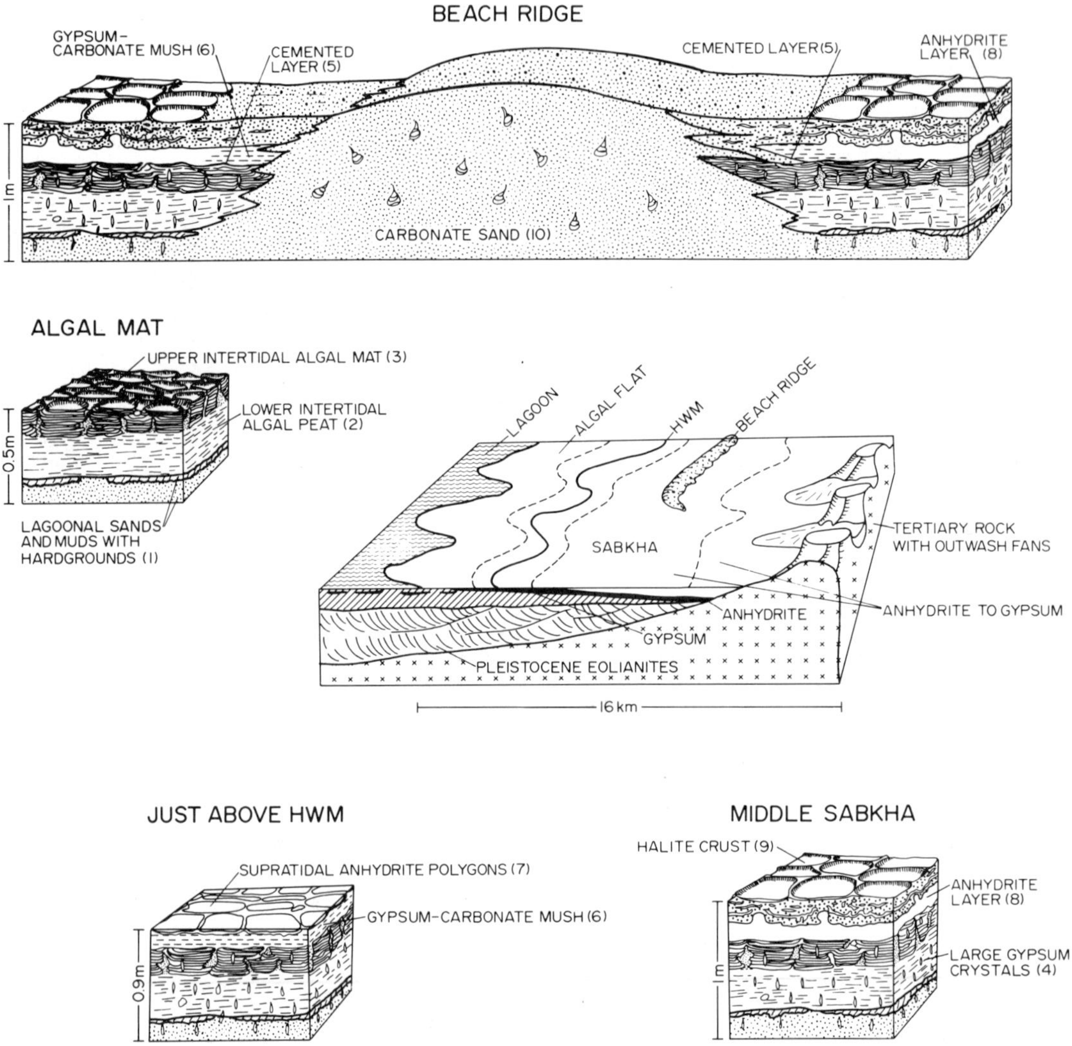

Figure 2.4 Vertical and horizontal distribution of the sabkha sediments based on the Arabian Gulf Sabkha (after Warren and Kendall, 1985). All the peripheral diagrams are keyed to the central block diagram. HWM, high water mark.

Subtidal Facies

Sedimentation along the Abu Dhabi coastline can be divided into two parts:

1. *Open marine sedimentation* occurs on the seaward side of the north-east trending barrier island system (Fig. 2.2). The islands are composed of Holocene and cemented Pleistocene oolite shoals separated by channels several kilometers wide and 7 to 10 meters deep (Purser and Evans, 1973; Loreau and Purser, 1973).

The seaward terminations of the channels are flanked by shoal areas of tidal ebb deltas, composed of oolitic sand. These Holocene deltas form lens-shaped sand bodies 2 to 10 km wide and 1 to 5 m thick. They form in wave- and tide-agitated areas where the ooids are stacked up by opposing onshore waves and offshore ebb-tidal currents. In the shoals the ooid grain size and cortex thickness decreases away from the channel levees. At a water depth of about 2 m an abrupt change occurs into a mixed ooid-bioclastic grainstone. In areas between the ooid shoals, patch reefs are growing immediately seaward of the Pleistocene barrier islands. When compared to the reefs of the open Indian Ocean, the coral fauna is sparse and restricted, dominated by genera of *Porites* and *Acropora* with *Platygyra, Cyphastrea,* and *Stylophora*. The depauperate reef biota is a response to the warmer and more saline waters of the Gulf. Many of the corals are dead or appear unhealthy and are coated by a flourishing coat of calcareous algae (*Lithothamnion, Lithophyllum,* and *Goniolithon*.). Beyond the Pleistocene islands, a broad extensive platform ramp is accumulating skeletal sands and passes offshore into progressively deeper and finer-grained carbonate silts and muds of the Gulf center. Sediment and porosity trends in carbonate shoals of the open Gulf are discussed in Chapter 6.

2. *Restricted marine sedimentation* occurs along the landward side of the barrier islands; this region includes the lagoon and salt flat sediments of the sabkha (Fig 2.3b). Landward of the barrier islands are extensive shallow-water lagoons filled with bioturbated packstones and wackestones. These lagoon sediments contain a diverse biota of many benthic species but relatively few phyla (Kendall and Skipwith, 1969). Molluskan sands and peneroplid foraminifers occur in the more open areas of the lagoon, whereas pelleted carbonate mudstones accumulate in the more protected reaches. Some of the pellets are hardened, a special adaption of the cerithid gastropods to prevent fecal contamination of their feeding area (B. C. Schreiber, pers. comm.). Hardening means that pellets are less liable to be destroyed by waves and currents and may not be obliterated during burial compaction. Major sediment components, such as mollusks, benthic forams, calcareous algae, and pellets are present in similar proportions to tidal flat sediments of the Bahamas and Florida, but in the lagoons of the Arabian Gulf bryozoans are more common, detrital eolian grains are common, as are gypsum and anhydrite crystals in the more saline areas. Tepee structures are common on the floor of the lagoon and can sometimes be seen upslope in the supratidal zone surrounding the lagoon (Kendall and Warren, 1987).

Intertidal Facies (Lagoonal)

The intertidal zone is divided into an upper and lower intertidal facies, although the lower intertidal facies of some workers is equivalent to the lagoonal facies of others (Fig. 2.4). What distinguishes the lower intertidal facies is the occurrence of an algal peat composed of the bioturbated remnants of an algal mat. The upper intertidal facies is a laminated algal mat forming across an extensive belt of algal flats up to 1 to 2 km wide (Fig. 2.3c). The mats accumulate in areas too hostile for the cerithid gastropods and grazing animals of the lagoon. Mats below the surface of the algal flat often consist of mm-laminated mudstones up to 50 cm thick composed of cyanobacteria interlayered with aragonite muds. Most of the literature on the sabkha gives the impression that the laminated algal mat is the only type of mat in the upper intertidal zone, yet in almost every area of mat visited by this author the laminated mat tended to occur in shallow channels (a few centimeters deep) separated by interchannel areas of pustular mat. The sedimentary signature of the pustular mat is a poorly layered organic-rich mush. Thus in an ancient equivalent, the intertidal zone need not be laminated to be algal in origin. This is important in constructing the type of vertical sequence that one should expect to see in an ancient sabkha.

Often the intertidal sediment is cross-cut by dessication cracks and contains lenticular [diagenetic(?)] gypsum crystals up to 2 mm long. Gypsum is especially common in areas of pustular mat. Aragonite, magnesite, and dolomite locally cement the surface sediments of the algal flats (Butler et al., 1982). When these algally bound intertidal muds are covered by prograding supratidal sediments, they are often intensely dolomitized in the shallow subsurface (McKenzie, 1981). Arabian Gulf sabkhas are one of two or three areas on the world's surface where Holocene dolomites are forming. They are the type area for the sabkha dolomites discussed in Chapter 5.

Low beach ridges composed of skeletal grainstones occur in many parts of the sabkha; actively accreting ridges often separate the intertidal from the supratidal (Fig. 2.3d). Major components are whole cerithid gastropods with lesser amounts of bivalve fragments and forams. Buried ridges have been trenched on the mid sabkha (Fig. 2.4; Warren and Kendall, 1985). They have not been observed previously on the Abu Dhabi sabkha as the tops of these older ridges have been deflated and the remainder of the ridge buried beneath a halite crust. Better-documented Holocene equivalents of the ridges within sabkhas are described from Kuwait by Picha (1978) and from northeast Qatar by Shinn (1973).

Supratidal Facies

Earlier work in the sabkha has subdivided the supratidal into three facies: lower, middle and upper supratidal (Fig. 2.4).

Lower Supratidal. The lower supratidal forms a belt just above the high water mark (hwm) up to 2.5 km wide and is characterized by a gypsum mush up to 30 cm thick. This zone is flooded once or twice a month during high spring tides. The gypsum mush consists of two parts, an upper unit and a lower unit. The lower part is composed of aragonite mud containing algal filaments and isolated gypsum crystals. The upper part is a thixotropic mush of small, mostly lenticular gypsum crystals up to 2 mm in diameter with local solution and reprecipitation of the gypsum as 1- to 1.5-cm-long lenticular crystals. Much of the mush grows as displacive gypsum crystals within the capillary zone, creating a penecontemporaneous bedded unit. Surficial gypsum mush is often altered to bassanite ($CaSO_4 \cdot \frac{1}{2}H_2O$) or anhydrite during hot, dry periods and then back to gypsum during cooler, humid periods.

Middle Supratidal. The middle supratidal is a belt of sediment up to 1.6 km wide characterized by sediments containing diagenetic nodular anhydrite. It is capped by a surface crust of ephemeral halite with thin patchy units of dry siliciclastic and carbonate sands blowing across it. The middle supratidal is flooded by lagoonal water on less than monthly intervals. On the lagoonal side of this zone, pockets of gypsum crystals dispersed through the near-surface eolian sands are diagenetically altered to anhydrite nodules and layers. In the subsurface the partial replacement of the gypsum mush by anhydrite has begun. Subsurface and surface anhydrite nodules increase in size and number in a landward direction until in some areas they eventually form interlocking polygons. Still farther landward the surface anhydrite is overlain by reworked quartzose eolianite, and the anhydrite polygons appear festooned and truncated in cross section. The truncation of the anhydrite polygons is a result of the continued subsurface precipitation of anhydrite lifting the overlying sediments above the capillary zone into the vadose zone, where they dry and deflate. This deflation surface is later covered by eolian quartzarenites. In the intertidal sediments below the gypsum/anhydrite mush and below the sabkha water table a widespread precipitation of displacive and pore-filling gypsum is taking place. The gypsum growing below the water table partially destroys the now buried intertidal algal laminae (Fig. 2.3e) The middle supratidal is a zone where aragonite muds are extensively dolomitized in the shallow subsurface (Butler et al., 1982; McKenzie et al., 1980).

Upper Supratidal. In the vicinity of Abu Dhabi the upper supratidal is up to 4.8 km wide and is flooded once every 4 to 5 years. According to the literature, the gypsum mush has been completely replaced by coalesced nodules of "chicken-wire" anhydrite in units up to 30 cm thick (Fig. 2.3f). The ongoing growth of anhydrite layers means that these units show thrust folds, ptygmatic and enterolithic folds, and diapir-like structures. In this zone the original algal lamination of the now buried intertidal unit has been partially or completely obliterated by the growth of anhydrite nodules and large gypsum crystals (6 to 25 cm in diameter). Anhydrite infills the internal molds of leached gastropods. In contrast to sediment in a more seaward position, the sediments deposited in the supratidal zone contain little or no dolomite.

The growth of displacive sulfates in the capillary zone above the water table raises the sabkha surface until the near-surface sediments leave the capillary zone and enter the vadose zone (Fig. 2.5). There they dry out and are blown away, leaving behind the characteristic erosion surface with thin deflation units capping the sabkha sequence. On the most landward side of the upper supratidal zone the anhydrite is being reconverted to gypsum by the slow seepage of continental

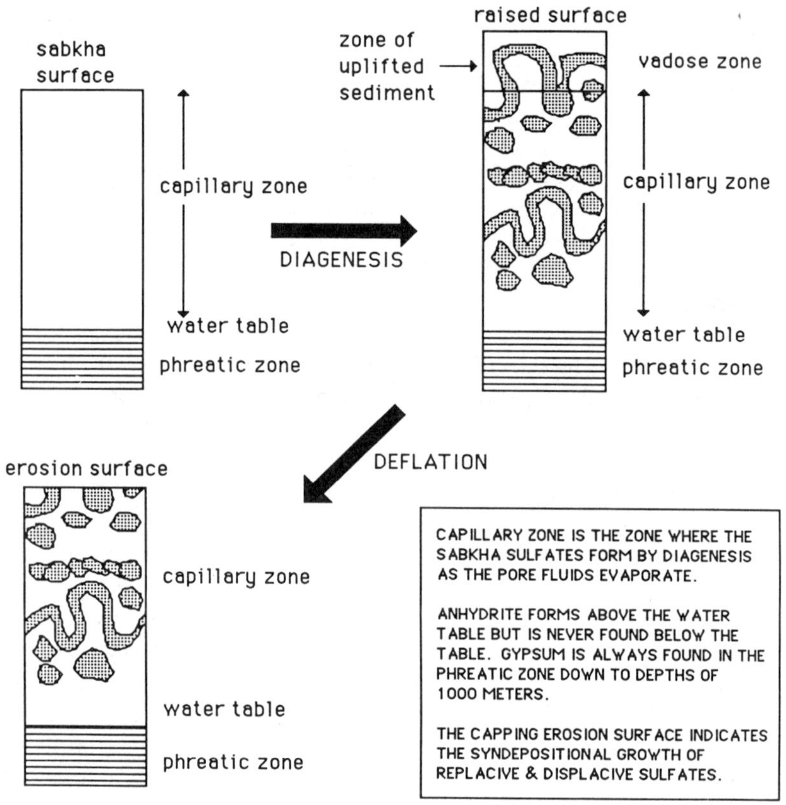

Figure 2.5 Sabkha hydrology showing the effect of displacive evaporite growth (stippled) within the sediment column. The increased volume of evaporites in the column lifts the near-surface sediments into the vadose zone, where they dry and are deflated to create a capping erosion surface.

groundwater into the sabkha. Displacive halite crystals can be found in the upper few centimeters of sediment in the upper supratidal, attesting to the infrequency of flooding in this region. These sediments are also covered by an ephemeral crust of halite cross-cut by pressure ridges.

Other Modern Sea-Margin Sabkhas

Gulf of Suez

Probably the next best understood area of sabkha deposition after the Arabian Gulf comes from the westem margin of the Gulf of Suez in the coastal strip extending from Abu Rudeiz to Ras Muhammad (Gavish, 1974, 1980). This area, named El-Qa, is a flat alluvial plain composed of alluvial silicate minerals derived from the Precambrian and granitic massif to the east. It contains some of the hottest and driest conditions in the Sinai. Carbonate components are less common in these sabkhas compared to the Arabian Gulf sabkha. Much of the matrix is siliciclastic and the evaporites are scattered as interstitial minerals; evaporites also are not as extensive as in the Arabian Gulf sabkha (Gavish, 1974). These sabkhas have not undergone as much diagenesis as sabkhas of the Arabian Gulf.

The sabkha environment extends from the shoreline to the base of the alluvial fans lining the faulted escarpments to the east. Subtidal sediments are mixtures of clastic fine-sands, derived from the alluvial input, and carbonate sands composed of skeletal fragments, pellets, ooids, and some intraclasts. Intertidal sediments vary from abundant fine-marine carbonates on the seaward side to coarse-grained clastics on the landward side. In many places there is an elevated sandy berm or a strip of pebbly beach rock which clearly defines the upper limit of the intertidal zone. The supratidal zone generally extends a few hundred meters inland, the distance being dependent on the the regional slope. The most landward extent of the sabkha is generally defined as the maximum extent of the diagenetic evaporite minerals. In places where the beach-berm is non-existent, stormwater occasionally covers large portions of the supratidal area. In such places a thin upper layer of halite may form over the top of the sabkha but generally disappears after a few weeks. The evaporite mineral distribution in a typical Gulf of Suez sabkha is very similar to the mineral suite in an Arabian Gulf sabkha. Like the Arabian Gulf, the Gulf of Suez sabkha supratidal is precipitating most of the evaporites within the supratidal zone. This forms a matrix-rich evaporite unit no more than 60–80 cm thick. Gavish (1974, 1980) attributes most of the ion supply to the sabkha to be via subsurface seepage of seawater after the capillary evaporation model of Hsu and Schneider (1973).

One of the most interesting differences between the Gulf of Suez sabkhas and the Arabian Gulf sabkhas is the distribution of the two sulfate phases (Gavish, 1974). In the Arabian Gulf sabkha anhydrite is the major sulfate phase. In the Gulf of Suez, gypsum is the dominant mineral in the recent sabkha and anhydrite is found in the more elevated older portions of the sabkha where it recrystallized from gypsum; thus in the Gulf of Suez sabkha, the anhydrite is a replacement mineral.

Mediterranean Coast, Egypt

A modern sabkha is forming along the Mediterranean coast near El Hammam, about halfway between Alexandria and El Alamein (West et al., 1979). There the coast consists of a series of ridges of carbonate beach and dune sands with all the ridges cemented into a hard limestone except the Coastal Ridge. Between the ridges are elongate interdunal depressions floored by sabkhas. The first sabkha depression immediately behind the coastal ridge of oolitic sand is more than 8 km long and a few hundred meters wide. Displacive nodular gypsum is growing in the carbonate-rich clastic sabkha of this most seaward depression. Landward, and to the south, beyond the Abu Sir limestone (Pleistocene), is another larger interdunal depression with halite-encrusted, algal-mat fringed lakes and sabkhas containing lenticular gypsum crystals and some small gypsum nodules. The matrix sediments in the sabkha are mostly brown, laminated, very fine quartz and carbonate sand, of both windblown and floodwater origin, with various quantities of gypsum. The displacive gypsum nodules are roughly spherical or elliptical, mostly between 1 and 4 cm in length. They are displacive as they push aside sediment laminae. Sometimes they form upward-bulging almost diapiric structures; incipient enterolithic structures. The gypsum nodules appear to only be forming in the capillary zone and form nodular evaporite units up to 60 cm thick.

The Egyptian sabkha is interesting for two reasons: first, it is an example of a matrix-dominated siliciclastic sabkha but in a coastal system locally dominated by carbonates; second, the nodular structure found in the sediments is a result of the growth of displacive gypsum not anhydrite.

Northern Sinai and Eastern Nile Delta, Mediterranean Coast

The northern Sinai coast is bordered for about two-thirds of its length by the Bardawil Lagoon, an evaporitic sabkha area (Levy, 1977, 1980). Sabkhas are of two types; *coastal sabkhas* (also called lagoon beach sabkhas) at present connected to the lagoon and flooded occasionally by seawater, and *inland sabkhas* cut off from the lagoon by sand dunes and forming under the influence of groundwater discharge with occasional flooding by rainwater. The largest documented coastal beach sabkha Sabkhat Nathal Yam is about 7 km long and extends about 1 km inland. To the south of the lagoon beach sabkhas, sand dunes extend inland to a distance of about 4 kilometers. The inland sabkhas form as small interdunal fills up to 1 km across along the coastal side of the dune field. Both types of sabkha are typified by siliciclastic matrices with displacive evaporitic salts and carbonate crystals growing in the capillary zone. This forms supratidal evaporitic units up to 60 cm thick. In the coastal sabkhas the sulfate mineral is always gypsum, no anhydrite was found in the sabkha. Magnesian-calcite, dolomite and halite were identified as minor components and halite can form crusts in the most restricted parts of the sabkha. The inland sabkhas were once connected to the lagoon but are now cut off from any surface connection, and most of the present water supply is via groundwater discharge into the sabkha depressions after rainy periods. The age of the inland sabkha sediment is not well known; some of the inland sabkha sediment may be late Pleistocene. There is a soft white sediment unit (20–30 cm thick) of calcian-rich dolomite in some of the inland sabkhas, beneath this is a sediment composed of well-preserved aragonitic gastropods embedded in a muddy calcitic groundmass. Radiocarbon dating shows the dolomite is older than 35,000 years and so may not be marine in origin.

THE MARINE SABKHA MODEL

An Arabian Gulf sabkha model explains many ancient prograding sequences of subtidal, intertidal, and supratidal evaporitic mud-flat sediments. The subtidal contains a restricted marine biota; often the intertidal is laminated by cyanobacteria, while the supratidal unit possesses a matrix of clastics (carbonates and siliciclastics) containing penecontemporaneous gypsum and anhydrite with a capping erosion surface (Fig. 2.4). However, many gaps exist in our understanding of modern sabkhas and at times the sabkha analogy has been applied to ancient successions without a full understanding of the implications of the Arabian Gulf sabkha hydrology and depositional setting (Warren and Kendall, 1985).

For example, many thick, ancient sequences of pure nodular anhydrite have been explained as stacked sabkhas deposited under conditions of slowly rising sea level. There are hydrological limitations associated with this hypothesis that complicate the issue. Sabkha sequences are almost always matrix-dominated; rarely, if ever, can the supratidal sabkha be evaporite-dominated. Sabkha evaporites grow displacively as nodules and crystals in the capillary zone of a mud flat composed of detrital siliciclastic and biogenic sediment. The thickness of nodular anhydrite in the supratidal unit is limited by the thickness of the capillary zone. Any evaporite sediments above the top of the capillary zone are blown away or dissolved (Fig. 2.5). This means that the supratidal evaporite phase in a single sabkha cycle can be no more than 1 m thick (Warren and Kendall, 1985). The only exception to a matrix-dominated ancient sequence was when the matrix was composed of mechanically deposited sulfate perhaps derived by storm or eolian transport from a nearby gypsum-depositing saltern platform. If such a succession was buried, the result would be a nearly pure sulfate unit composed of sabkha cycles, each 1 to 2 meters thick, interlayered with saltern sulfate units many meters thick. No modern examples of this style of evaporite deposition exist.

Anhydrite in the modern sabkha is stable only in the capillary zone above the water table; gypsum is the sulfate mineral that occurs below the water table (Fig. 1.3a). If ancient sabkhas were stacked one atop the other by a rising relative sea level, then as the capillary zone sediment subsided below the water table, the nodular anhydrite would reconvert to gypsum. Once below the water table the near surface sediment of the stacking sabkha would be gypsiferous, not anhydritic, and would not reconvert to anhydrite until buried to depths of 1000 ft or more. If calcite pseudomorphs formed in the sabkha at this shallow burial stage they would mimic gypsum, not anhydrite. Once buried deep enough, the gypsum would reconvert to anhydrite, and if there was not a substantial flowage of the sediment, it would duplicate the original trilogy. But now the nodular anhydrite of the sequence would be a burial-stage anhydrite, not a syndepositional phase.

Even within a single sabkha cycle the evolution of the hydrology as the sabkha progrades affects the preservation potential of the sabkha. As Arabian Gulf sabkhas prograde, the water table on the landward side of the sabkha falls and deflation occurs (Patterson and Kinsman, 1981). Thus in some ancient sabkhas the supratidal sediments on the landward side of the sabkha could have been removed completely by deflation. The cemented layer atop the subtidal and intertidal sediments probably stopped deflation at the top of the intertidal. Ancient intertidal—subtidal carbonate cycles containing early dolomites and nodular anhydrites in the Tansill Formation of the Permian Basin are probably a result of this process (Warren and Kendall, 1985). The delicate balance between deflation of the sabkha surface and the position of the water table means that the supratidal zone of a sabkha sequence possesses a maximum width and thickness dependent on the hydrology. The subsurface extent of a completely preserved ancient sabkha may represent only a portion of what was a much wider and thicker supratidal succession. In this type of setting any subtidal and intertidal gypsum that formed syndepositionally below the crust would have been converted to nodular anhydrite by burial to depths of more than 500 m. The result is a shoaling sequence of open marine to restricted lagoonal carbonates with anhydrite pseudomorphs after gypsum. The sequence is capped by a thin erosion surface with clasts of upper intertidal sediment.

Reservoir Occurrence: Coastal Sabkhas

Potential reservoirs in a coastal sabkha area lie not in the evaporite-dominated section of the sabkha supratidal but in the associated subtidal—intertidal grainstones and dolomitized mudstones. The evaporite-plugged supratidal acts as a seal to these underlying porous units. Sabkha-associated reservoirs form in a number of settings:

1. In the beach-barrier system separating the open marine from the restricted marine of the lagoon
2. In beach ridges which become encased in the evaporites of the mid-sabkha
3. In tidal deltas which cut the barrier islands, and in tidal channel sandy infills
4. As dolomitized muds of the lagoons
5. As buried desert dunes beneath a sabkha cap

The barrier islands and beach ridges form shoestring sands parallel to the paleoshoreline and depositional strike. The tidal deltas form sand lenses perpendicular to the depositional strike, while the tidal channel sands infill meandering channels trending down the depositional dip. Dolomitized peritidal mudstones created by fluxing brines follow the diagenetic trends discussed in Chapter 5. At its simplest the porous dolomitized mudstones were deposited as lagoonal muds paralleling the paleoshoreline with the best diagenetic permeability created along the interface between the sabkha and the marine.

CONTINENTAL SABKHAS

Continental and marine sabkhas often show similar vertical sequences. A marine sabkha is a near coastal salt flat dominated by marine-derived brines and processes; a continental sabkha is an inland salt flat dominated by continental brines and processes (Fig. 1.9). Both environments receive water via either subsurface or overland (storm-induced) flow and may be associated with eolian dune fields and intermittently submerged interdunal corridors. Excellent descriptions of continental sabkhas can be found in Hunt et al. (1966), Amiel and Friedman (1971), Eugster and Hardie (1978), Hardie et al. (1978), and Handford (1982).

Continental sabkhas, like coastal sabkhas, are characterized by the displacive growth of evaporites in the capillary zone (Fig. 2.5) and are associated with the margins of ephemeral saline lakes. Hardie et al. (1978) observed that continental sabkhas have been little studied but were able to distinguish three major depositional complexes in continental saline lakes:

1. Alluvial fan-ephemeral saline lake
2. Ephemeral stream floodplain-dune field-ephemeral saline lake
3. Perennial stream floodplain-perennial saline lake

The *alluvial fan-ephemeral saline lake* depositional complex occurs in fault bounded, orogenic basins in tectonically active areas. Inflow into this setting is predominantly from groundwater and occasional storm runoff. This setting is probably the best documented type of saline lake complex, especially in the U.S. Basin and Range. Due to the importance of groundwater seepage supplied from the adjacent ranges, this setting is one that often contains extensive continental sabkhas. Deep Springs Lake and Bristol Dry Lake in California and many other Basin and Range playas in the United States as well as Sabkha Yotvata in Israel are documented examples of this setting (Fig. 2.6).

Sabkha Yotvata is a continental sabkha with an aerial extent of 45 km^2 and infills the entire 10 km width of the southern Arava Valley (Amiel and Friedman, 1971). It is located some 45 km north of the Red Sea and 110 km south of the Dead Sea in the Great Rift Valley, the transform depression separating the Jordanian and Palestinian Plates. To the east and west the sabkha is bordered by steep-sided valley margins; to the east the valley wall rises up to 1300 m above the valley floor while to the west it rises up to 600 m. The valley floor is only 75 m above sea level. Such sheer valley walls have led to the development of an extensive bajada which interfingers with the sediments of the sabkha. In the valley to the north of the sabkha is an extensive dune field extending some 40 km up the valley. To the south is another smaller dune field which partially covers the southernmost sediments of the sabkha. Matrix sediments are furnished to the sabkha by a series of short ephemeral streams (wadis) on the bajada and from the dunes by northerly winds. Matrix grain size in the sabkha shows a bull's-eye pattern of distribution with fine to coarse sands and pebbles about the margin and silty to sandy clays in the center. North and south the bull's-eye is complicated by the presence of dune sands (Fig. 7 in Amiel and Friedman, 1971). The displacive evaporite minerals of the sabkha grow in the capillary and phreatic zones; they also follow a bull's-eye pattern. Gypsum dominates the periphery and halite the central portion of the sabkha (Fig. 2.6a). The sabkha water table is as deep as 4 m about the periphery

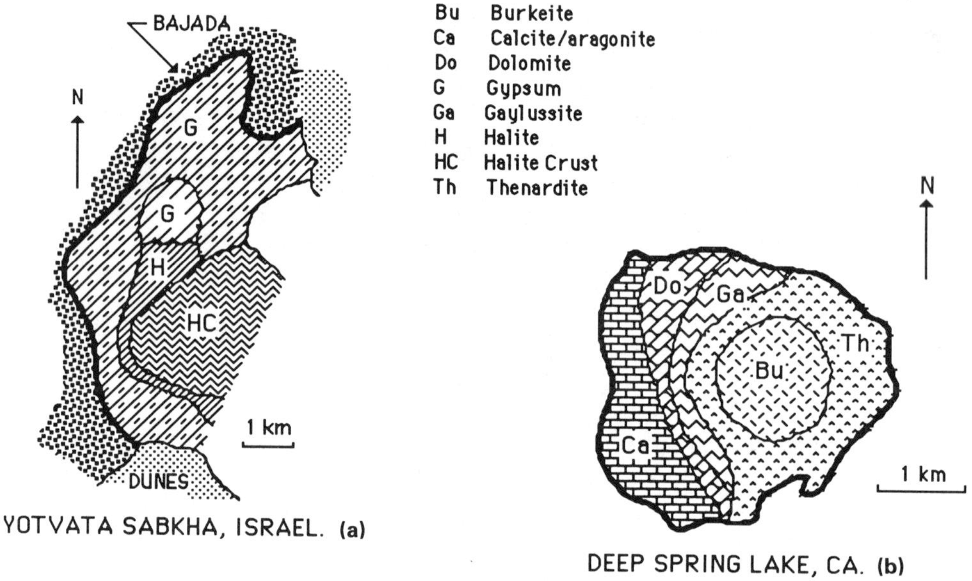

Figure 2.6 Surficial geology of Sabkha Yovata, Israel, and Deep Spring Lake, California (Amiel and Friedman, 1971; Eugster and Hardie, 1978). Note the off-center bull's-eye pattern of sediment distribution and the saline mineral suite in Deep Spring Playa.

and as shallow as 1 m in the central part of the sabkha. Hence the puffy ground and halite crust occur in the center of the sabkha, where displacive near-surface evaporites are growing in the capillary zone and distorting the sabkha surface.

Deep Springs Lake is a small playa in eastern California just east of Owens Valley (Fig. 2.6b). Holocene faulting has dropped the basin floor to 1500 m in the southeastern corner to form a 5-km^2 playa with a 1-km^2 saline lake near the center. The playa is fed by a series of springs whose waters are in contact with igneous and metasedimentary rocks. The center of the playa is occupied by soluble, mainly Na-rich salts— thenardite, burkeite, trona and halite—deposited as a bedded crust which is deformed into overthrust pressure ridges. Dolomite forms in the more peripheral, less saline areas. The playa is surrounded by the distal sediments of a series of alluvial fans with carbonate cemented springs at their feet. Vertical sequences from this setting are dominated by alluvial fan and sand flat (distal fan) facies. Death Valley, California, is probably the best known playa of the Basin and Range setting, yet Death Valley is a very steep-sided and unusual lacustrine depression. Its floor is some 24 m below sea level, the lowest continental locality in the western hemisphere. In many areas of Death Valley the coarse-grained debris flows of the alluvial fan abut the salt crusts of the brine pan. Sand flats, so obvious in other Basin and Range playas, are poorly developed or absent in Death Valley due to its steep valley walls (Fig. 2.7a and b).

The *ephemeral stream floodplain–dune field–ephemeral saline lake* complex is found in tectonically stable areas associated with arid-zone subtropical high-pressure belts. Continental sabkhas occur in this lacustrine complex but to a lesser extent than in the alluvial fan–ephemeral saline lake complex. Most of the sediments in this setting are composed of evaporite-free sand flats or mud flats. In a vertical sequence, ephemeral braided stream floodplain sequences are the dominant sedimentary facies. Inflow is by storm runoff, with storm events often spaced years apart. The dune fields in such complexes are derived by eolian reworking of the floodplain and lake sediments during the dry periods between storms. The subtropical position of this complex means that vertical sequences contain evidence of Pleistocene climatic fluctuations. Many lake complexes in southern central Australia (Lake Callabonna, Lake Eyre, Lake Frome, Lake Gairdner, and Lake Torrens) were freshwater to brackish water lakes some 35,000 years ago, yet today are dry lake beds with only short-term episodes of filling. Quaternary climate has been very important in controlling the dominance of the various sedimentary processes. This should be taken into account when taking a model based on this type of sabkha into ancient lacustrine settings.

Lake Callabonna is rather interesting in the context of rapid climate change. Some 30,000 years ago the lake was a freshwater lake and was surrounded by grassy plains. This vegetation supported huge herds of *Diprotodon,* a hippopotamus-sized marsupial, the largest marsupial known to exist. Yet some 16,000 years ago the lake dried up so rapidly that it caused the extinction of vast numbers of these marsupials. Today, the fossil carcasses can be found beneath thousands of mud-covered mounds littering the lake surface.

Lake Frome in South Australia is another example of the ephemeral-stream floodplain complex (Fig. 2.8). Lake Frome is a large playa approximately 100 km long and 45 km wide with an area of 2700 km^2. It lies at the center of a closed drainage basin of some 400,000 km^2 and is fed mainly from the west by a series of short, relatively steep gradient streams from the Flinders Ranges some 40 to 50 km to the west. The lake contains many islands mainly in the south, where the lake floor is at its lowest, some 2 m below mean sea level. These islands are the erosional remnants of earlier Pleistocene sediments and are composed of gypsiferous quartz sands atop a clay surface. Around the edges of the lake are quartzose spits and bars, beach deposits, and delta fans, as well as Pleistocene transverse and longitudinal desert dunes. Delta fans have formed at the mouths of the larger ephemeral streams and have been built by coalescing ephemeral streams. Stream channels are generally straight to gently curved with few point bars. Organic rich muds fill the lowest parts of the channels, while coarser sediment and gravel bars

(a)

(b)

(c)

(d)

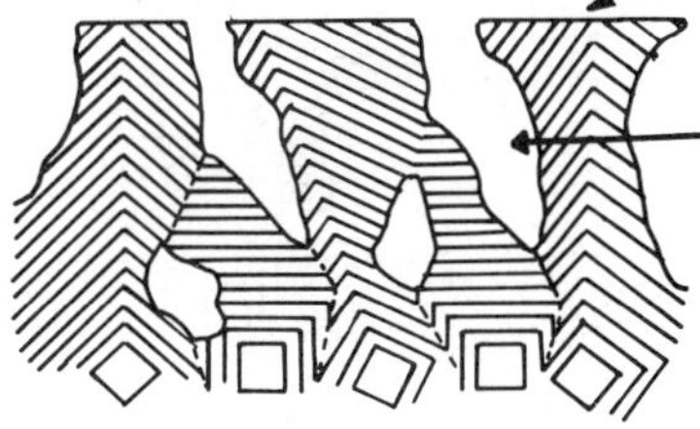

(e)

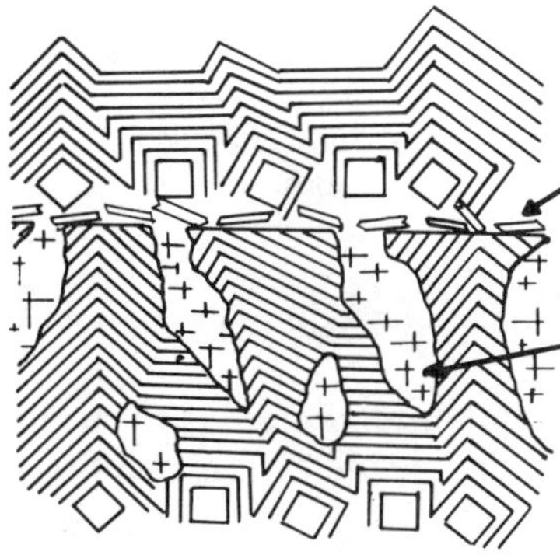

(f)

Figure 2.7 Continental sabkhas: (a) view of salt crusts in Death Valley Sabkha showing alluvial fans in contact with crust; (b) Devil's Golf Course, Death Valley, showing fretted surface created by capillary growth interspersed with dissolution (ridges are 15-30 cm tall); (c) gypsum enteroliths due to the displacive growth of modern gypsum in Late Pleistocene laminated gypsum, Salt Flat Playa, west Texas (spatula handle is 15 cm); (d) pressure ridges of halite forming in fresh halite crust, Deep Springs Playa, California; (e) salt crust of aligned chevrons truncated by freshening during the flood stage; (f) thick salt unit broken by detrital couplets. [parts e and f after Shearman, 1970].

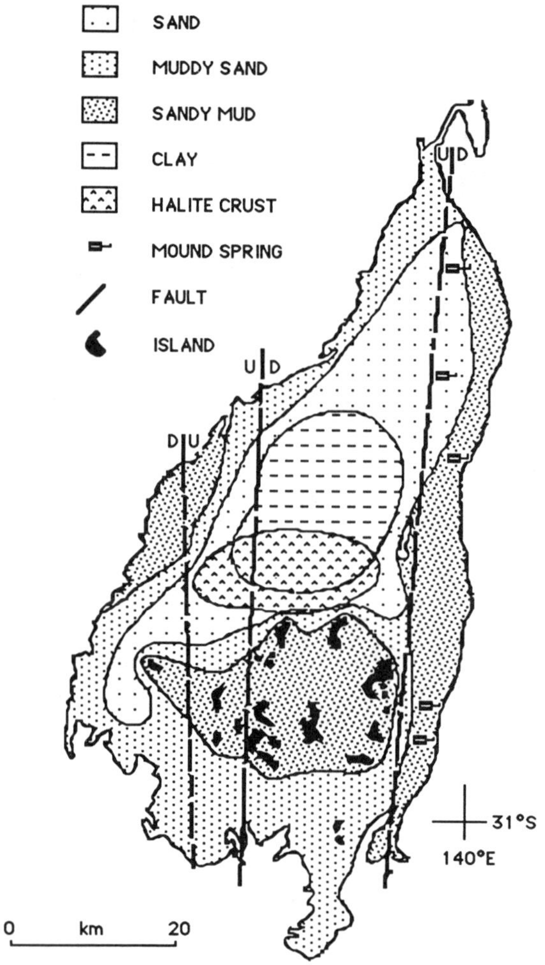

Figure 2.8 Surface sediments of the Lake Frome Playa, South Australia (after Draper and Jensen, 1976). Note the bull's-eye pattern of sediment and association of carbonate spring mounds with a north-south-trending fault on the eastern side of the playa.

are found across the remainder of the channel depression (Draper and Jensen, 1976).

Similar fine-to-coarse sands and lesser amounts of gravels form the bulk of the delta sediment. The interchannel areas on the delta are composed of reworked eolian sands. Marginward salt pans are developed as large flat areas peripheral to, but sometimes within, the fan. Further out into the lake the channels die out into the sands and muds of the lake basin. The bulk of the lake sediment is a quartzose sandy mud or mud with lesser evaporites. Often, such sediments are organic rich and wispy laminated. Displacive gypsum plates and prisms crosscut the sediments in many areas. In the center of the lake is a halite crust up to 20 cm thick and composed of halite rafts and chevrons. The crust is ephemeral and dissolves during times of flood. The lake also displays large active spring mounds along its eastern side. The mounds are composed of varying proportions of detrital sediment and carbonates with authigenic aragonite, calcite, and dolomite cements forming an algal tufa bindstone. The mounds mark the southwestern outflow zone for some of the confined aquifers of the Great Artesian Basin of Australia.

Perennial stream floodplain–perennial saline lake depositional complexes occur in several tectonic settings but usually lack extensive sabkhas. Any sabkhas that are present are localized muddy embayments about the lake shoreline. Most perennial saline lakes such as the Dead Sea and the Great Salt Lake are characterized by steep slopes surrounding the lake; this allows saline water to pond in an arid climate. A lack of sabkhas is often the result of these steep basin sides with the associated lack of an extensive capillary zone. Water inflow is more permanent and localized to rivers and a few arroyos; it is not diffuse about the lake periphery, and lake levels do not fluctuate as rapidly as in ephemeral lakes. Sedimentary facies fringing the lake are commonly old lake-bottom deposits, not depositionally active sand or mud flats. The Dead Sea and Great Salt Lake are modern examples of this type of setting (Figs. 3.6 and 4.1). Lacustrine sediments located peripheral to the perennial lake are "sabkharized" by long-term falls in the lake water level. The matrix originally formed in a subaqueous environment but has been subaerially exposed by a relative lowering of the lake water table, usually induced by late Pleistocene climatic changes. The fall of the water table into previously subaqueous sediments moved sediment into the capillary zone, where they were subject to the processes of capillary evaporation. The deposition of secondary nodular and enterolithic sulfates and large displacement halite crystals in the new capillary zone in effect "sabkharizes" the originally subaqueous lake sediments.

A good example of this process can be seen in the late Pleistocene sediments of Salt Flat Graben in west Texas, where enterolithic gypsum, displacement halite, and replacement halite are growing today in laminated gypsum-dolomite sediments that are at least 5000 years old (Fig. 2.7c; Warren and Hussain, 1984). Large modern displacement halite crystals can be found in the laminated Pleistocene carbonates of the Lisan Formation exposed in mudflats about the modern Dead Sea (Gornitz and Schreiber, 1981).

Sedimentation Styles

Continental sabkhas in both types of ephemeral saline lakes are sites of fluctuating water levels within the sediments as well as sites of ephemeral subaqueous deposition when storm waters flood the sabkha surface. In core it is almost impossible to differentiate the two types of ephemeral lake sabkha, so they are presented as a single section (Fig. 2.9). About the edges of some playas the inflow waters deposit coarse-grained meander belts and braided streams which lose their channel identity as they flow out into the playa. Once in the playa proper they dump their bedload as an extensive sand sheet about the playa strandline. As the playa alternately floods or dries, the position of the lake shoreline migrates back and

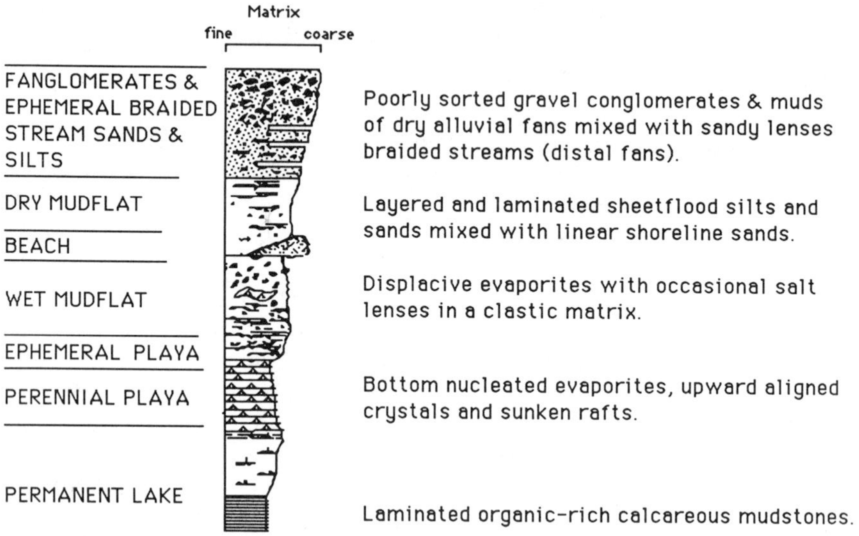

Figure 2.9 Idealized vertical sequence from a playa dominated by sabkhas. In a prograding setting the sequence will coarsen upward.

forth across the basin. During flood stages this reworks fluvial sands into shoreline sands. During dry stages the sands are reworked by eolian processes.

Owens Lake and Searles Lake, California, are good examples of lakes with this type of changing inflow. The strandline is characterized by gravelly sands deposited as fluvial and eolian sands, with each depositional package no more than 1 to 3 m thick. Back upslope toward the highlands the channels pass into arid-zone alluvial fanglomerates and debris flows of the proximal fan. In other playas the sands flow into the lake as a series of coalesced sandy deltas. Lake Frome, South Australia, is a good example of this type of inflow. Gilbert-style deltas form during the pluvial water-full stages in a playa lake when the streams flow directly into the lake waters. At such times the lake waters are relatively fresh and well mixed. Many modern playas were perennial water bodies in the pluvials of the Pleistocene (Lake Eyre, Lake Frome, Death Valley); sediments deposited in that type of setting are discussed further in Chapter 3. Some playas were still ephemeral in pluvial periods (Bristol Dry Lake, California), but the style of deposition on the fans changed from the "dry fan" type forming today to a more humid or "wet fan" type in pluvial times, when there was more surface moisture on the landscape.

For much of its Holocene depositional history a playa was dry or drying, and evaporites were deposited as bedded crusts in the lower areas of the basin. Episodes of subaqueous evaporite deposition were short-lived and the resulting deposits only centimeters thick. Usually, the surface of the playa was emergent, a site of groundwater discharge and evaporation, best called a continental sabkha.

After a flood the drying lake sediments often sat in the capillary zone above saline phreatic waters. Gypsum, anhydrite, and halite crystallized in the near-surface sediments as the capillary pore waters evaporated, in much the same way as they do today in the Arabian Gulf sabkhas (Fig. 2.5). Such intrasedimentary crystals grew either by pushing aside the surrounding sediment or by poikilitically enclosing the matrix. Crystals that pushed aside the surrounding sediment probably grew more slowly than did those enclosing matrix (Kastner, 1970). The growth of displacive crystals tended to destroy any primary sedimentary structures and replace them with a chaotic mass of crystals in a randomly oriented matrix. Such crystals have formed displacement halite and chaotic muds in Bristol Dry Lake, California, and the salt flats around the edges of the Dead Sea (Handford, 1982; Gornitz and Schreiber, 1981). Displacement gypsum is growing in the mud flats of Salt Flat Playa, Texas, and about the edges of the brine ponds of Lakes Eyre, Frome, Gairdner, and Torrens in South Australia. Displacive evaporites will grow in arid settings wherever the capillary zone intersects the playa surface and can often be seen as puffy ground on the playa surface. Hardie et al. (1978) used displacive evaporite growth to distinguish between two types of mud flats in an ephemeral lake setting:

1. A *dry mud flat*, lies closer to the basin edge. It is characterized by the preservation of sedimentary structures formed under alternating upper and lower flow regime conditions. Vertical sequences dominated by mud settling layers deposited during poststorm settling, but they also contain gravel layers, laminar beds, and antidunes formed under upper flow conditions as well as climbing ripples. Such units are sometimes called storm couplets.
2. A *saline mud flat,* where all the sedimentary structures are destroyed by the growth of evaporite crystals, often in the capillary zone but also from phreatic waters. Dolomite also forms as a diagenetic replacement in mud flats around ephemeral lakes such as Deep Springs Lake; sometimes these dolomites can even form capillary crusts which are overthrust into tepees as in Salt Flat Playa in west Texas (Warren and Hussain, 1984).

Thick salt crusts form in areas on the playa where ponded surface waters evaporated to dryness or where groundwater returns to the lake surface and evaporates. Often these areas are the topographically lowest parts of the playa. Crusts can attain thicknesses of 30 cm or more. Continual addition of new salt crystals from evaporating pore waters causes an increase in the volume of the crust. The thickness of the associated capillary fringe limits continued vertical growth of the crust, so the crust attempts to expand horizontally. The force of crystallization creates a biaxial stress in the crust that ultimately forms cracks which grow into overthrust salt polygons (pressure ridges), knobs, and salt-lined fractures. Once initiated the cracks become sites of groundwater seepage and can attain a relief that is sometimes measured in meters (Fig. 2.7d). Overthrust ridges rarely retain the simple inverted V cross section of the initial break. Continual salt growth and dissolution in the fissure deforms and anneals the salt into various knobs, ridges, and pinnacles.

During dry periods eolian dust is trapped by capillary adhesion on the moist, rough surface of the salt crust. During wet episodes the storm waters flood the sabkha, dissolve the upper part of the salt crust, and at the same time carry in a new load of clastic material. Evaporation during the next dry period creates a new salt layer atop the dissolution surface. The knobbly surface seen on many playa salt crusts is the result of continual deposition and fretting of a growing salt layer. The halite in the initial crust is often composed of upward-pointing subaqueous halite chevrons. Freshening episodes often plane off the upper part of the chevrons as undersaturated waters eat into the crust along the inter-chevron boundaries. The end result of a freshening episode covering the salt crust with brackish water is a porous salt crust with an irregular truncated upper surface. Capillary evaporation then infills the porosity with clear halite (Fig. 2.7e). Continued dissolution and reprecipitation of halite and other soluble salts in the vadose and capillary portions of the crust destroys the initial aligned chevron texture.

Each salt layer in a thick salt unit deposited as part of an ephemeral salt lake is in large part derived from the recycling of an older salt layer. Layers are separated and crosscut by mud partings composed of detrital material carried in by floodwaters and eolian sediment deposited on the emergent salt surface. The final buried sediment is a thick bedded-salt composed of randomly aligned recrystallized evaporites forming units which are meters thick, layered, and crosscut with eolian or detrital impurities (Fig 2.7f). This type of bedded salt contains randomly aligned crystals due to continuous dissolution and recrystallization. It contrasts with bedded salt deposited in a continuously subaqueous and/or phreatic environment where even after burial the crystals in the crust still retain a growth aligned fabric of upward pointing large gypsum crystals, chevron halite, or trona (Figs. 3.2b and 3.3).

The recrystallized textures found in the bedded salt units of most ephemeral lakes are a direct result of hydrology in the capillary zone. Flooding events in ephemeral lakes are relatively rare; most evaporite minerals in ephemeral lakes are moved about the lake flat as soluble load in shallow groundwaters, not as mechanically reworked sediment; and as water always flows to the lowermost point or the position of the lowest potentiometric head, the groundwaters in ephemeral lakes gradually seep into the lowermost parts of the lake. On its journey to the lowermost portion of the lake the basinward flowing groundwater can come near or to the surface a number of times (resurging groundwaters). In such areas are found the displacive salts and ephemeral crusts of the saline mud flat. In early stages of formation these capillary crusts will contain metastable salts even along the more-landward saline mud flats. Less stable salts are quickly leached from the crust as it is elevated by capillary growth or subjected to a light rain. The less-soluble mineral fraction is then left behind in the crust, often to form a carbonate crust such as the dolomitic crusts of Salt Flat Playa, west Texas (Warren and Hussain, 1984). Brines derived from the partial dissolution of such crusts are usually simple chemical solutions containing only one or two major solute species (e.g., NaCl, Na_2SO_4, Na_2CO_3). When they precipitate later as surface or near-surface crusts, layers, and nodules, they will form simple mineral suites by evaporation as the brine resurfaces on the lower parts of the playa floor. This process is a natural mineral fractionation very similar to processes used to obtain the more saline brines from solar evaporation pans.

Even if marginward crusts survive dissolution at the surface and are buried, they can later be dissolved in the shallow subsurface whenever the basinward-flowing groundwaters are undersaturated with respect to the relevant mineral phase. This process adds further to the chaotic textures forming in the saline mudflat. Farther out in the basin, in the central brine pan area, the upward movement of groundwater reprecipitates evaporites in the capillary zone and a new capillary crust at the lake surface. The process of fractionation can begin all over again. Only the most soluble salts are carried into the lowermost portions of the lake basin. As a consequence of this continual recycling of material, the groundwaters become increasingly more saline toward the basin center—so saline that gypsum and halite become stable mineral phases and accumulate as bedded and displacive units in the lowest parts of the basin. Even here the continual recycling of brine through the capillary zone destroys the initial chevron texture and leaves a massively recrystallized unit. Basinward brine flow explains why continental ephemeral lakes develop well-defined bull's-eye patterns with the most soluble salts accumulating in the lowermost portion of the lake floor surrounded by saline carbonates of the strandline.

Vertical Sequence

Continental sabkhas are transitional up-dip into continental facies such as alluvial fans and ephemeral braided stream floodplain sediments. Similar continental sediments can occur on the landward side of marine coastal sabkhas, and vertical sequences from continental sabkhas and marine coastal sabkhas can be almost identical, especially when a lack of preserved body fossils exists in the sequence. However some important differences occur. Facies belts tend to be concentric—symmetric or asymmetric "bull's-eyes" in continental settings, and along-strike continuous belts in coastal marine settings (Figs. 2.1, 2.4, and 2.6). The following description is an artificially constructed ideal vertical sequence for a continental sabkha based mainly on the work of Hardie et al. (1978) and my own work. In the two major saline lake depositional associations (ephemeral lakes) showing widespread continental sabkhas, the following facies tracts are common:

alluvial fan or
ephemeral stream → sand flat → dry mud flat → saline mud flat → salt pan
floodplain ← ephemeral saline lake →
←——— CONTINENTAL SABKHAT ———→

Additional sedimentary facies (*dunefields*—composed of either terrigenous sands, mud pellets, or evaporite grains; *springs*—travertine or tufa mounds; and *shoreline features*—deltas, beach ridges, or spits) may be interspersed within the general facies tract at any of several positions; they should be considered as modifications of the model rather than an integral part of it. The sand flat, the dry mud flat, and the saline mud flat may together be considered the continental sabkha environment.

Given that this is the depositional setting of most continental sabkhas, an idealized section can be modelled by "Walther's Law" (Fig. 2.9). Assuming a prograding sequence the features of the continental sabkha are:

1. Overall coarsening-upward grain size profile
2. Upward increase in density of root mottling and rhizoconcretion development
3. Upward increase in preservation of primary sedimentary structures (less disruption by intrastratal displacive evaporites)
4. Upward transition from zones dominated by halite, trona, or other soluble saline minerals; to gypsum; to alkaline earth carbonates (more magnesium-rich carbonates should be lower in the section)
5. General cyclicity throughout the section, induced by the periodic storm events (deposits of each facies often exhibit "storm couplets" in both detrital and evaporitic facies)
6. Upward transition from accretionary bottom- or surface-nucleated evaporite fabrics at the base of the sequence to displacive or poikilitic intrastratal fabrics to surficial crusts (if preserved) higher in the sequence

Vertical changes in evaporite mineralogy in core are not always those predicted from simple brine concentration and progradation within a bull's-eye depositional setting of more saline minerals passing up into less saline. The preserved sequence depends on the direction of brine flow and changes in playa hydrology with time as well as changes in the position of the evaporite depocenter (lowest point) within the basin. Sabkhas and brine pans in tectonically active areas can move around the lake as the tilt of the basin floor changes over time. Wherever more saline brines are upwelling, the less soluble salts can be buried beneath more soluble minerals. However, the more soluble salts (such as the bittern salts) are not usually precipitated at the surface and are often preserved beneath the less soluble salts due to downward brine flow from mineral fractionation. The dissolution of near-surface salts during a flooding event can create dense concentrated intrasediment brines in the following emergent dry period. Such dense brines sink through earlier crystal accumulations, displacing the lighter, less concentrated brines, which in turn emerge at the surface, there causing dissolution. It is from the descending, dense brines that the permanent more saline salts accumulate as early diagenetic replacement minerals, such as polyhalite, dolomite, and so on (Table 1.3). These minerals often pseudomorph the crystal shape of the mineral they replace. Density mixing of subsurface brines during emergence of the playa surface also encourages replacement of metastable by stable minerals as well as the recrystallization of earlier formed halite, so contributing to the early diagenetic lithification of the salt deposit (Hardie, 1984).

In cores taken from alluvial-fan-associated sequences, the sabkha unit passes up into fanglomerates. In ephemeral stream settings it passes up into braided stream units (Fig. 2.9). In both cases with a prograding continental sabkha system, most of the sabkha matrix is supplied from the hinterland. This means that the matrix in a single sabkha cycle shows a coarsening upward trend. Coastal sabkhas grow by lateral accretion as sediment is carried from the lagoon up onto the salt flat; this means that the matrix in a marine sabkha fines upward. The contrast between fining-upward and coarsening-upward may be one feature which can be used with wire-line logs to distinguish continental from marine coastal sabkhas. Similarities between marine and continental sabkhas can far outweigh the differences especially in areas of siliciclastic coastal sabkhas. The processes depositing displacive evaporites in the capillary zone are essentially the same; often, only the initial ionic composition of the waters and the humidities of the overlying air are different. Today this means that it is rare to see bittern salts accumulating in a near marine setting (Kinsman, 1976). In most coastal sabkhas the calcium sulfate minerals are much more commonly preserved in the sediment column than halite. Bittern salts are probably more likely in continental sabkhas, even so, high levels of $CaSO_4$ minerals or halite are far more common than bittern salts in the capillary zone of modern continental sabkhas.

Evaporite minerals forming in the sabkha can sometimes distinguish continental sabkhas and marine coastal sabkhas (Table 1.1a). The chemistry of inflow waters into a continental setting are often determined by bedrock weathering reactions in the source area. The myriad of more soluble salts such as trona, pirssonite, burkeite, and hanksite are, if present, the true signature of a continental setting because they reflect the greater proportional abundance of Ca^{2+}, Na^{+}, HCO_3^-, or SO_4^{2-} relative to these ions in seawater. A number of early diagenetic reactions also occur in the shallow subsurface which are perhaps more indicative of the fluctuating hydrological and geochemical conditions of continental settings (Table 1.3). Carbonates may alter to gaylussite in sodium-rich brines; gypsum may dehydrate to bassanite, then anhydrite; gypsum may react with sodium-rich brines to form glauberite; mirabilite formed in a cold climate or in the winter months may dehydrate to thenardite (Kendall, 1979, 1984). The early conversion of gypsum to anhydrite is thought to be more common in continental settings than coastal ones due to the greater temperature extremes and lower relative humidities (Kinsman, 1969, 1976). The occurrence of some other minerals, such as zeolites, authigenic feldspars, magadiite, and chert in continental sabkhas, reflects the reaction of brines with volcanic glass or silica-rich groundwaters. Their presence results from the common association of continental sabkhas with extensional tectonism, hydrothermal activity, and volcanism. They are a common but not integral part of a continental sabkha succession.

Working with cuttings, as is often the case in ancient evaporites, means that the bittern salts dissolve in the drilling mud before they ever reach the surface. In many ancient evaporites, it is the larger scale depositional setting and the presence or absence of marine fossils which are often the two most useful clues in separating the continental and marine sabkha settings.

Reservoir Occurrence: Continental Playas

Few reservoirs have been documented in continental sabkha or arid lacustrine settings, probably because few geologists think in terms of a continental setting when interpreting evaporite-associated reservoirs. The following discussion is conjectural, based on sand, gravel, and dolomite trends in modern continental sabkhas as well as reservoir types in producing lacustrine fields in Angola, the Green River Basin, the Gulf of Suez, and China.

Potential reservoir facies in most continental playas do not occur in the saline mud flats (sabkhas) or ephemeral brine pan areas; rather they are usually found in the siliciclastic apron about the playa edge. As in a coastal sabkha, the evaporite capping of the continental sabkha can act as a seal to any underlying sands or dolomitized mudstones. The type of reservoir will vary according to the nature of water inflow and sediment reworking about the playa edge. In some playas the surface inflow forms coarse-grained meander belts and braided streams which lose their channel identity as they flow out into the playa. Once in the playa proper they dump their bedload as an extensive sand sheet about the playa strandline. The position of the lake shoreline migrates back and forth across the basin; this process can rework the fluvial sands into shoreline sands during flood stages. During dry stages the sands are reworked by eolian processes. The strandline is characterized by gravelly sands deposited as fluvial and eolian sands with each depositional package no more than 1 to 3 m thick. Back upslope toward the highlands, the channels pass into arid zone alluvial fanglomerates and debris flows. In other playas the sands flow into the lake as a series of coalesced sandy deltas. Gilbert-style deltas, formed in a playa lake during the high-water stages when the lake waters are well mixed, along with other delta types can also contain sandy reservoirs in their topset and upper foreset beds.

The arid setting of most playas means that eolian conditions alternate with sabkha, shoreline and fluvial conditions. The porous facies about the playa edge are the result of the wetting and drying of the playa margin. This forms a heterogeneous gravelly-sand reservoir with variable porosity and permeability. Correlation of individual pays in such reservoirs in ancient counterparts is difficult. The best method when using wire-line logs is to correlate the shales and muds which separate the reservoir sands. The muds are either the lacustrine muds of the high-water stage in the lake or the sabkha muds of a low-water stage. In either case they form a widespread marker with what was originally a near-horizontal upper surface. Stream inflow to many continental sabkha/lacustrine depressions is radial, so the sand-filled channels trend radially beneath these marker beds into the evaporite-dominated basin center. Dune sand trends are controlled by the prevailing wind directions at the time of deposition and may trend at any angle to the paleodip. Dolomitized mudstone trends are poorly documented in continental sabkhas and cannot be modeled at this time.

On the larger scale the most important control on potential reservoir occurrence is probably the depositional gradient of the area separating the brine pan from the proximal zone of the alluvial fans. If the gradient is steep, as it is in most perennial lakes and some ephemeral lakes such as Death Valley, extensive reservoir facies are unlikely. The siliciclastic apron is dominated by debris flows and poorly sorted muddy sediments. The best chance for a reservoir facies in such settings is probably along the lake shoreline or in areas of eolian reworking. If the gradient is lower, as it is in most ephemeral lake settings such as Bristol Dry Lake in California and Lake Eyre and Lake Frome in Australia, there can be an extensive sand flat. Thus the distal end of alluvial fans and the sand flat in this setting along with eolian reworked sequences can form potential reservoir facies. Depositional topography and facies characteristics can be traced in high frequency seismic lines; these observations can be used to help define prospective areas before the drill enters the ground.

The semiarid setting of many playas means that the reservoir is prone to early, often syndepositional, calcite cementation. In addition, many potential lacustrine reservoirs are open ended on the up-dip side. In their original deposited geometry the porous units of the sand flat and the distal alluvial fan often lack an up-dip seal. To create a trap may require uplift or faulting to induce a dip reversal. The fault-prone setting of alluvial fans/ephemeral lakes in rift basins, transform, and foreland areas means that lake deposits in these areas are prospective for hydrocarbons. Often, the result of such faulting is a series of one-sided or half grabens with each rotated block showing some degree of dip reversal. The rapid subsidence and burial of these tectonically active areas means that most buried lacustrine sand bodies are broken up by steep, near vertical faults related to the uplift of the adjacent highlands and the subsidence of the lacustrine mud flats. Reservoirs will be broken up into smaller blocks within this setting and truncated by steep faults. Drilling targets are usually small in area, but stacked traps are common with the faults creating good vertical migration pathways for the hydrocarbons. The heterogeneous nature of the reservoir facies and the lack of predictable connection between individual sands and gravels means that recovery rates are low and water flooding not very efficient.

Carbonate sand bodies formed as lacustrine sand bodies are also potential reservoirs (also Chapter 3). Such sands grow in shoal water areas as oolitic, pisolitic, and skeletal grainstones. Unlike siliciclastics, carbonates do not require a hinterland source and can form in shoal waters anywhere in the lake basin. Carbonate sand bodies can form shoreline sands with shoestring geometries as in

Great Salt Lake (Fig. 3.5) but can also form shoal-water buildups above active horst blocks and diapirs anywhere in the lake. The nonmarine shell-hash reservoirs of Cabinda in West Africa lie above horst blocks and are sourced by lacustrine dolomitic shales (Brice et al., 1982).

ANCIENT SABKHAS AND RESERVOIRS

Probably the first ancient sabkha to be recognized was in the lower Purbeck (Uppermost Jurassic) of England (Shearman, 1966). Since that time many carbonate—sulfate sequences have been interpreted as ancient sabkhas; examples include the Visean of Eire (West et al., 1968), the Upper Devonian of Western Canada (Fuller and Porter, 1969), the Middle Carboniferous of the Canadian Maritime provinces (Schenk, 1969), the Jurassic of the Arabian Gulf (Wood and Wolfe, 1969), the backreef strata of Guadalupian and Leonardian age in the Permian Basin (Kendall, 1969; Ramondetta, 1982), the Smackover/Buckner Formations, and their equivalents of the Gulf of Mexico (Budd and Loucks, 1981). All of these examples are dominated by a carbonate matrix which may pass laterally and up-dip into redbeds. Permian continental sabkha complexes occur with siliciclastic matrices and a great deal of halite, both as interbeds and sporadic displacive crystals, in the Lower Clear Fork Formation of the Palo Duro Basin in west Texas (Handford, 1981b). The following discussion of ancient sabkhas concentrates on those examples that contain economic hydrocarbon accumulations.

San Andres Formation, West Texas

Geology

The San Andres Formation (Leonardian-Guadalupian) extends from Texas to as far west as Arizona and Utah. Subsurface studies on the San Andres have focused on two localities: the Permian Basin of southeast New Mexico and west Texas, which includes the North and Northwest Shelves, and the Central Basin Platform; and the Palo Duro Basin in the Texas Panhandle (Fig. 2.10). Past studies were concentrated on the North and Northwest Shelves or the Central Basin Platform, as these areas contain very productive hydrocarbon trends; as of 1982, 80% of the oil produced in the Northwestern Shelf was taken from reservoirs within the lower San Andres (Ramondetta, 1982). The San Andres Formation is also the most productive unit in the whole of the Permian Basin of west Texas. Recently the San Andres of the Palo Duro Basin has become an area of interest as a potential radioactive waste repository (Bein and Land, 1982). Unlike the producing dolomitic intervals of the other three regions, the San Andres in the Palo Duro Basin is composed primarily of salt, a relatively impermeable and self-annealing mineral. Most of the carbonate intervals in the region are nonproduc-

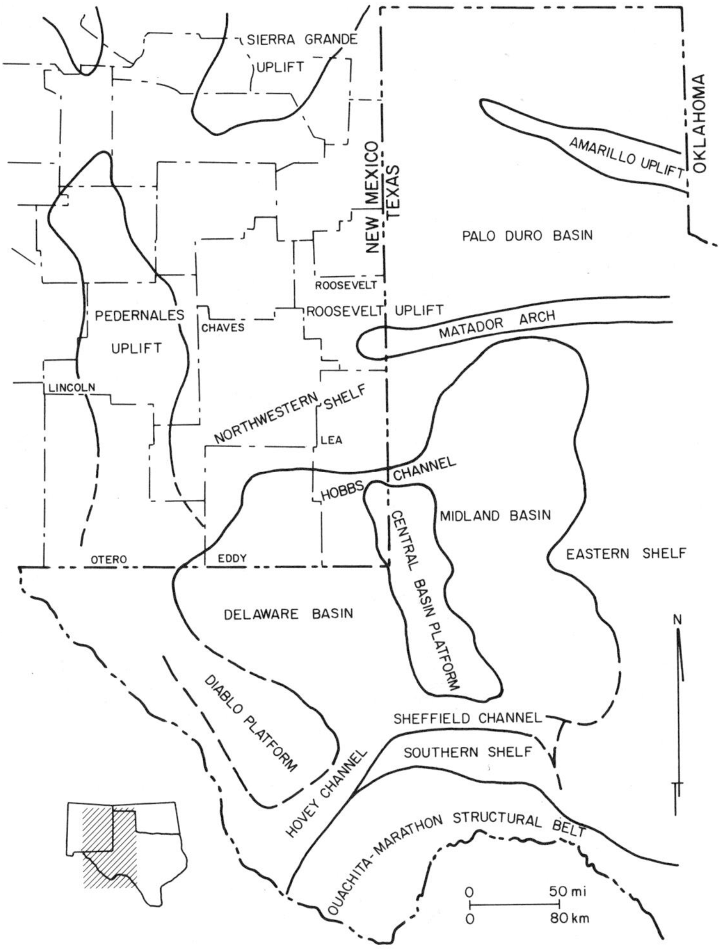

Figure 2.10 General setting of the Permian Basin, west Texas during San Andres time.

tive, as they are plugged by salt. The Palo Duro area is discussed further in Chapters 3 and 4.

The more porous San Andres units of both the Central Basin Platform and the Northwest Shelf are usually interpreted as progradational sequences or regressive sequences which were deposited after an initial transgression at the beginning of San Andres time (Fig. 2.11). Several stacked cycles occur within the overall San Andres regression (e.g., P_1 to P_5). These cycles consist of the following shallowing-upward units:

4. An evaporitic cap of laminar or nodular anhydrite interpreted as supratidal coastal sabkhas by many authors (e.g., Chuber and Pusey, 1972). *Note:* The

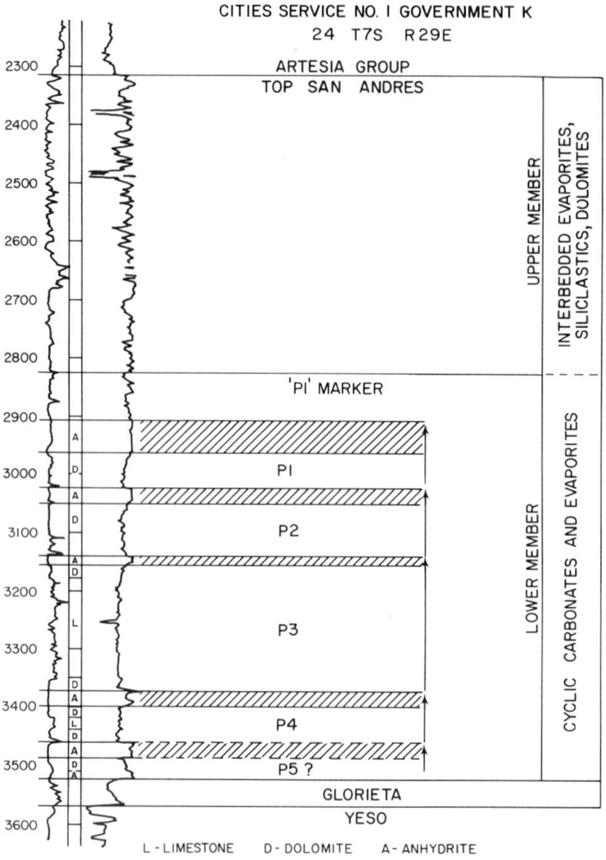

Figure 2.11 Typical wire-line signature of the San Andres Formation on the Northwestern Shelf, Permian Basin, west Texas shows Upper and Lower San Andres units; members in the lower unit are much more readily discernible as widespread cycles (from Elliott and Warren, *in review*).

dominance of sabkha in all lower San Andres reservoirs is questioned in Chapter 3.

3. Shallow lagoon and/or intertidal sediments (often dolomitized)
2. A thicker unit of shallow outer shelf sediments (often limestone)
1. A thin basal transgressive unit

Chuber and Pusey (1972) subdivided these cycles into strongly agitated shelf-edge and minimal-flow-strength back-shelf cycles. The cyclical nature of the San Andres is typical of Permian deposits around the world and is often explained by the glacial-eustatic sea level changes induced by the Permian Ice Age.

Hydrocarbon Occurrence

San Andres exploration is now taking place in a very mature province and San Andres oil is being recovered using secondary and tertiary recovery techniques. San Andres reservoirs are characteristically slow but prolific producers of oil and gas, usually at relatively shallow depths. Traps are a result of several factors, including stratigraphy, diagenesis, structure, and hydrodynamics. The most important trapping mechanism is usually stratigraphic, but combination traps are also common (Fig. 2.12). The most common trap is a regional up-dip porosity pinchout from porous dolomite into the lower-porosity dolomites and

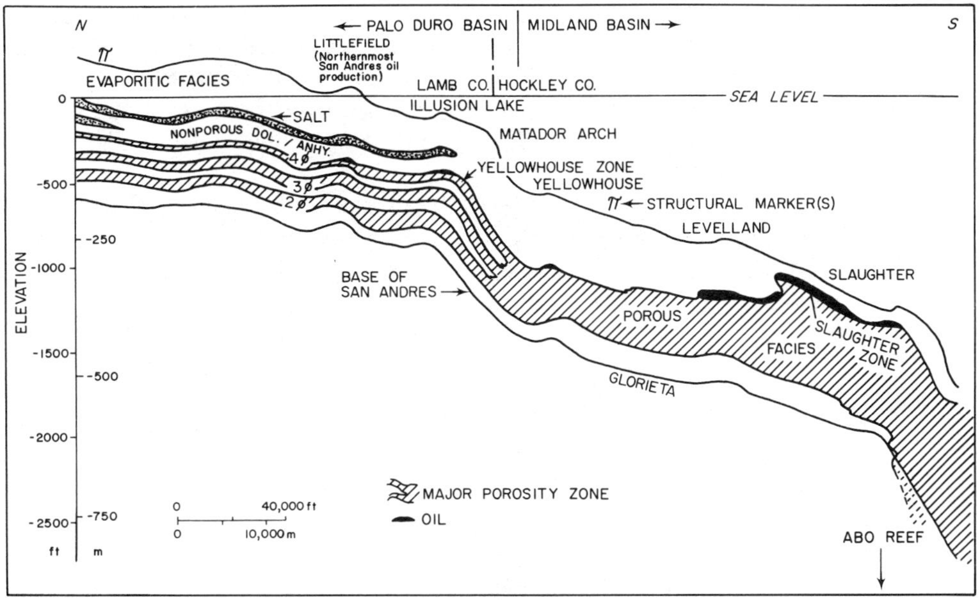

Figure 2.12 Schematic of hydrocarbon distribution on the Northwest Shelf Permian Basin, west Texas (after Ramondetta, 1982). For locality see Fig 2.10.

supratidal evaporitic sediments (Dunlap, 1967; Gratton and LeMay, 1969; Zaaza, 1978; Pitt and Scott, 1982). Such pinchouts are intimately related to the cyclical nature of San Andres sedimentation and the progradational basin architecture during San Andres time. Within several shallowing-upward cycles the more porous subtidal and intertidal facies interfinger with tight supratidal facies, creating an efficient reservoir-trap mechanism. The reservoir units are usually subtidal and intertidal dolomitized limestones with good intercrystalline porosity, while the nonporous evaporitic supratidal sediments typically serve as the seal. The source for the hydrocarbons is the underlying basinal shale of Wolfcampian age (Ramondetta, 1982). Migration probably occurred via vertical fractures upward into the reservoirs of the San Andres. Oil column heights greater than structural closure confirm the import of stratigraphic traps within the San Andres, even in areas dominated by combination traps.

Some structural control on oil accumulations is commonplace in San Andres reservoirs; in fact, stratigraphic/combination traps are the most efficient oil-trapping mechanisms in the San Andres. Draping and differential compaction over paleotopographic highs are common structural features, and earlier highs often influenced subsequent deposition. Kumar and Foster (1982) cited San Andres production over a bryozoan mound where the shoaling of sediments above the mound resulted in the deposition of the porous lithologies of the reservoir. These porous carbonates are surrounded by nonporous mudstones. Ramondetta (1982) cited San Andres reservoirs as associated with differential compaction along an abrupt shelf margin/basin transition in the Yellowhouse–Illusion Lake–Littlefield trend (Fig. 2.12). Similar features are seen in the Levelland–Slaughter trend. In the Levelland–Slaughter trend, freshwater diagenesis took place in these highs during low sea-level stands and produced episodes of secondary porosity enhancement (Todd, 1976).

Hydrodynamics within the Permian Basin also influenced trapping configurations. Eastward-flowing groundwater created eastward-dipping oil-water contacts which further complicated an already complex trap. Groundwater flushing has also reduced oil quality in many shallow reservoirs.

Ordovician Red River Formation, Williston Basin, U.S.A.

Geology

The Williston Basin is located on the southern edge of the Canadian shield, straddling the U.S.-Canadian border and centered on Saskatchewan and North Dakota. The basin itself is circular, about 800 km across, and is filled with 4600 m of Phanerozoic sediments (Fig. 2.13). Ordovician carbonates were deposited in a shallow epeiric sea and show evidence of shoaling cycles containing environments ranging from shallow marine to supratidal (Roehl, 1967). Lithologies include bioclastic and argillaceous limestones, dolomites, and evaporites. Ruzyla and Friedman (1981, 1985) studied the Ordovician Red River Dolomite in the Cabin Creek Field on the Cedar Creek anticline in Montana on the southwest side of the basin. They concluded that the Red River in that area was deposited in a sabkha setting, and most of the following discussion is based on their work.

Sediments. The Red River Formation is 90 to 180 m thick. The lower part is a nonporous marine limestone, and the upper part is 40 to 50 m thick and is composed of three cycles (A, B, C), each consisting of a limestone-dolomite couple deposited in environments ranging from subtidal to supratidal (Fig. 2.14). Subtidal lithologies are composed of mottled dolomitic biomicrite wackestones that contain fragments of echinoids, brachiopods, bryozoans, trilobites, mollusks, ostrocods, and gastropods, with lesser amounts of solitary corals and stromatoporoids. These units are poorly bedded and heavily bioturbated; they were laid down in a quiet-water marine setting. The intertidal sediments are also dolomitic

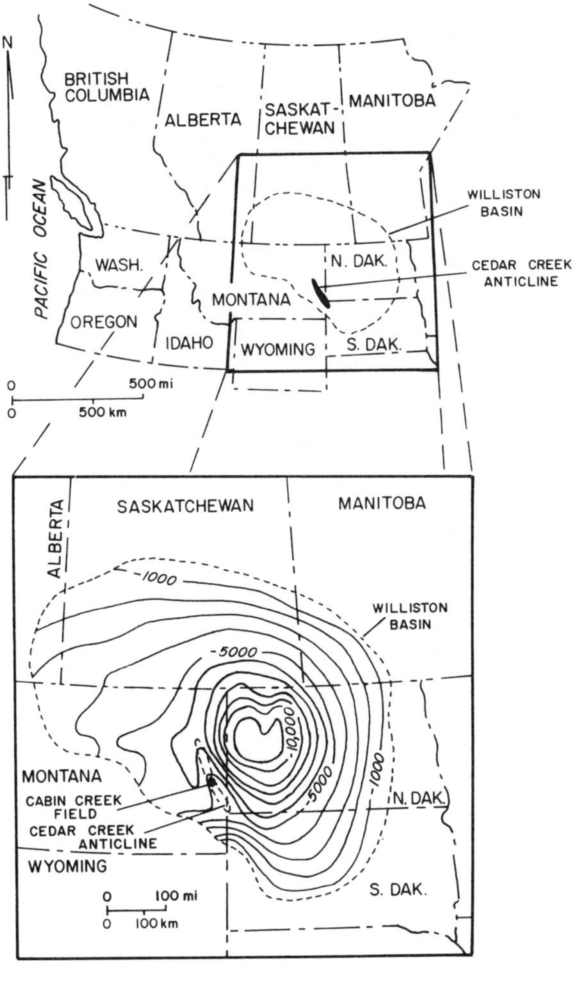

Figure 2.13 Location map showing outline of Williston Basin, Cedar Creek Anticline, and Cabin Creek Field (based on Ruzyla and Friedman, 1985). Contour interval is 1000 ft.

biomicrite wackestones, but they were deposited under conditions of greater flow strength. The deposits contain graded beds, cross-beds, ripple marks, imbricated shells and conglomerates. They also contain allochems characteristic of an intertidal setting, such as oncolites, ooids, intraclasts, and peloids. Supratidal sediments in the Red River Formation include wavy-laminated micritic dolomites and organic-rich mudstones. They contain evidence of restricted and at times subaerially exposed conditions in the form of desiccation cracks, erosion surfaces, intraclast breccias, solution collapse breccias, and nodular anhydrite. The supratidal sediments and some intertidal sediments show evidence of supratidal leaching as they contain empty gypsum and anhydrite molds, corroded anhydrite

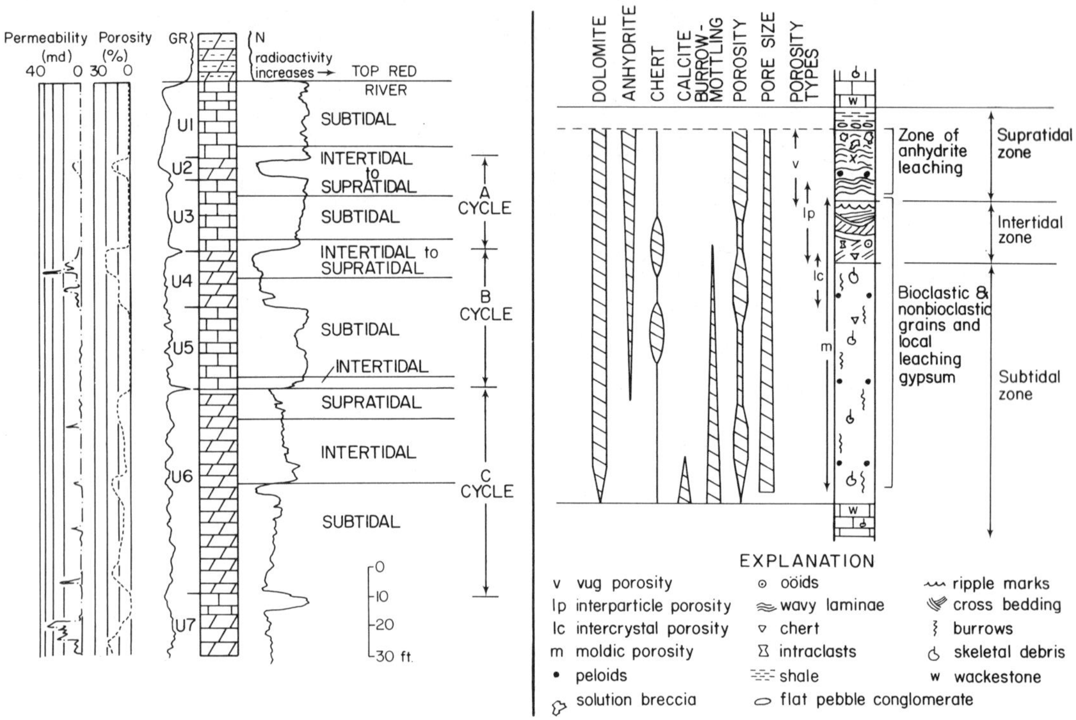

Figure 2.14 Typical columnar section showing wire-line signature, porosity, and permeability plus idealized interpreted sequence. Note that the base of the dolomite is transitional down into limestone (based on Ruzyla and Friedman, 1985).

crystals and solution collapse breccias. The early micritic dolomites of the supratidal best retain the original depositional fabrics.

Dolomitization. The Red River Dolomite is related to two periods of diagenesis, an early penecontemporaneous sabkha stage which precipitated penecontemporaneous dolomite in the supratidal sediments, and a later brine reflux stage which dolomitized portions of the underlying intertidal and subtidal sediments (Chapter 5). The first dolomite, a sabkha dolomite, was caused by the arid sabkha setting; in fact, dolomitization of the supratidal sediments is almost complete. Such dolomites are finely crystalline and occur in the wavy-laminated or peloidal mudstones. The second dolomite, a coarser-grained brine-reflux dolomite, occurs in the intertidal and subtidal sediments immediately below the supratidal dolomites. The proportion of dolomite decreases downward from the base of each supratidal unit due to the sinking of dense hypersaline waters which originated in the supratidal zone. The brine-reflux dolomites selectively replace skeletal fragments and matrix; locally, the descending brine selectively dolomitized burrows.

Hydrocarbons and Porosity

Porosity in the Red River Dolomite is restricted to the dolomitized units, which in turn are controlled by the depositional setting, specifically the occurrence of hypersaline and fresher waters during deposition. Interparticle porosity is best developed in the intertidal sediments between peloids, intraclasts, ooids, and skeletal fragments. Intercrystalline porosity is best developed in dolomitized mudstones and occurs as the spaces between individual dolomite rhombs. Porosities up to 25% (13% average-dolomite) and permeabilities up to 142 md (7.9 md average-dolomite) occur in areas with good intercrystalline porosity. In addition, vuggy and moldic porosity occur throughout the Red River Dolomite wherever the sequence has been leached by fresher waters. Production is from the porous zones in the three dolomitized units in the Red River; all dolomites controlled by a sabkha depositional system. The average depth to pay across the field is 2750 m, with a 15.2 m net pay thickness, a productive area of 30.5 km^2 (7620 acres), and an estimated 224 million barrels of oil in place.

Lower Clear Fork Formation (Permian), Texas

Geology

The Middle Permian (Leonardian) evaporites, carbonates, and redbeds of the lower Clear Fork Formation in the Palo Duro Basin of the Texas Panhandle show many features which are also found in modern continental and coastal sabkhas (Handford, 1981b). The Clear Fork Formation in the Palo Duro Basin is not productive but has been thoroughly investigated as a potential repository for radioactive wastes (Fig. 2.10). The Clear Fork was deposited in a near-coastal salt flat complex with both nonmarine and marine elements (Fig. 2.15). The major facies include:

1. Laminated red siltstones and mudstones passing into coarse-grained arkosic sandstones in the northwest Texas Panhandle and New Mexico. Lobate geometries, and sedimentological features such as small braided channel fills, caliche zones, and sheet-flood laminae indicate deposition in wadi, desert-alluvial, eolian plain, and dry mud-flat environments. This facies passes basinward into
2. Chaotic halite mudstones composed of isolated skeletal halite hoppers up to centimeters across floating in a mudstone matrix. Deposition of this unit was in a saline mud flat where the near-surface sediments were soaked with an evaporating brine. It passes basinward into
3. Massive bedded halite and laminated anhydrite. The wavy-laminated anhydrite-dolomite units were deposited as algal mats in an Arabian Gulf–style setting. Other laminated units are composed of poorly developed crystals of halite which pseudomorph vertically-aligned swallowtail gypsum; this form

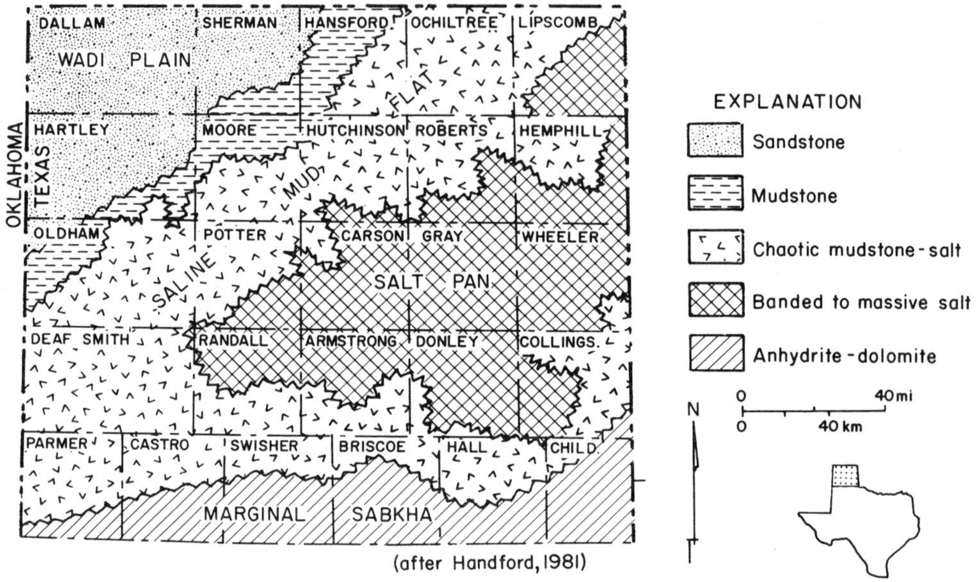

Figure 2.15 Distribution of major facies belts in the Palo Duro Basin during Clear Fork time (after Handford, 1981b).

of gypsum grows only in a subaqueous environment (Warren and Kendall, 1985). Such gypsum was deposited in shallow-water salt pans scattered over the salt flat. The bedded halite, often associated with the gypsum pseudomorphs, contains upward-aligned chevrons and was also laid down subaqueously in the same salt pans. These facies in turn pass basinward into

4. Nodular, nodular-mosaic, and massive anhydrite and dolomite. The dolomite is often shaley at the base and grades upward from burrowed mudstones and skeletal wackestones to cross-laminated oolitic or pellet dolomite packstones-grainstones and laminated dolomite mudstones with algal laminae, intraclasts, fenestral cavities, and nodular anhydrite. This shoaling sequence is interpreted by Handford (1981b) as marine passing up into marine marginal and sabkha environments. It is the most basinward of the facies types and passes south into the marine limestones of the Midland Basin.

This grouping of sabkha facies is not truly continental nor marine (Fig. 2.15); it is a setting that passes from continental to marine depending on relative sea level and the rate of sediment yield. During the early stages of deposition after the initial transgression at the base of the lower Clear Fork, a carbonate–anhydrite sabkha was constructed in the northern Panhandle along a terrigenous sediment dominated coastline. Rapid sedimentation and slow rates of subsidence in the northern Panhandle allowed the sabkha to prograde across a very shallow marine inner shelf. As the sabkha prograded, the up-dip areas no longer received marine water and sedimentation virtually ceased (stage 1, Fig. 2.16). Southward in the Palo Duro Basin subsidence rates and perhaps water depths increased, causing the sabkha progradation to slow down, and eventually the shoreline position became almost static. Roughly equal rates of subsidence and sedimentation

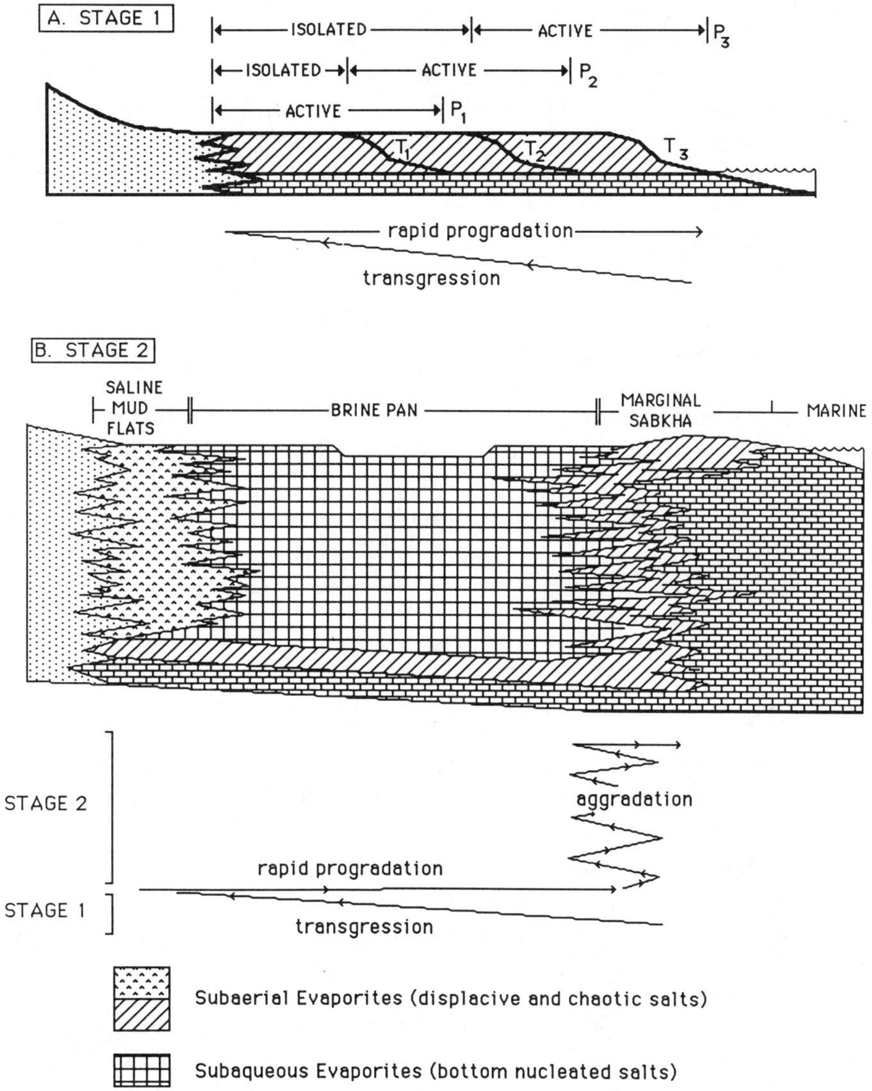

Figure 2.16 Two stage development of evaporite deposition in the Lower Clear Formation (after Handford, 1981b). Stage 1: initial transgression followed by construction of a sabkha which prograded through time ($T_1 \rightarrow T_2 \rightarrow T_3$) to a final position T_3; Stage 2: subsidence leads to aggradation and the development of an extensive brine pan behind a marginal sabkha.

formed an aggradational, or climbing upward stratigraphic pattern (stage 2, Fig. 2.16). Continued subsidence in the up-dip areas caused periodic flooding of the salt flat pans by brines from the inner shelf. The result was widespread deposition of bedded subaqueous halite directly above the marginal-sabkha anhydrite–dolomite facies.

To explain the impoundment of brines in salt pans during stage 2, Handford provided a rather interesting explanation of the origin of the brine filled depressions first documented by Kendall (1979). He postulated that the pans were formed by the growth of displacive nodular anhydrite in the surrounding sabkha. This raised the surrounding sabkha surface and created local depressions or shallow ponds which eventually were filled with subaqueous evaporites.

The Upper Minnelusa Formation, Wyoming

Achauer (1982) described interdunal sabkhas from the Upper Minnelusa Formation (Permian), Rozet Field, Powder River Basin, Wyoming (Fig. 2.17). The section he studied came from the Upper Minnelusa from the Minnelusa–Opeche contact to the top of the "Mel" marker (Fig. 2.18). This is a widespread marker easily recognized on gamma and sonic logs. At the time this succession was deposited, the area was a broad marine shelf adjacent to an extensive eolian dune field. Conditions on the shelf alternated between open marine and eolian deposition.

Geology

The Upper Minnelusa in the Rozet Fields is composed of a number of shoaling cycles. A complete cycle (Fig. 2.18) is composed from top to bottom of

3. Supratidal facies of anhydrite with "chicken-wire" texture and/or a capping erosion surface with rip-up breccias. These evaporite units are no more than 1 m thick.
2. Intertidal facies of algal-laminated or layered dolomicrite which may be desiccated or slightly brecciated.
1. Subtidal facies of peloidal dolomicrite or dolomitized packstone containing fusulinid or crinoid fragments.

The succession fits the classic sabkha trilogy and the anhydrite units are less than 1 m thick. Complete cycles are often interrupted by exceptionally clean, well-sorted quartzarenites deposited in a variety of subjacent settings including eolian, beach, and very shallow subtidal (foreshore) environments.

Detailed mapping of the subtidal dolomite units, especially the lowermost unit directly above the "Mel" marker (Dolomite 1), further refines the sabkha

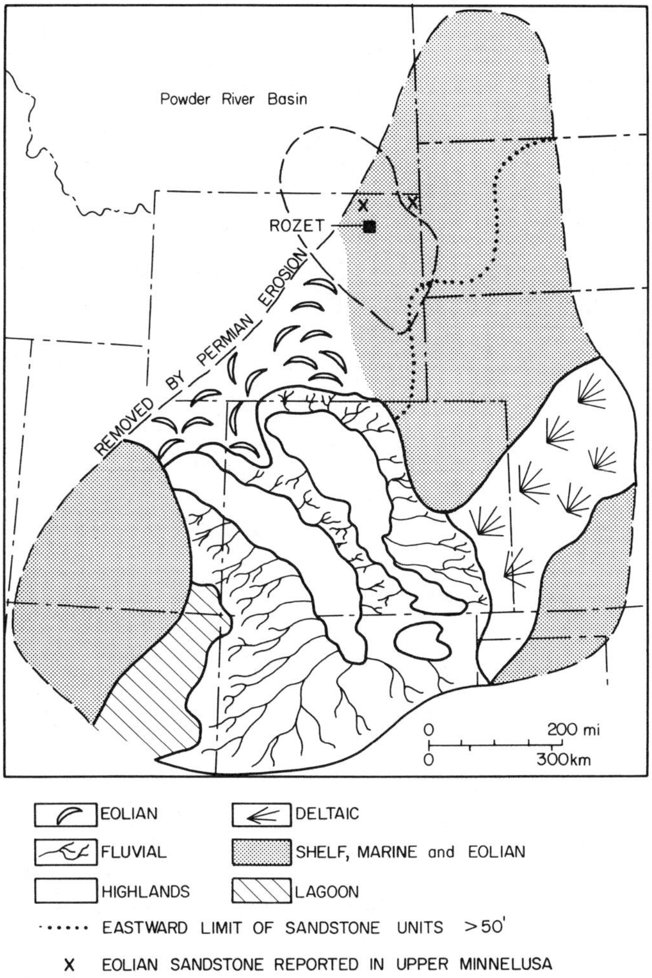

Figure 2.17 Paleoenvironmental reconstruction for the northern Rocky Mountain region during Upper Wolfcampian time (after Achauer, 1982). X shows documented locations where marine marginal and eolian deposits interfinger in the subsurface.

interpretation. The dolomite forms a series of narrow northwest trending channels separated by quartzarenites (Fig. 2.19). This trend is an eolian sabkha trend, with the sabkha dolomites forming in the lows between a series of parallel crested desert dunes. The depositional setting was an area adjacent to an open marine sabkha. It was similar to a region south of Abu Dhabi where the modern coastal sabkha of the Arabian Gulf passes landward into interdunal continental sabkhas. Often, the full sabkha "trilogy" beginning with the subtidal and ending in the supratidal is not found in the Minnelusa, as the cycles are periodically interrupted by episodes of quartz sandstone deposition. The reservoir facies in this area is the quartz sandstone facies wherever it has not been plugged by anhydrite.

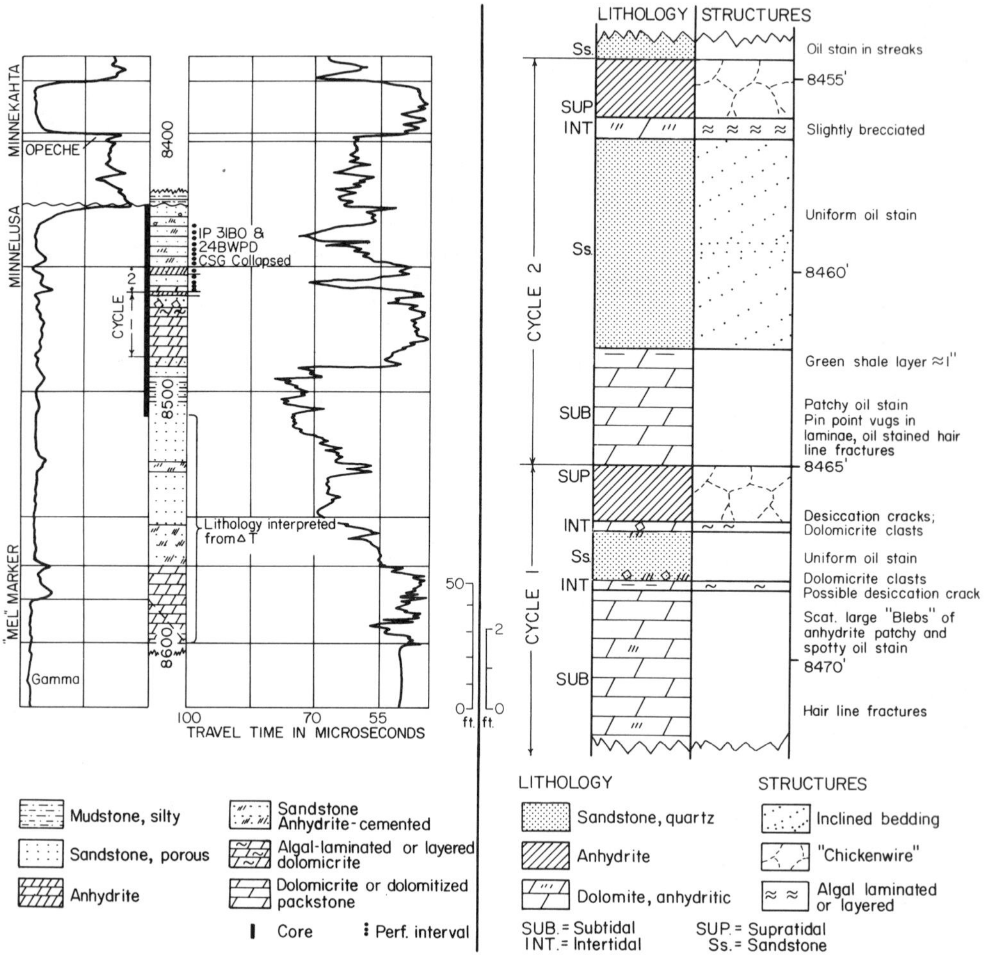

Figure 2.18 Gamma ray-sonic log profiles through the Upper Minnelusa Formation, the Opeche Shale, and the Minnekahta Dolomite (after Achauer, 1982). Also shows an expanded graphic log and interpretation of the two depositional cycles documented on the wire-line log. Logs are from the Arco Peter Svalina No. 15 well, NENE, sec. 25, T50N—R 70 W, South Rozet Field, Wyoming.

Rift Basins

Ephemeral-water evaporites are common deposits in the early stages of rifting in an arid climatic setting. Rifts occur in tensional settings and are characterized by sedimentation in grabens or half grabens. They may be intracontinental as in the case of the East African and Baikal Rifts. Then they are filled with alluvial and sabkha/lacustrine sediments (sometimes evaporitic). Alternatively rifts may be mid-ocean ridges, but in this case the sediments are marine pelagics veneering

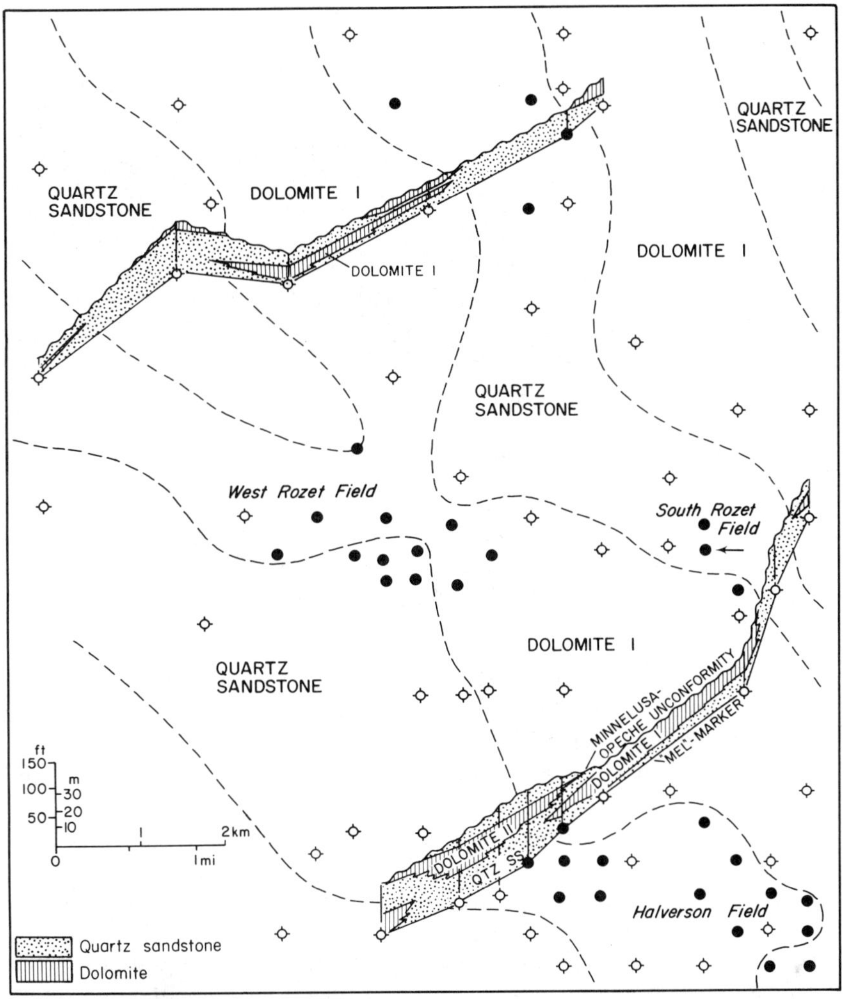

Figure 2.19 Distribution of the Dolomite 1 member hung from the "Mel Marker" (after Achauer, 1982). Note the elongate interdunal trends of the sabkha dolomites; arrow points to Arco Peter Svalina No. 15 well.

oceanic crusts, not evaporites (e.g., the Mid-Atlantic Ridge). Most importantly for this discussion of ephemeral water evaporites are situations where the rift is an incipient ocean like the Gulf of Suez or the Danakil depression; then the continental sediments can be encased and sealed by sea-marginal evaporites (Angelier, 1985). In a sea-marginal rift setting the rises and falls of tectonically active transform offsets and associated volcanism often form a zone of restriction to inflowing seawater so allowing thick sequences of basin-wide evaporites to be precipitated (Fig. 2.20).

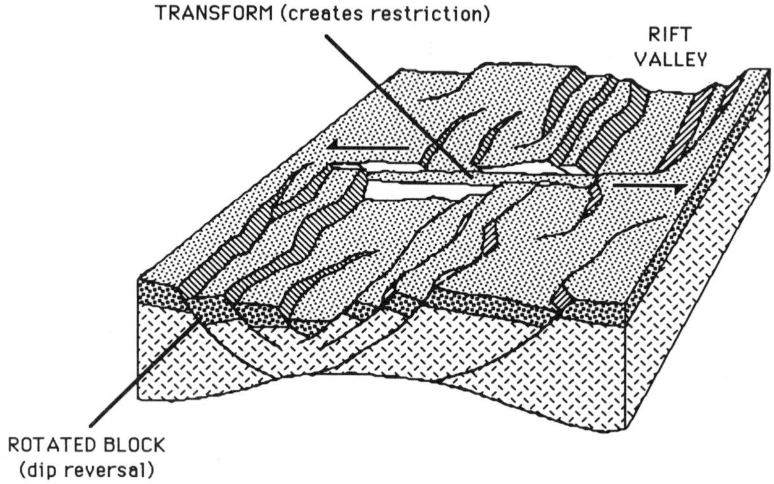

Figure 2.20 Rift setting showing rotated fault blocks dipping away from the basin axis. Many continental rifts are offset by transforms which act as zones of restriction to water inflow as the sea begins to enter the rift depression. This generates a series of offset restricted basins sequentially filled by marine evaporites (e.g. the Proto-Atlantic).

The present southwestern margin of the Red Sea is floored by a series of rotated fault blocks which tilt away from the axis of spreading to form the Danakil Alps and Depression, a Pliocene rift that is still evolving and accumulating ephemeral water evaporites (Hutchinson and Engels, 1970; Lowell and Genik, 1972). On the deep asymmetric west side of the Danakil depression at least 1000 m of virtually pure, Quaternary halite has been drilled (Holwerda and Hutchinson, 1968). It is a modern example of relatively pure ephemeral-water salt accumulating in a narrow restricted trough, a consequence of continental breakup. The evaporites probably precipitated during periodic incursions of the Red Sea into the Danakil depression. The modern depression is only 40 km from the Red Sea and has a large area below sea level to a maximum depth of 117 meters. The depression was covered by seawater as recently as some tens of thousands years ago (Tazieff, 1970; Lalou et al., 1970). It would take only relatively minor tectonic subsidence in the sill region to flood it once again. Geologically this environment is very active with ongoing faulting, igneous intrusion, volcanism, salt flow and hot springs. Seismically the area is also very active with evidence of recent faulting in alluvial fans and in recent volcanics.

An earlier evaporitic episode in the Red Sea rift is marked by up to 7.55 km of Oligocene and Miocene evaporites mixed with siliciclastics and volcaniclastics. Volcanic sills across the narrower part of this marine trough periodically restricted the inflow of marine waters into this earlier depression and so initiated evaporite deposition. The relative purity of the evaporites deposited during this stage was aided by the arching associated with early rifting which carried terrestrial runoff and the associated siliciclastics away from the rift axis. From reflection of seismic data the deepest part of the Red Sea basin, west of the axial rift, is estimated to contain at least 5000 m of bedded evaporites. The Amber Well penetrated 3340 m of almost pure halite in this basin, but this is not a true thickness as it has been influenced by salt flow.

Tilted half-graben fault blocks covered with continental deposits and sealed beneath marine marginal evaporites are found in the subsurface of the continental shelves along both sides of the Atlantic. They formed along the incipient rift of the proto-Atlantic and have been used to date the time of rifting in different areas (Rona, 1982). Jurassic evaporites and diapirs are present off the coast of Nova Scotia and Newfoundland and in the Gulf of Mexico. Lower Cretaceous evaporites occur off Brazil and West Africa from Nigeria to Angola. Thus the proto-Atlantic rift opened from north to south. Rona went on to note the evaporites of the Atlantic region only occur in those parts of divergent margins that were oriented more or less perpendicular to the oceanic fracture zones. Presumably because the sheared margins such as north Brazil did not open into sizable basins until oceanic circulation was established in the main rift of the proto-Atlantic.

In addition to active linear rift seas opening into oceanic realms some basin-wide evaporites accumulate in aulocogens. Aulocogens are the failed arm of a triple point junction, failed in the sense that one arm typically stops spreading after a few million years, often before or just after the development of true oceanic crust in the rift center. The Viking-Central Graben of the North Sea during Triassic-Cretaceous is a good example of an aulocogen floored by Permian rift evaporites (Sclater and Christie, 1980). The North Sea Viking-Central Graben actually shows several periods of rifting ranging from the Permian to the Cretaceous (Zeigler, 1980) with evaporites deposited during and after the earlier Permian stage of rifting (Glennie and Buller, 1983). The Paleozoic Mississippi Embayment and the Proterozoic Adelaide "geosyncline" in South Australia are other examples of evaporitic aulacogens (Von der Borch, 1980).

Ephemeral-water evaporites form during the early rift stage and occupy characteristic positions in the sediment infill of a rift basin; areas that were the topographically lowest points on the continental land surface. Often these were located near the central axis of the continental rift. The intracratonic setting means most of the evaporitic sediments below the basin-wide sea-marginal evaporites will be subaerial or shallow-water continental. They overlie and are interbedded with other continental deposits of the early rift and underlie marine deposits of the spreading margin. Thick basin-wide shoal-water and ephemeral-water marine evaporites were deposited as sea-marginal salts when the continental rift opened up to the sea; these thick, tight evaporites often seal potential reservoir sands deposited as fluviolacustrine sediments. Of course, not all the basin-wide evaporites will overlie continental deposits; some will overlie the basaltic lavas and dykes of the spreading rift. Brine depths during evaporite deposition in a marine-margin rift basin will depend very much on the topography of the basin when marine waters first start to seep into the basin depression. However, the requirement that more water leaves the basin than enters it, means most rift evaporites are extremely shallow water or sabkha deposits. Evaporitic sediments in Cabinda, Angola fit this pattern, as do the sediments of the Newark Series in the eastern USA (Brice, et al., 1982), and the Gulf of Suez (Abdine, 1981).

Doming and extension within rift basins leads to a series of rotated fault blocks with associated dip reversals (Fig. 2.20: Mitchell and Reading, 1986; Warren, 1986b). Dip reversal prevents hydrocarbon leakage on what would otherwise be an open up-dip end to the rift fill and has created many evaporite-sealed reservoirs in the Gulf of Suez, the circum-Atlantic basins including Angola, and some of the lacustrine basins in Bohai region, China. Historically most of the hydrocarbons in evaporitic and other rift basins have been discovered in sand reservoirs on the up-thrown side of extensional blocks and encased in evaporites or shales. Most of the production comes from the highest part of a tilted, listric faulted block. Some of the hydrocarbons present in a rift setting are probably sourced by organic-rich sediments of rift-valley lakes (Warren, 1986a). In the Gulf of Suez there are evaporitic source rocks in Miocene rift evaporites and subjacent potential reservoirs in the siliciclastic fans of the rift margin (Richardson and McArthur, 1986).

The Suez graben is a well-documented example of a hydrocarbon-rich rift setting. All of the oil fields in the Suez basin lie along a trend parallel to the northwest-southeast axis of the graben. Essentially all the oil discoveries are concentrated along three major pre-Miocene uplifted blocks that are second-order structures within the first-order rift. During the Miocene period, step-faults tilted the blocks toward the Gulf. The new highs, generated by this faulting, were the source areas for new deposits and were suitable for reef growth and lagoonal sedimentation that today act as reservoir and source for some fields (Ahmed, 1972). Reservoirs in the Suez basin range from Paleozoic through Miocene. Most of the oil is trapped in pre-rift and rift strata on the higher parts of tilted extensional blocks beneath a basin-wide evaporite seal (e.g., Ramadan Field; Fig. 2.21). The principal reservoirs in the Gulf are the lower Miocene sandstones within the Kareem and Rudeis Formations, the pay zones of the largest Egyptian fields. The pre-Miocene Nubian sandstone is also an important reservoir rock, especially in the cluster of fields in the offshore central production belt (Soliman and El Badry, 1970). Recently oil was discovered on the down-thrown side of a major normal fault in the October field West Abu Rudeis trend (Fig. 2.21; Abdine, 1981). The discovery well GS173-1 tested 4900 BOPD from a 30 m thick oil zone in the basal Nukhul sandstones and conglomerates; sediments presumably eroded from an adjacent uplifted fault block. This occurrence has important implications for oil occurrence within evaporite-sealed rift basins as previous exploration models have concentrated on the highest portions of uplifted blocks. The biggest problem in exploring the Gulf of Suez and extending geological interpretations down to the sub-salt sediments, has been seeing through (seismically) the basin-wide Miocene salt.

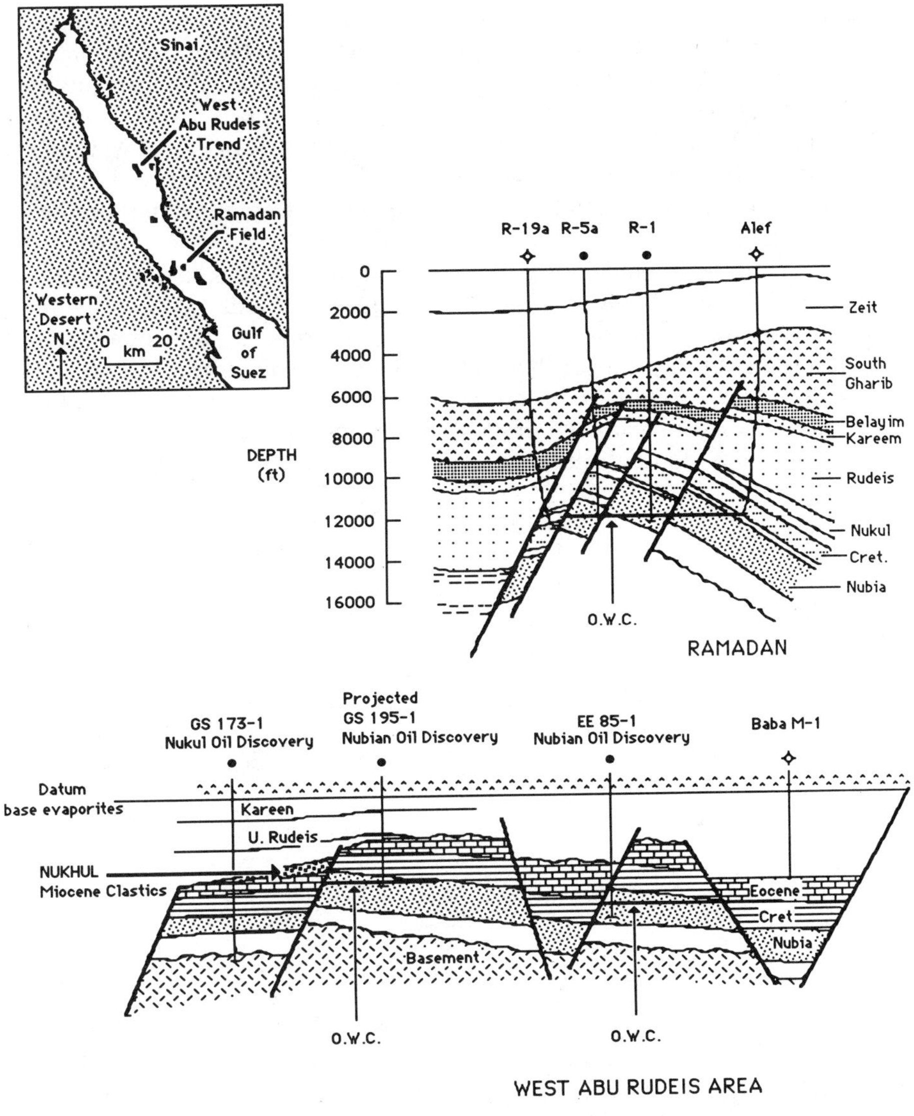

Figure 2.21 Gulf of Suez showing location of some of the major rift-related oil fields and cross sections from Ramadan and West Abu Rudeis fields (after Abdine, 1981).

SUMMARY

Sabkhas form shoaling sequences in both continental and coastal mud flats. Most of the evaporites in a sabkha are deposited in the capillary zone as intrastratal displacive nodules and crystals. Vertical sections from ancient coastal and continental settings can be distinguished by evaporite mineralogy, the presence of body fossils in the lower part of the cycle and the geometry of the salt flat (Figs. 2.4 and 2.9). Idealized sabkha sequences are shoaling cycles with displacive evaporites in the upper part of the salt flat. Each cycle is capped by an erosion surface, the result of displacive crystal growth raising the surface of the wet mud flat into the vadose zone. Prograding marine sabkhas often fine upward, while prograding continental sabkhas tend to coarsen upward (this can sometimes be seen in wire line logs).

Grainstone and sandy reservoirs in marine sabkhas are in the subtidal and intertidal sand bodies sealed by evaporites. Sandy reservoirs in continental sabkhas are the infilled ephemeral stream channels and dune sand sheets dispersed about the edge of the ephemeral lake. Dolomitized sabkha mudstones can form reservoirs with good intercrystalline porosity. At this stage the controls on dolomite distribution are not well understood, but in marine sabkhas the dolomitization seems to occur just landward of the sabkha shoreline (Chapter 5).

CHAPTER 3

Shallow Water Evaporites: Lakes, Salinas, and Platforms

INTRODUCTION

Shallow-water evaporites have accumulated in both marine and continental settings. Until the 1970s most subaqueous textures were documented in ancient halite-rich basins (Dellwig, 1955) and few documented examples of modern subaqueous deposition existed. Then studies were published of Quaternary subaqueous evaporites in Baja California, southwestern United States, Sicily, South and Western Australia and the Middle East (Table 2.1). This work is leading to a new understanding of the significance of laminar and bedded evaporites (both sulfate and halite). In the last decade many ancient evaporites that were previously interpreted as sabkha sequences were reevaluated and often found to be combinations of a number of depositional settings. Ancient evaporites are not simply sabkha or subaqueous; rather, they are combinations of depositional settings, evolving over time from one setting, say sabkha, into another, perhaps subaqueous environment. This chapter gives criteria useful in determining a shallow-water origin for ancient evaporites. The term *shallow water* must be applied wisely to ancient evaporite sequences and should be used only when sufficient evidence is present in core or outcrop. With the reference models available at the present time the distinction between subaqueous and subaerial evaporites is readily made (Warren and Kendall, 1985). It is much more difficult to assign water depth in ancient subaqueous evaporites, especially in sulfate-rich evaporites where gypsum converts to anhydrite with burial and destroys much of the original texture.

Two types of modern shallow subaqueous evaporite basins are known, those fed mostly by marine-derived waters and those fed mostly by continental waters. Ancient basins fed mainly by marine waters often contain evaporite salterns laid down in areas as diverse as sea-marginal, platform, or platform margin, and were deposited on both ramps and shelves. At times saltern evaporites formed basin-wide evaporites (Chapter 1; Warren, 1986b). Modern seawater-fed subaqueous evaporites have only been found in sea-marginal settings. Basins fed by continental waters contain subaqueous evaporites laid down in modern and ancient lacustrine settings. Inflow waters enter marine and continental basins as either surface inflow across washover flats, inlets, and in rivers, or they enter the basin via the subsurface as resurging groundwaters. In a marine basin the waters can be modified by rainfall or riverine input. Unless they form in a near-coastal position, continental basins possess no marine input and rely on meteoric groundwaters and surface runoff to renew water supply.

The hydrology of all shallow-water evaporite basins changes with time. Marine-fed basins are less susceptible than continental basins to short-term changes (weekly to annual) in both water level and salinity. Marine basins are connected to the isochemical reservoir of the ocean so that as evaporation removes water from the basin, seawater will flow or seep into the basin to replace it. This replenishment mechanism is called evaporative drawdown. Because seawater chemistry has changed little since the end of the Precambrian, most marine evaporites are deposited in a readily predicted mineralogical suite (Chapter 1). Short term or seasonal variations in weather patterns can affect the mineralogy of evaporites deposited by affecting the salinity of the seawater-derived brine. A sudden increase in rainfall or humidity will decrease brine salinity in the basin and often produce laminae composed of less saline minerals. Similarly, a storm can wash in unaltered seawater to a basin, so generating a pulse of less saline minerals. Such periodic freshening of an evaporite basin often led to a situation of density stratification of the water body with less dense seawater or meteoric water overlying a much denser brine. This situation existed until the surface waters concentrated to where the densities of the upper and lower brine masses were equal and the brine body mixed. Density stratification in these ancient platform-size water bodies meant that the bottom waters stagnated for long periods (months to years) and may explain the organic rich nature of some ancient evaporitic carbonates and evaporites (Kirkland and Evans, 1981; Warren, 1986a).

Coastal and saltern evaporite basins are also affected by longer term hydrological changes (hundreds to thousands of years), often controlled by shifts in relative sea level. A transgression raises the regional water level flooding the basin with fresh seawater; this prevents or restricts the deposition of evaporite minerals. A regression can lower the regional water level to where the basin is isolated from marine waters; this change may convert it to a continental evaporite basin or stop evaporite deposition. Slow falls in the regional water level also reduce the hydrodynamic head supplying seawater to groundwater-fed coastal basins. This may increase the salinity of waters in the basin and change the basin from a carbonate/gypsum depositing basin to a halite depositing one.

Continental basins and their hydrologies are even more at the mercy of climate than coastal basins. They receive their waters almost exclusively from meteoric sources. Changes in weather patterns or long-term humidity may mean a period of perennial lacustrine deposition or complete desiccation. If the rate of water supply to a perennial lake basin is reduced for many seasons, the lake may change from a perennial to an ephemeral lake with the possible production of a deflation surface in the basin center surrounded by a fringe of eolian evaporite and siliciclastic dunes. This transition is seen in many of the world's Quaternary playas (Chapter 2).

MARINE BASINS

Modern Subaqueous Sulfate Deposits

Most modern shallow-water evaporite deposits are forming in coastal lakes or salinas (Warren and Kendall, 1985; Kushnir, 1981; Arakel, 1980). Coastal salinas are found around the edges of the Mediterranean, the Black Sea, the Red Sea and along the southern and western coast of Australia. The best documented examples are the South Australian salinas where sequences of subaqueous gypsum up to 10 m thick have formed in the last 6000 years (Warren, 1982a, b; 1983).

Salinas of South Australia

The salinas occur as isolated lacustrine depressions within surrounding Quaternary calcareous coastal dunes. They are filled with Holocene bedded evaporites and carbonates. Solutes are supplied by resurging marine-derived groundwaters. Seawater plus some meteoric dune water seeps into the basin as the surface waters of the salina evaporate to levels below the nearby ocean (evaporative drawdown). In plan view the sediments in South Australian salinas show a bull's-eye pattern consisting of a carbonate rim surrounding a central zone of laminated gypsum (Fig. 3.1). Individual salinas measure up to 20 × 12 km with centers filled by as much as 10 m of laminated, subaqueous gypsum. The carbonate unit of the lake edges is a distinct unit from the more central gypsum; the two are often separated by a sharp, often nearly vertical contact. The $CaSO_4$ phase in the basin is always gypsum, not anhydrite, as the temperatures are not high enough, nor the basin arid enough, to deposit anhydrite. The South Australian coastal zone experiences Mediterranean style weather with cool, wet winters and hot, dry summers. Annual rainfall is 250 to 600 mm/yr and annual evaporation is

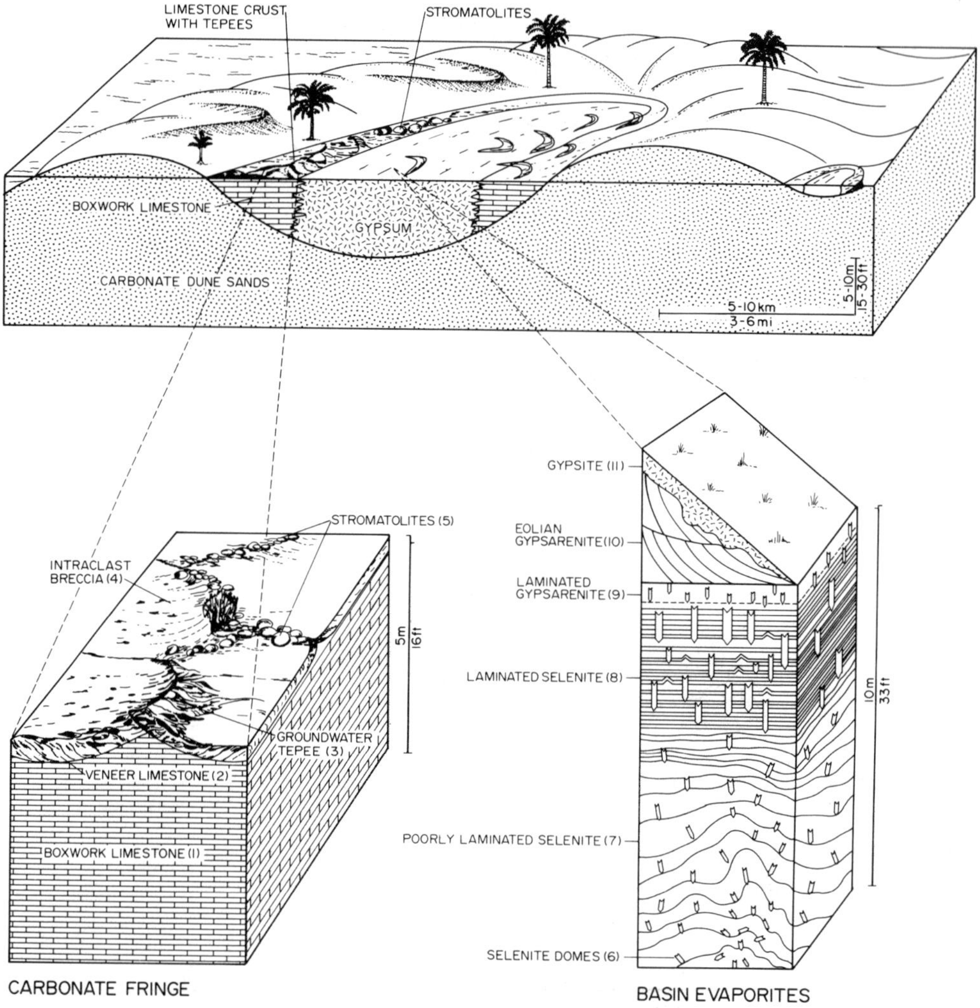

Figure 3.1 Depositional facies in a modern subaqueous setting, based on the salinas of South Australia (after Warren and Kendall, 1985). The facies can be subdivided into a central subaqueous gypsum facies surrounded by a carbonate facies.

1500 to 2250 mm/yr. Winter temperatures are 10 to 20°C and rarely fall below freezing; summer temperatures are often in the 30s but can climb into the 40s.

Much of the salina gypsum is coarse grained and bottom nucleated; pore infilling gypsum is rare. The South Australian salinas are the only documented occurrence of naturally occurring, widespread, Holocene subaqueous gypsum deposition (Warren, 1982a). The gypsum is usually >90% pure and is laid down as a characteristic shallowing-upward evaporite unit (Fig. 3.1). At the base of the succession are massive, poorly layered domes of coarse-grained gypsum (selenite). The elongate gypsum crystals in this zone show little or no preferred orientation, and carbonate impurities are distributed randomly through the gypsum (Fig. 3.2a) Higher in the section the degree of lamination increases and the dome amplitude decreases until the domes pass into horizontally laminated selenite, where individual laminae appear to crosscut large upwardly aligned gypsum crystals (Fig. 3.2b). These large crystals at first appear secondary and their Miocene equivalents in the Sicilian Basin were once interpreted as secondary gypsum after anhydrite (Ogniben, 1957). Studies of identical crystals in the South Australian salinas have shown that the gypsum crystals (30 to 50 cm long) are primary, not secondary crystals. They grow with their long axes perpendicular to bedding, an effect of crystal impingement and growth alignment. Carbonate laminae form by the precipitation and accumulation of aragonite peloids during the spring and early summer. The peloids are pellets from ostracods and brine shrimp mixed with the micritized remnants of algal tubules. At first the peloids mantle the underlying gypsum and often are captured by a cyanobacterial mat covering the gypsum. Peloids are subsequently encased by the upward poikilitic growth of the large gypsum crystals during the summer and fall. The central zone of coarse-grained gypsum, punctuated by carbonate laminae, is in turn overlain by a mm-laminated, sand-sized gypsarenite accumulation. The upper part of these horizontally laminated gypsarenites is often reworked into wave-oscillation ripples. Laminated and rippled gypsum is in turn overlain by a thin, massive, poorly bedded gypsarenite unit deposited under seasonally vadose or subaerial conditions. Capping the whole succession is a unit of cross-stratified eolian gypsum and, in areas stabilized by vegetation, a pedogenic crust of gypsite (silt-sized) has formed atop both lacustrine and eolian sediments (Fig. 3.2c).

The carbonate unit forming about the edges of South Australian salinas is composed of a boxwork limestone unit (4 to 5 m thick) overlain by a fenestral limestone sheet less than 1 m thick (Fig. 3.1; Warren 1982b). Boxwork limestone is a diagenetic unit created by the cannibalization of earlier lacustrine carbonate by inflowing "fresher," but still saline, marine-derived groundwaters. The "boxwork" contains algal structures, evaporite pseudomorphs and other highly altered bedded fabrics dependent on the preexisting sediment types (Fig. 3.2d). Some areas of the limestone sheet above the boxwork limestone consist of subaerial algal tufas or subaqueous mm-laminated stromatolites and algal mats (Fig. 3.2e). Other areas consist of fenestral capillary crusts crosscut by tepees (Fig. 3.2f). The stromatolites grew as current-aligned domes up to 40 cm tall in the continuously subaqueous parts of the lake edge or as mats in areas of very shallow and sometimes ephemeral surface water along the lake shoreline (von der Borch et al., 1977; Warren, 1982b). In some areas a little farther out in the lake, the stromatolite domes cap tepee structures. Like stromatolites the tepee structures are confined to the capping carbonate unit and form an extensive fenestral limestone sheet up to 60 cm thick. Tepees form readily in areas of resurging groundwater, where seasonal changes in the groundwater head encourage tepee growth (Warren, 1982b). Most of the limestone sheets caught up in the tepee structures are

(a)

(b)

(c)

Figure 3.2 Salina sediments: (a) coarse-grained gypsum in the domal core at the base of the gypsum sequence, New Lake, Kangaroo Island (pen for scale); (b) primary coarse-grained gypsum which has poikilitically enclosed aragonite laminae, New Lake, Kangaroo Island, South Australia; (c) eolian gypsum sand overlain by ''gypsite'' soil, Cooke Plain, South Australia; (d) boxwork limestone, Marion Lake, South Australia; (e) tepee structure, Deep Lake, South Australia; (f) slabbed mm-laminated stromatolite taken from top of tepee, Marion Lake, South Australia.

capillary crusts composed of fenestral lime mudstones containing stromatactic-like voids, pisolites, laminar cements, and internal sediment.

Lake MacLeod, Western Australia

Lake MacLeod is a halite-filled salina 120 km long and 40 km wide lying just to the north of Shark Bay (Logan, 1981; Handford et al., 1984). The surface of the lake now lies well below sea level and is fed by evaporative drawdown from the nearby Indian Ocean. To the north, and at the surface, the lake is filled by a massive, 5 m thick gypsum layer. To the south over much of the salina there is a massive halite unit which is 7 m thick in some parts of the lake. Both the gypsum and the halite overlie a thin unit of lagoonal carbonate. Gypsum deposited on the lake floor today is a gypsarenite with less than 10% carbonate.

Seawater flows into the lake as a series of carbonate precipitating springs fed from fractures and caves in underlying calcareous coastal dunes. When the water first enters Lake MacLeod it has normal marine salinity. Similar spring-fed normal marine waters are thought to have filled the lake when some 6000 years before present, the Australasian Sea rose to a level that was within a meter of its present level. Since then the sea has been at its present level and has supplied water to the lake by a combination of evaporative drawdown, brine reflux, and spillover. Concentration of this water has filled Lake MacLeod with up to 15 m of evaporites. Initially the water flowing into the lake was concentrated to where it precipitated a carbonate unit across the floor of the lake (Logan, 1981). As the limestone accumulated it created an aquitard which reduced the volume of brine seeping out through the lake floor. Salinities in the lake increased until a gypsum unit was laid down across the lake floor. The gypsum formed an even more effective aquitard than the carbonate unit. Salinity of the lake brine then increased to where halite could precipitate and accumulate in the lake particularly at its southern end. Carbonate dominated the evaporite mineral suite at the northern end, and passed rapidly south into gypsum. This relatively thin band of sulfates passed farther south into halite which dominated the mineral suite over much of the lake during this early depositional phase. As the playa filled with sediment, the groundwater head drawing water into the lake was reduced, the perennial water body became a playa. Meteoric input became important at this late stage in the depositional fill much in the same way it did in the South Australian salinas (Warren, 1982a). Now the salinity of the surface waters was once again reduced to where gypsum was the stable mineral phase and the Ibis Gypsite unit was deposited across the lake. Local springs continued to feed the lake throughout its Holocene history, carbonate was deposited adjacent to the springs and gypsum and halite filled the salina in an asymmetrical fashion away from these point water sources (Handford et al., 1984).

The dominance and preservation of a thick halite unit in Lake MacLeod is due to the arid setting of this lake compared to the gypsum-filled lakes further south. In the southern Australian examples, meteoric water played a role throughout the depositional history and became increasingly important as the lakes filled with sediment. By contrast, in Lake MacLeod, the arid conditions meant seawater was always the dominant inflow. The lack of substantial meteoric input during the depositional history allowed the lake waters to remain in the halite precipitation field for extended periods of time (Warren, 1983b).

In general, independent of the degree of aridity, the inflow zones in modern coastal evaporite lakes are marked by carbonates, whereas the areas of brine ponding are defined by subaqueous gypsum and/or halite deposition. In arid settings the more saline minerals dominate the sediment section while in semi-arid settings the assemblage becomes less saline toward the top of the section and the margins are thick carbonates. The dominance of boxwork limestone in many South Australian salinas and its absence in salinas in more arid settings may be due to the relative importance of climate. If so the presence of a thick boxwork unit about an evaporite-filled lake may well be an indicator of a semi-arid versus arid setting (Warren, 1983b).

Gypsum is not only deposited as coarse-grained gypsum (selenite) in these Holocene coastal lakes, but salinas filled with sand-sized gypsum are also common. In fact it is the presence of beds of coarse-grained gypsum (selenite) which makes the southern Australian salinas unique among modern gypsum deposits. The more usual form of gypsum is as sand-sized detrital accumulations. Such deposits occur in many Australian salinas, in lagoons about the Red Sea, and as was discussed in Chapter 2, in continental playas (Table 2.1; Kushnir, 1981; Arakel, 1980; Warren, 1982a, 1983). In subaqueous and some ephemeral brine ponds the gypsarenite is deposited as laminated units composed of sand- and silt-sized gypsum disks and needles or cleavage fragments. Often, this clastic gypsum is deposited as cross-laminated and algally bound units. Basal scour surfaces and intraclast breccias show that these clastic evaporites were subjected to periodic high-energy storms.

Modern Subaqueous Halite Deposits

Halite also accumulates in shallow water and does so in three main forms: detrital halite, crust halite, and displacement halite (Arthurton, 1973; Gornitz and Schreiber, 1981). When detrital halite is laid down, it is composed of fragments of surface-grown hopper rafts and small individual halite crystals. Halite usually precipitates by the evaporative concentration of waters in a brine pond but may also form as a result of brine mixing (Raup, 1970). Detrital halite units are often ripple cross-laminated when deposited, and ripples composed of halite ooids (halolites) have been observed in very shallow waters of the southern Dead Sea (Weiler et al., 1974). Detrital (mechanically deposited) halite is rarely documented in the rock record, probably because it is a type of halite which with its initial high permeability is susceptible to dissolution and reprecipitation as clear, coarse-grained, secondary halite.

Halite crusts are the best understood type of halite deposition. Sabkha

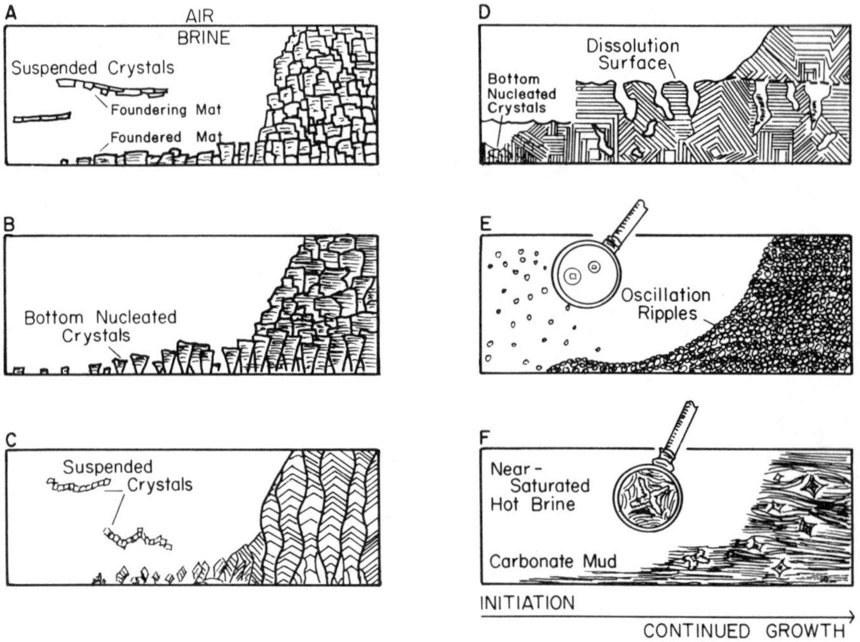

Figure 3.3 Depositional characteristics of halite deposited under shallow water conditions (after Gornitz and Schreiber, 1981).

crusts were discussed in Chapter 2. Crusts are common in many evaporite brine ponds (Shearman, 1970); they have been grown in the laboratory (Arthurton, 1973; Weiler et al., 1974), and their equivalents have been seen in ancient deposits (Wardlaw and Schwerdtner, 1966; Lowenstein, 1982). Crusts form by

1. Foundered, horizontally aligned rafts (mats) of halite crystals coated by bottom-formed overgrowths (Fig. 3.3a).
2. Bottom-nucleated crystals growing *in situ* appear similar in cross section to formation 1. Horizontal lineations in the crystals are formed by fluid inclusions and indicate successive growth stages (Fig. 3.3b).
3. Foundered rafts of halite crystals showing corner-oriented or randomly aligned crystals which develop chevron-shaped lineations from edge-oriented crystal growth (Fig. 3.3c).
4. Bottom-nucleated crystals, growing *in situ* and showing corner-oriented or randomly aligned crystals. Chevron-shaped growth lineations similar to formation 3. Cycles of flooding and dissolution followed by reconcentration yields repeated erosion surfaces (Fig. 3.3d).
5. By accumulation of, and overgrowth upon, detrital halite particles (Fig. 3.3e). Often, such accumulations are rippled when deposited.

Flow strengths in a brine body often vary during the deposition of a halite crust, so most crusts deposited subaqueously are composed of a combination of the various halite types. For example, rafts require relatively quiet water conditions to grow; when the water is disturbed they will sink to the bottom or be washed onto the strandline (Fig. 3.4a), simultaneously moving bottom waters can be depositing oscillation ripples. When the waters quiet down, the deposition of rippled bedforms stops, and rafts or bottom-nucleated chevrons once more accumulate on the floor of the brine pan.

The most common texture in ancient halites deposited under shallow water conditions is bedded halite, formed by the superposition of subaqueous crusts. Each crust is separated from the adjacent crust by films or thin beds of detrital carbonate, sulfate, or terrigenous sediment. They are identified by the presence of vertically aligned chevron halite (Fig. 3.4b; Kendall, 1984). Halite is a cubic crystal, and, because edges in such crystals grow much more rapidly than faces, the result is a growth-aligned texture of elongate halite chevrons with coigns forming upward directed crystal apices (Figs. 3.3c and d). Freshening of the overlying waters (implying shallow water) or occasional subaerial exposure of the uppermost surface of the crust leads to partial dissolution of the crust. During freshening or exposure, the crystal coigns are rounded or the chevrons truncated. The contact areas between adjacent crystals in the crust can be partially dissolved to form voids (Fig. 3.3d). Subsequent concentration of the pore fluids infills the cavities with clear halite (Fig. 2.7f). Continually subaqueous halite crusts are dominated by vertically aligned chevrons; during deposition the relatively stable brine salinities meant that it was less susceptible to recrystallization than are halite crusts deposited in sabkhas. There the crust was deposited as subaqueous aligned chevrons but ended up extensively dissolved and recrystallized in the capillary or vadose zone of the sabkha (See Chapter 2).

The presence or absence of the fluid-filled inclusions in halite crystals within subaqueous crusts can be used as an indirect indicator of water depth at the time of crystal deposition (Kendall, 1984). Inclusion-rich layers in a zoned halite crystal forms under conditions of supersaturation and rapid crystal growth. Inclusion-free crystals with better crystal shapes form under conditions of slower crystal growth. Consequently, a halite unit composed of numerous alternations between inclusion-rich and inclusion-poor halite layers indicate that rapid changes in brine concentration and brine stability occurred during deposition of the halite unit (Fig. 3.4b). This can occur only under shallow brine conditions. Thus the layering in chevron halite indicates shallow-water precipitation and contrasts with the clear halite crystals of a deep-water halite deposit described from the Silurian in the Michigan Basin (Chapter 4). In the Southern Basin of the Dead Sea, halite crusts grow in quiet-water, shallow-water areas while ripple-laminated detrital halite forms in the reworked shallows where crusts are broken up by wave action (Weiler et al., 1974).

Displacement halite is composed of skeletal halite cubes which grow displacively in the bottom sediments (Fig. 3.3f). When the skeletal development is

(a)

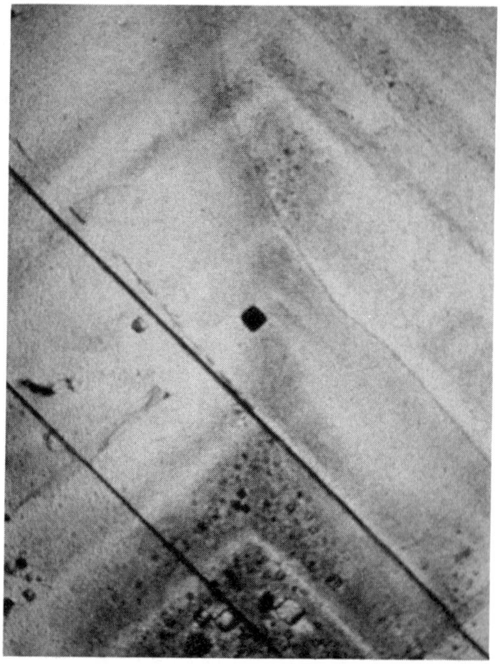

(b)

(c)

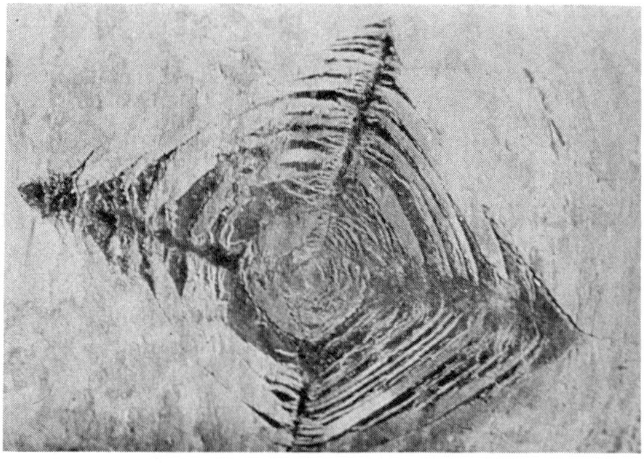

(d)

Figure 3.4 Halite morphologies: (a) stranded rafts of halite, Lake Eliza strandline, South Australia; (b) inclusions outlining lamina in upward-aligned chevron in San Andres Formation salt, Palo Duro Basin, Texas Panhandle, lamina about 0.1 mm thick; (c) modern displacement halite from Dead Sea mud flat (courtesy of B. C. Schreiber); (d) ancient displacement halite (pagoda halite) from Zechstein of England (courtesy of D. Smith).

extreme, it is called "pagoda halite" to describe its highly dispersed growth form, which in two dimensions often resembles the silhouette of a Thai temple (Fig. 3.4c and d). Displacement halite has been grown in the laboratory (Arthurton, 1973), and has been observed in the evaporitic carbonate mud flats of the Dead Sea (Gornitz and Schreiber, 1981) and in ancient sequences (Southgate, 1982; Parnell, 1983). To form, displacement halite needs supersaturated brine within the mud matrix, and so many workers postulate that it usually grows below subaerial mud flats of playas and sabkhas. Thus abundant skeletal halite in an ancient sequence indicates strandline conditions. In contrast, based on his study of a lower Jurassic lacustrine shale at Holyoke, Massachusetts, Parnell argued that pagoda halite can grow subaqueously. Even so, the sequence he studied was deposited under shallow-water conditions in shales that show evidence for periodic emergence. This is not to say that all halite cubes enclosed by a displaced mud matrix are basin edge indicators. Ray Garber (pers. comm. Nov., 1985) found clear centimeter-sized halite cubes growing displacively in the deep water sediments of the Dead Sea. However, these crystals are perfect cubes and lack the tiered edge of pagoda halite (Fig. 4.2b).

Other Holocene Salina Evaporites

Solar Lake, Gulf of Elat

The Northern Sinai coast of the Gulf of Elat is characterized by steep rifted-margin topography and a clearly defined fringing-reef (Aharon et al., 1977; Gavish, 1980). In most areas the mountainous granitic and metamorphic massif meets with gulf waters to produce a narrow steep shore-zone composed of slump blocks, coarse beach-rock plates and narrow reef flats. Locally, wadis extend as fan deltas into the gulf to produce a wider, less steep coastal plain. Small brine ponds or salinas have developed in depressions on this narrow coastal plain. Probably the best known of these is Solar Lake located immediately north of a large fan delta (Wadi Murah) and 18 km south of Elat. Solar Lake is a meromictic brine pond, about 140 m long and 65 m wide, with a maximum depth of 5 m. It is separated from the open gulf waters by a porous beach-barrier, some 60 m wide and up to 3 m high, composed of coarse sand and pebbles of igneous and metamorphic rocks with abundant molluscan shell debris.

The main supplier of water to the pond is evaporative drawdown and the associated seepage of gulf waters into the pool during high tide. Minor amounts of seawater may also enter the pond by splashing over the barrier during strong winter storms. In summer, evaporation exceeds inflow raising the total dissolved solids to 165 gm/l. In winter the inflowing seawater and minor amounts of rainwater form an upper water layer of relatively low salinity. This results in stratification of the water body across a well-developed halocline and a corresponding thermocline. The pond is unusual in that its waters are heliothermal with bottom water temperatures reaching 57°C in the winter and spring, and year-round surface water temperatures of 20 to 28°C. During stratification the lower layer absorbs heat, convection is inhibited, and the bottom layer becomes anoxic with relatively high levels of H_2S in the hot bottom waters. By summer, the evaporation of the upper water body causes a weakening of the density stratification and finally the overturn of the water body. At that time the water level in the pond is as much as 0.5 m below sea level. With overturn the water body becomes homogeneous and aerobic; it also cools down by convective heat loss to a consistent 28°C.

The sediments of the pond reflect the mode of water supply and the associated density stratification of pond waters. The margins of the pond are covered by thick algal mats, the slopes are composed of a white crust of large bottom-nucleated gypsum crystals and aragonite while the deep pond bottom is covered by black mud and loose gypsum mush (Lyons et al., 1984). The algal mats are cross-cut by well developed polygonal cracks giving the mat a "lily-pad" appearance. The cracks form during the late summer and fall when the mat surface is subaerially exposed. Buried beneath the algal and gypsum layers of the pond is a carbonate layer with *Cerithium* gastropods indicating deposition during the initial restriction of the pond. Carbonate laminae in the algal mats and in the gypsum crust are composed of authigenic aragonite, Mg-calcite, dolomite and calcite with the dolomite forming as a replacement of the aragonite (Friedman et al., 1973; Aharon et al., 1977). The bulk of the authigenic dolomite occurs in the gypsiferous sediment not in the algal mats and has an aragonite precursor. It is thought the dolomite is the result of a throughflow of hot hypersaline brines with a Mg/Ca ration increased by the precipitation of gypsum, a form of brine reflux dolomite (see Chapter 5).

Ras Muhammad Pool, Southern Sinai

At the most southern tip of the Sinai there is a small peninsula called Ras Muhammad mainly composed of tectonically uplifted Pleistocene reefs (Gavish, 1980; Kushnir, 1981). At the termination of the peninsula is a 1 km long tidal inlet with a small brine pool at its end called Ras Muhammad Pool; the pool is an oval feature about 100 to 170 m in diameter and not more than 50 cm deep. It is separated from the inlet by a porous barrier some 250 m wide composed of reef material and carbonate sand. Most of the water feeds into this pool as a series of seawater springs located around the pool's edge. Even in this very arid area the pool never dries up although the extent and depth of water in the pool varies with the seasons. A constant recharge by seawater always maintains some water in the pool. During summer and fall, the water volume in the pond is reduced by high evaporation and salinities in the pond reach 300 per mille; in winter and spring the pond expands and salinities fall to 100 per mille.

Sediments in the pool are composed of alternating layers of gypsum mush and algal mats; there is no primary or diagenetic formation of anhydrite in this pool (Kushnir, 1981). The thickness of the gypsum layers ranges between 0.3 and 6 cm and averages 1.8 cm. Porosity of the gypsum layers ranges between 64 and 69 percent. The gypsum in the layers is mainly prismatic gypsum (crystals up to 6 mm long) associated with minor amounts of aragonite, calcite, quartz and feldspar. In the mid-portions of those cores taken through the algal sediments some of the algal layers contain gypsum crystals which appear to have grown displacively within the algal layer. The continued growth of gypsum in the algal layer eventually destroys the lamination. After the decomposition of the algal material the evidence of a former algal origin is ultimately destroyed. Only the carbonate grains captured by the algal mat remain to indicate its former presence. The final fabric as revealed in the lowermost layers of the cores (about 80 cm deep) taken by Kushnir is a laminated gypsum mush with thin interlaminations of carbonate crusts and ooids with very little identifiable algal material.

In the supratidal sediments around the pool evaporites are accumulating. Surface sediments are cemented by a halite crust while beneath the crust and above the water table are concentrations of anhydrite. At the level of the water table there are diagenetic accumulations of gypsum and carbonate minerals.

Hutt and Leeman Lagoons, Western Australia

Hutt and Leeman Lagoons are located in the northern part of the Perth Basin, Western Australia (Arakel, 1980). Like the salinas of Southern Australia and Lake MacLeod, they are shallow water evaporitic basins that formed during the last 6000 years as a result of seawater ponding in a Pleistocene coastal-dune terrain. Both lagoons are filling elongate interdunal hollows between late Pleistocene and Holocene calcareous beach dune barrier complexes. Hutt Lagoon extends 24 km in a north-south direction and is up to 3 km wide; Leeman Lagoon is 28 km long in a north-south direction and up to 3 km wide. The lagoons have filled with sediments in four successive stages each with its own characteristic association of facies (Arakel, 1980).

1. *Open marine embayment stage* represented by a sheet of lithoskeletal grainstone incorporating skeletal remains and seagrass leaf-sheaths.
2. *Marine lagoonal stage* formed as an upward fining marine sediment sequence deposited during a time of restricted open-marine circulation. This was the result of the accretion of coastal barrier-beach and dune complexes at the entrances to the lagoons. In Hutt Lagoon there was simultaneous progradation of alluvial units of the Hutt River.
3. *Evaporitic pond stage* followed the complete surface isolation of the lagoon from the sea. It was characterized by a sequence of gypsum deposited in the lagoonal brine pond as layered and laminated sediments. Most of the gypsum is composed either of bottom-nucleated gypsum crystals or of reworked

gypsarenites and less common gypsrudites. The finer-grained muds form interlayers within the gypsum; they are composed of micrite pellets in Leeman Lagoon and calcareous clays in Hutt Lagoon. Gypsum layers are white and about 5 mm thick, while the muds are dark gray laminae 4–15 mm thick. Localized subaerial exposure resulted in the reworking of some evaporites toward the lagoon center. A pellet intraclast grainstone formed about the edges of Hutt Lagoon and now completely covers the gypsum-rich units of Leeman Lagoon. Arakel (1980, p. 1313) also mentions the occurrence of hemipyramidal anhydrite and nodular gypsum in this veneer. This is a rather unusual occurrence of anhydrite given the semi-arid climatic setting of this coastal region of Western Australia. Both the nodular gypsum and the anhydrite are diagenetic phases forming under vadose, probably capillary zone, conditions.

4. *The pond-playa stage* is the contemporary stage. Mineral precipitation today is restricted to ephemeral halite crusts in ponded waters and the vadose precipitation of $CaSO_4$ minerals in the evaporitic lagoonal sediments.

CONTINENTAL BASINS

Modern Continental Lacustrine Basins

Subaqueous continental deposits form on the floor of a perennial salt lake or on the bottom of brine pan during the flood stage in an ephemeral lake (Chapter 2). Perennial lakes and pans usually form in the topographically lowest region of a closed basin. For example, Lake Magadi is located just south of the equator in a low point in the Gregory Rift Valley of Africa at an elevation of 600 m above sea level. Lake Magadi is the only saline lake in the East African Rift Valley where bedded evaporites have accumulated (Eugster, 1980). To maintain a perennial brine body in such an evaporitic setting requires continuous water inflow to the lake either as river inflow or as spring seeps. In Lake Magadi the perennial inflow is from springs. Perennial surface brines are relatively rare in continental salt lakes; ephemeral salt pans are far more common (Chapter 2). Whenever present, a perennial brine lake is at the mercy of the local climate. Lake water level and salinity are subject not only to seasonal but to long-term fluctuations, often a direct response to climatic or tectonic changes. In the short term the volume of inflow water increases after a storm; in the long term a change to a more humid climate increases the volume of water in the lake depression. Both scenarios flood the lake with fresher water and often form a density-stratified water body with a dense bottom brine overlain by fresher surface water. Tectonic uplift also affects deposition; it can convert what was formerly a hydrologically open basin into a closed basin, or vice versa. It can also change the position of the topographically lowest point in the lake.

Lacustrine Carbonates

Alkaline earth carbonates and evaporites usually accumulate in perennial lakes as bull's-eyes complicated by spring mounds forming carbonate outliers well out into the basin center evaporites. Evaporative carbonates and salts are accumulating today in Mono Lake (California), Great Salt Lake (Utah), Lake Natron (Tanzania), the Dead Sea (Israel), and many other continental perennial lakes (Figs. 3.5 and 4.1). Some lacustrine carbonates form as spring precipitates and

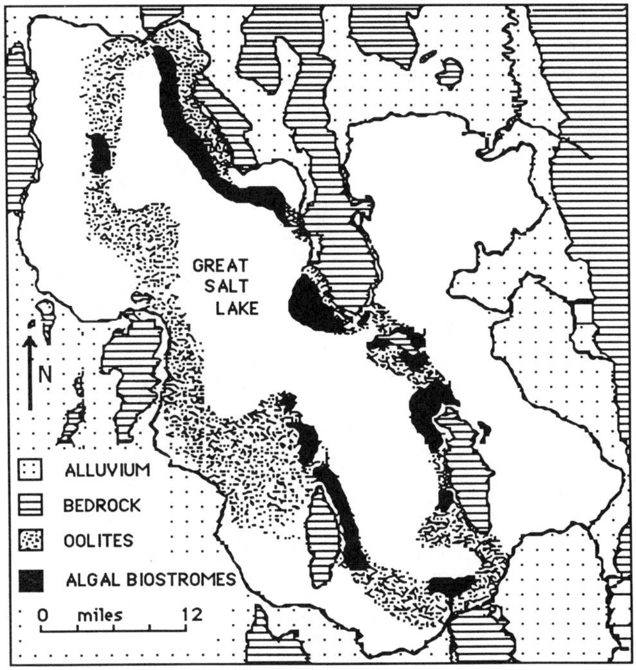

Figure 3.5 Sediments in Great Salt Lake, Utah (after Eardley, 1938). Note the rim of shoal-water carbonates around the subaqueous center of laminated carbonate.

soil encrustations in the salt flats surrounding the perennial water body (Chapter 2). During periodic floods they are broken down and carried into the lake as clastics. Fine-grained carbonates can also crystallize in the water column and then settle to the lake bottom as part of a laminated lacustrine sequence.

Laminated sediments in the central area of Great Salt Lake, Utah, are in large part composed of brine shrimp fecal pellets (Eardley, 1938). Such pellets are a mixture of aragonite and detrital matter, mainly quartz and clays. The brine shrimp in Great Salt Lake is *Artemia gracilis,* a filter feeder; thus the fecal pellets in the bottom sediment represent a sampling of the mineral matter suspended in the water column implying aragonite is precipitating in the water column of Great Salt Lake. Carbonates can also precipitate about the lake shoreline as spring mounds, algal mounds, and stromatolites (Fig. 3.5). In Great Salt Lake the algal mounds belong to an earlier, less saline period (Halley, 1977). Oolite shoals form as sand waves along the shore of Great Salt Lake. Individual ooids are aragonite and show a radial crystal structure. When and how the individual ooids grew is not well understood (Sandberg, 1975). In Mono Lake the spring tufas form pinnacles and needles up to 3 to 4 m tall as well as extensive pavements about the lake shoreline. Pisolites grow in a near-shore position in Lakes Magadi and Natron, where the mixing of Ca—HCO_3 waters with alkaline brine has formed a pisolite pavement (Eugster, 1980). Similar calcite pisolite pavements are growing on the edge of Pastos Grandes in the Andean Altiplano more than 3000 m above sea level (Risacher and Eugster, 1979). There Holocene pisolites are 1 to 200 mm in diameter and grow in shallow, hot-spring-fed stagnant pools as CO_2 is removed from the resurging groundwaters by degassing and algal photosynthesis.

Lacustrine Evaporites

Lake Magadi is one of a few salt lakes where shallow-water evaporite deposition is well documented and has become the modern counterpart for ancient continental bedded evaporites. The lake is off the main rift at the bottom of a steep-sided valley and is the lowest point in the African Rift. It is located 2° south of the equator. Magadi is fed by both sheet flood from ephemeral streams and by perennial springs (often thermal) flowing over and through a surrounding volcanic terrain. The springs form perennial lagoons located about the edge of the lake. The lagoons are the only permanent surface water in the lake; unlike the more central trona depositing areas they never dry up. Magadi lagoons contain a thriving fish population of *Tilapia alcalica,* a fish that survives at temperatures of 39°C and at pH values up to 10.5. Inflowing waters to the lagoons are enriched in sodium and bicarbonate ions from the surrounding volcanics. These basinward-flowing surface waters are concentrated as they move out into the lake. First they precipitate alkaline carbonates and then deposit trona along with other sodium bicarbonate minerals.

Trona ($NaHCO_3 \cdot Na_2CO_3 \cdot 2H_2O$) forms subaqueously as a bottom nucleated crystal pavement in the more central areas of the lake. This part of the lake desiccates in the dry season, subjecting the upper part of the trona pavement to subaerial exposure. The lake experiences two rainy seasons, so that the texture of the trona deposit reflects the marked seasonality of the lake climate. The longer rainy season (March–April) usually provides enough runoff to flood the trona-forming areas of the lake with water up to 1 m deep. Each flooding event supplies new Na^+ and HCO_3^- to the lake basin by dissolving the efflorescent trona crusts on the mud flats surrounding the central pan. Throughout the dry season (May–October) evaporation is intense and the level of the surface brine drops, until by June–July the lake center is dry. Trona begins to form when the lake brine is no more than a few centimeters deep (Eugster, 1980). As evaporation continues to dry up the surface water, trona forms in two modes: as rafts and as bottom-nucleated, upward-pointing blades, both similar textures to the subaqueous halite discussed earlier in this chapter (Fig. 3.3). The rafts contribute only a minor amount of trona to the pavement but are important as they inhibit further evaporation and raise the temperature of the underlying brine through a greenhouse effect. Temperatures of 60°C have been measured below the trona rafts. Winds blowing across the lake often break up the rafts and drive the fragments into heaps about the edge of local brine pools. As the brine level continues to fall, the upward-growing bottom-nucleated trona breaks through the rafts, which in turn become new sites for bottom-nucleated deposition. Finally, the brine level falls below the surface of the newly formed trona crystals and the primary depositional cycle is complete. Trona continues to crystallize from the interstitial brines, reducing the porosity of the trona unit. Each season a trona layer 2 to 5 cm thick forms, but the average accumulation rate for the Holocene accumulation rate of the Lake Magadi trona is at most 0.4 cm/yr. Something like 75% of the trona in the lake is recirculated each year by dissolution and reprecipitation (Eugster, 1980).

The present trona pavement is rock hard, composed of an interlocking meshwork of growth-aligned trona crystals. Even when the lake center dries, the brine level stays very close to the surface of the pavement. During exceptionally dry periods it may drop 1 or 2 cm into the trona layer. Whenever the surface of the layer is exposed during the dry season it can be covered with a considerable amount of windblown material. The trona pavement and the underlying trona deposit, which is 75 km^2 in aerial extent and is 7 to 40 meters thick. The deposit is made up of a succession of trona-detrital couplets. Light-colored trona layers are made up of rosettes of upward-pointing, growth-aligned trona crystals. They are separated from each other by thin bands of trona colored by dark, mainly windblown dust. The presence of continual interstitial brine seems to prevent the complete recrystallization of trona crystals within the accreting crusts. Each layer retains some of the growth alignment of the originally deposited crystals. Expansion polygons or pressure ridges, several meters in diameter, cut up the trona into a series of overthrust saucers with small brine pools in the center of each polygon. Pressure ridges form as a lateral component of compression is created by the sideways expansion of near vertically aligned trona crystals jostling for space in the growing crystal pavement.

Trona in Lake Magadi is mined using a dredging barge. Currently the deposition of trona in the lake is so rapid that at present removal rates Magadi trona is a renewable resource; mined trona is replaced each year with a new crop. If for any reason the dredging barge breaks down for more than a day or two, the trona crystallizing about the base of the barge freezes the barge in place. Trona has to be physically removed from the base of the barge before operations can continue.

Lake Magadi is also an area of silica diagenesis where sodium silicates of the High Magadi beds are altered to cherts (Eugster, 1969, 1980). The High Magadi beds were deposited 9000 to 10,000 years ago during a lake high stand when the lake waters were some 13 to 14 meters above the present level. The sediments of this time were mostly bedded volcanic muds and cherts deposited as laminated mudstones containing abundant fish scales. Three sodium silicates have been documented in the High Magadi beds: magadiite—$NaSi_7O_{13}(OH)_3 \cdot H_2O$, kenyaite—$NaSi_{11}O_20.5(OH)_3 \cdot H_2O$, and makatite—$NaSi_2O_3(OH)_3 \cdot H_2O$. Of these, only magadiite occurs in large quantities. Magadiite formed both as a lake-wide precipitate in the Late Pleistocene when the lake was density stratified and as an authigenic mineral precipitated during shallow diagenesis by the interaction of alkaline brines with siliceous materials such as volcanic glass (Jones et al., 1967). Even today, as the spring waters emerge about the edge of the lake, they precipitate hydro–sodium–aluminosilicate gels, which with burial convert to analcime, erionite, and magadiite.

Magadiite is an important actor in lake diagenesis, as it is a precursor of bedded chert. When a magadiite horizon is traced from depth (1 m) to the surface, the conversion magadiite → surface chert is seen. There are two mechanisms converting the magadiite to chert:

1. The leaching of sodium by dilute surface runoff (Eugster, 1969);

$$NaSi_7O_{13}(OH)_3 \cdot H_2O + H^+ \rightarrow 7SiO_2 + Na^+ + 5H_2O$$

2. The spontaneous crystallization of chert from magadiite. In this process sodium is expelled even in the presence of brines; it does not require the fresh water of process 1, and better explains the occurrence of calcite-filled trona casts in the chert (Hay, 1968, 1970)

Transformation of magadiite to chert involves a 25% reduction in volume, as shown by shrinkage cracks and surface reticulations in the resulting cherts. The distinctive appearance of chert derived from magadiite is a useful feature in identifying equivalent cherts in ancient lacustrine beds. Eugster (1969) described another impressive result of this shrinkage, namely chert polygons up to 50 m in diameter with bounding ridges up to 2 m high. Cherts after magadiites are recognized by their soft-sediment deformation features, crystal casts of trona sprays, and shrinkage cracks. To date, magadiite has been found only in alkaline lacustrine settings. In addition to Lake Magadi, magadiite has been found in Lake Chad in western central Africa, Alkali Valley in Wyoming, and Trinity County in California. Ancient Magadi-type cherts have been found in many ancient lacustrine basins, in the Jurassic of Wyoming, the Proterozoic Callana Beds of the Adelaide geosyncline in Australia, and in Precambrian banded iron formations worldwide (Eugster and Chou, 1973).

Siliciclastic sedimentation about the lake margin is not well documented in the African Rift Lakes, yet to geologists interested in hydrocarbon accumulations these poorly studied marginward settings are probably the most important within the lacustrine system (potential reservoirs). Some general characteristics of continental lacustrine clastics have been described in Chapter 2. Lake Magadi is surrounded by a clastic apron, but it has not been described sedimentologically.

Lake Natron, another rift valley lake, is located some 30 km southwest of Lake Magadi. It contains more standing water than Magadi and during the wet season typically contains water 1 to 3 meters deep, although it can reach 10 m locally. In lagoonal sediments about the lake edge the muds contain up to 10% TOC (Ron Surdam, pers. comm.). Modern surface inflow into Natron is from one perennial stream, Uaso Nyiro, yet no sediment from the river actually reaches Lake Natron. Riverine sediment is deposited in an adjacent swamp before reaching the lake. The swamp was created by a gradient reversal induced by an uplifted horst block at the lake margin. The Uaso Nyiro dumps its traction and suspended load into the swamp. Water continually spills out of the swamp into Lake Natron as clear sediment free water. The effect of this swamp on bedload implies that mineralogical changes in ancient lacustrine units are not always a response to climatic changes; mineralogy can change due to tectonics.

Two other rivers flow toward the shores of Lake Natron, the Peninj and Malambo Rivers. Today, the rivers are ephemeral; surface water sinks into the riverine sediments before it ever reaches the lake shoreline. Some 9000 to 10,000 years ago during a high water stage, these rivers, and the Uaso Nyiro, deposited fan deltas along the edge of Lake Natron. Two hundred kilometers north of Magadi is Lake Bogaria, a fresher-water perennial lake which possesses more perennial inflow than either Magadi or Natron. Lake Bogoria is located in an asymmetric rift valley. Fan deltas are forming along the steeper gradient eastern side of the lake and Gilbert-style deltas on the less steep western side. An important observation based on the African Rift valley lakes is that perennial continental lakes do not often deposit evaporites at the same time that they deposit thick fluviodeltaic successions. Episodes of fluviodeltaic deposition and thick evaporite deposition are separated in time and space—a restriction inherent in the lake hydrology. It is almost impossible to form evaporites when the lake waters are diluted by fresh water carrying siliciclastics into deltas. The same inverse relationship is true for the fan deltas and evaporites in the Dead Sea (Chapter 4). However, the time interval resolvable in the siliciclastic-to-evaporite transition in the African Rift Lakes is less than a few thousand years. In ancient lacustrine settings such as the Eocene Green River Formation, the two sedimentation styles may appear syndepositional.

A transition of depositional styles from subaerial salt flats to perennial lakes is related to changes in lake water levels. In the Pleistocene water level changes are often a reflection of climate. The effects are easily demonstrated in the Basin and Range Province of southern California (Fig. 3.6). Today the saline lakes of the area (except Mono Lake) constitute a series of ephemeral lakes surrounded and capped by extensive sabkhas. In the Pleistocene these same lakes formed a chain of lakes stretching from east of the Sierra Nevada to Manly Lake (now Death Valley). During the highest water stage the water level in these lakes stood as much as 200 m above the present playa floor. Searles and China Lake coalesced to occupy some 1000 km^2, about 10 times the present area. What is more, these lakes were at times brim full, with Pleistocene rivers extending from one

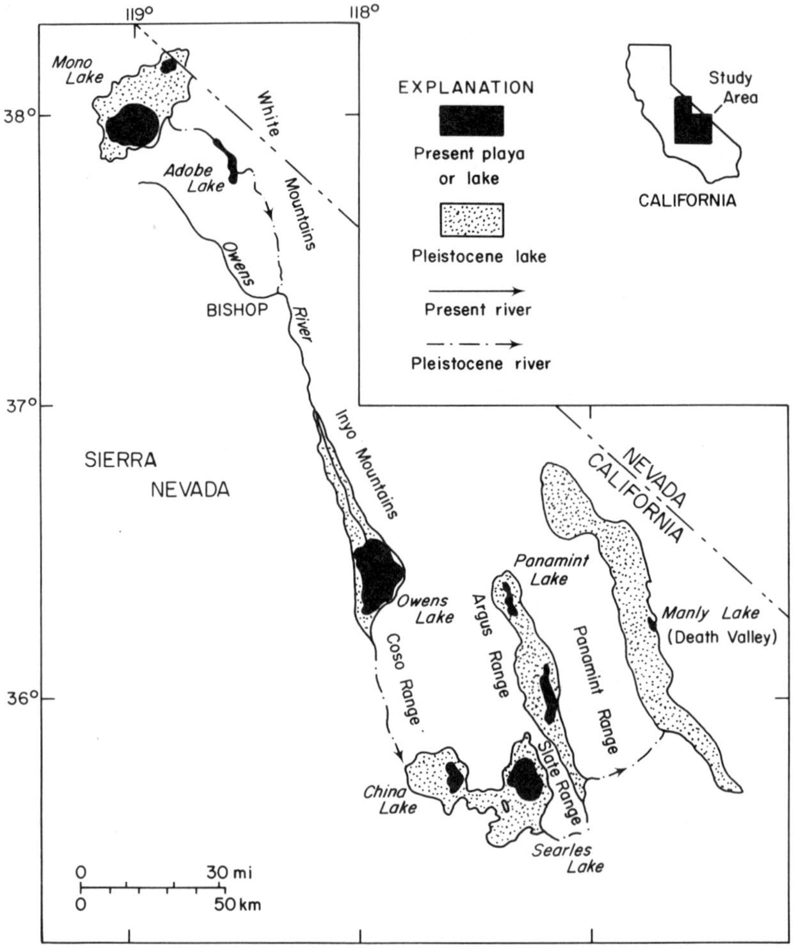

Figure 3.6 Playas in the Basin and Range of California (after Smith, 1979). The playas, including Death Valley (Lake Manly), may have formed an interconnected chain of lakes during the Pleistocene, yet today, with the exception of Mono and Adobe Lakes, are isolated closed basins. All the modern brine pans are far smaller than their Pleistocene equivalents.

basin to the next. Deep cores in the lakes show that most lakes are filled with alternating units of muds and salts. The saline layers can often be correlated with times of glacial retreat and the mud units with times of glaciation (Eugster and Hardie, 1978; Smith, 1979). Bischoff et al. (1985) in a study of Searles Lake in California found that the sediment-to-climate correlation does not always hold. The Pleistocene paleoclimate record of this midlatitude playa does not mimic the marine climatic record.

Vertical Sequence and Hydrocarbon Potential

The mineralogy in a vertical section formed in a continental perennial lake is dependent on the chemistry of the inflow brines (Fig. 1.2); however, the textural changes up-section are independent of mineralogy (Fig. 3.7). The base of the section is often a clayey or muddy sequence plugged with evaporites; this allows the waters to pond and subaqueous evaporites to accumulate. Evaporites deposited under perennial water conditions are often composed of bottom-nucleated halite chevrons, giant gypsum crystals, or aligned trona blades and rosettes. Individual crystals are filled with brine inclusions, a result of times of rapid crystal growth in shallow brines. The upper surface of crystals in the laminae and atop the larger scale evaporite beds can be truncated and covered by detrital material mixed with organic material, reflecting freshening of the overlying waters and the associated planktonic blooms. Above and sometimes below the evaporites are laminated carbonate units deposited either as marginal lagoons and mud flats or as anoxic basin center carbonates. The latter were formed during high water stages in the lakes, times of relative freshening above a perennial halocline. The sediments are often detrital/carbonate—evaporite couplets laminated and layered on a scale of millimeters to centimeters. Siliciclastic sand bodies cut across the laminated carbonate units and were deposited either as fan deltas or as river-dominated deltas. Sand-thicks in the deltaic sediments are perpendicular to the lake shoreline and trend radial to the basin. These sandy beds pass up into, or are laterally equivalent to, siliciclastic mud flats, continental redbeds, dunes, stream sands, and in some tectonically active areas the distal sediments of the bajada (Fig. 2.9).

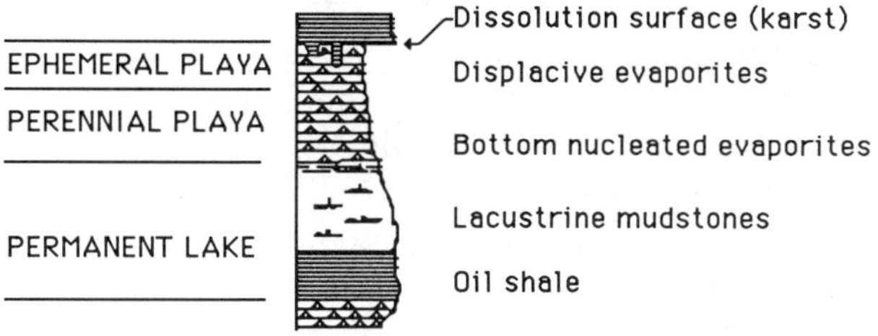

Figure 3.7 Vertical sequence from a perennial continental salt lake.

In rift settings, most of the potential reservoir sands in perennial lakes (fluviodeltaic) contain significant percentages of volcaniclastics. Under burial diagenesis these mineralogically immature sands are susceptible to grain collapse and pore plugging, features detrimental to reservoir properties. However, rift settings also possess high geothermal gradients, and this can partially allay the effects of mineralogical immaturity by the early generation and migration of hydrocarbons. Lacustrine muds in rift settings are oil prone, especially continental rifts which later develop into marine evaporite-filled basins (Warren, 1986a, b). In such settings the reservoir and the source are often subjacent to the seal. If the lacustrine source was derived in large part from continental organic material and was worked over by bacteria and other microorganisms before final burial, the resulting oil possesses a high wax content and an associated high pour point. All these features affect recovery detrimentally, in many lacustrine-sourced reservoirs it is less than 20%, usually in the range 10 to 14%.

Even so, saline lakes in rift valley settings probably make better exploration targets than perennial saline lakes or playas in other continental settings. In a nonrift perennial lake the groundwater circulation is relatively shallow, mainly in the porous units about the lake margins. Evidence for shallow circulation is seen in the spring mounds and seeps along the toes of the alluvial fans surrounding the lake. This rapid shallow circulation leads to widespread calcite cementation in the porous sediments of the alluvial fan and sand flat. Second, the lake easily becomes a dried-up playa, and drying oxidizes any organic matter (proto-source rocks) in the central lake muds. On the other hand, rift valleys are the topographically lowest parts of the local landscape, so tend to be water saturated environments for extended periods. Complete drying of the organic-rich lake muds is rare. The high salinity pore waters tend to be anoxic, and the rapid burial rates associated with rifts quickly isolate potential source rocks from near-surface oxidizing conditions (Warren, 1986a). Fluid circulation in a rift setting is deep and

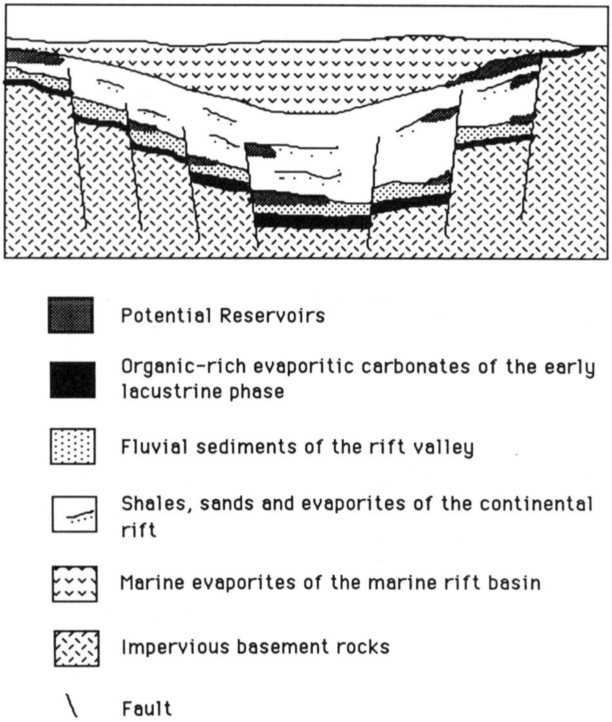

Figure 3.8 Occurrence of lacustrine source rocks (black) near the base of a rift-fill sequence later filled by sea marginal basinal evaporites (after Warren, 1986a). Potential reservoirs are the fluviodeltaic sands about the basin edge. Dip reversals and normal faulting create the trap.

thermally driven. The faulting associated with rifting provides vertical migration pathways for fluids containing early oil. The emplacement of this early oil prevents any further loss of porosity by shallow burial cementation even when the matrix is mineralogically immature (Fig. 3.8).

So far the chapter has concentrated on some of the better known modern examples of shallow subaqueous evaporite deposition. There are many other variations in subaqueous deposition but to understand them one must go to the ancient strata where shallow water evaporite deposition was more widespread and diverse.

ANCIENT EVAPORITES

Compared to sabkhas ancient subaqueous evaporites are not yet well documented in the geologic literature. This is the result of three factors:

1. Until the last decade no reliable modern or late Tertiary models existed for subaqueous deposition
2. Subaqueous evaporites tend to be relatively pure massive units (Warren and Kendall, 1985); they are easily picked on wire-line logs and rarely if ever cored in a drilling program
3. Relatively pure subaqueous evaporites, especially gypsum, are subject to early recrystallization, flowage, and intense burial diagenesis which leave only indistinct relicts of the original texture (Chapter 1). This burial overprinting is not as important in sabkha sequences, where the matrix tends to be more rigid, as it is composed of less mobile carbonates and siliciclastics (Warren and Kendall, 1985)

Even when much of the original depositional texture has been obliterated, it is becoming increasingly apparent that shallow-water subaqueous evaporite deposition played a major role in many ancient evaporite basins. Subaqueous evaporites have been described from the Late Miocene (Messinian) of the Mediterranean, the Ferry Lake Anhydrite in the Gulf of Mexico, the platform sulfates of the Permian Zechstein Basin, the ramp evaporites of the Palo Duro Basin, and other ancient giants.

Models of Ancient Marine Shallow-Water Evaporites

Studies of ancient marine evaporitic sequences show that evaporites were deposited on a scale and in settings that do not occur today. For example, when the San Andres evaporites were deposited on the Palo Duro ramp in the Texas Panhandle, one could have probably walked seaward in ankle to knee deep brine for 100 or more kilometers. With enough drinking water and a few camels, one could have walked across the seafloor of the Mediterranean some 5.5 million years ago. At that time the Mediterranean was a huge desert—playa complex fed by marine and continental waters. No modern equivalents exist in scale or depositional diversity to these types of ancient depositional settings.

Many ancient shallow-water sulfates were deposited in shelf, ramp, and basin settings either as shallow-water bottom-nucleated crystals or as mechanically deposited, laminated clastics. Extensive saline flats, hundreds of kilometers wide, existed in some basins, yet the flats were covered by brines only a few tens of centimeters deep. Some authors call these flats tidal, but many of the transitions from subaqueous to subaerial on such flats were probably not tidal but related to storms and associated wetting, freshening, and drying episodes. Laminated sediments were deposited on these flats both as algally bound storm deposits and as subaqueous deposits similar to those already described from the Holocene of southern Australia.

In the former case, laminae were formed wherever subaerial evaporitic flats were flooded by sediment-laden stormwater. Once the storm subsided, these sediments, mainly gypsarenites and pelleted carbonates, were captured by extensive algal mats which covered the surface of the salt flat. The predominant sedimentary structures were mechanical traction features such as graded and reverse-graded beds or rippled layers subsequently bound by algal structures. The graded beds and the rippled layers record waning flood conditions. The reverse-graded beds record either episodes of storm surging (increasing energy levels) across the tidal flat or episodes of brine dilution that induce gypsum recrystallization and slower crystal growth in the upper part of the laminae. In ancient sequences the shallow-water conditions are demonstrated by the occurrence of carbonates within the cross-bedded evaporites. Often, such interbeds contain micritic domal stromatolites, bird or dinosaur footprints, or pellets from fossil brine shrimp and ostracods.

In ancient times the other predominant type of shoal-water sulfate deposition was as aligned, coarsely crystalline, selenitic gypsum. It formed wherever the flat was covered by a few centimeters or more of brine. Such coarse-grained gypsum was deposited as single crystals, clusters, crusts, and superimposed beds. Some of their equivalents occur in the Holocene salinas of southern Australia, but there is a far greater variation in crystal form in pre-Quaternary sequences. The best development of these shallow subaqueous sulfates is in the Late Miocene (Messinian) evaporites of the Mediterranean. There, most of the gypsum has not been buried deep enough to convert to nodular anhydrite.

Mediterranean Evaporites and the "Messinian Salinity Crisis"

Some 5.5 million years ago (Messinian or late Miocene) the Mediterranean was an area of basin-wide evaporite deposition where an evaporite unit up to 2 kilometers thick was laid down (Fig.12). Desiccation and restriction was brought about by the collision of Africa with Eurasia. The Mediterranean Basin is a syn- or post-orogenic basin, the exact details of reconstruction for this time are still disputed. Dewey et al. (1973) argued for a Miocene collision of North Africa with the Iberian Peninsula along the line of the Betic Cordillera. At the time it converted several oceanic straits into collisional sutures in the Alpine-Appenine region. Laubscher and Bernoulli (1977) criticized the Dewey scheme and documented evidence of compressive tectonism from Cretaceous time onward. Milnes (1978) suggested that the final stages of continental underthrusting locking the land masses together was completed by the early Oligocene. Biju-Duval et al. (1978) showed the picture was probably a little more complex in the various Mediterranean sub-basins which contain Late Miocene evaporites as the sub-basins differ in age and genesis. The Western, Tyrrenhenian, and Aegean Basins are relatively young (Oligocene to Holocene) peripheral basins, whereas the Eastern Mediterranean is a remnant of an early Mesozoic ocean with its southern continental margins preserved. The Black Sea is thought to be a retro-arc basin possibly as old as Late Cretaceous. Whatever the tectonic setting of the various Mediterranean sub-basins, they all show a similar Late Miocene to Holocene fill. Some 5.5 million years ago all the sub-basins (except the Black Sea) were filled with thick (1–2 km) sequences of shallow-water with minor deepwater evaporites deposited during a 200,000 year period (Cita, 1982). All the evaporites were underlain by deep-water pre-Late Miocene sediments and covered by deep-water Plio-Pleistocene clastics.

In the 1970s the Mediterranean was a site of intense geological controversy. DSDP Legs 13 and 42A had been drilled in 1971 and 1975, a total of 22 sites on the floor of the Mediterranean Sea. In thirteen of these holes gypsum or anhydrite and/or halite of Messinian age (Late Miocene-5.5 million years ago) was recovered from sediments now buried beneath 2000 m of ocean water. At the end of Leg 13 three of the on-board party of the *Glomar Challenger* (Hsu et al., 1973) were convinced that the evaporites of the deeper areas of the Mediterranean subbasins had been laid down in relatively shallow to ephemeral water. They postulated further that at the time the Mediterranean was a deep desiccated basin whose floor was some 2000 m below sea level, yet it was covered by shallow-water desert lakes and sabkhas (Fig. 1.15a). This radical conclusion stirred up a major controversy among many land-based geologists who saw the Mediterranean as a classic deep-water evaporite (Fig. 1.15b) and were dubious as to the significance of the nonuniformitarian conclusions of the DSDP shipboard party. Leg 42A followed, while at the same time many outcropping Messinian strata on the Italian Peninsula, Sicily, and Spain were reevaluated (Vai and Ricci Lucchi, 1977; Schreiber, 1977; Schreiber and Decima, 1978) and found to be relatively shallow-water deposits. When the consensus report of Leg 42A was published (Hsu et al., 1977), the interpretation was almost unanimous that the Messinian evaporites of the Mediterranean were deposited in shallow water, interspersed with the occasional deeper water deposit. However, not all Mediterranean geologists believe the evaporites are shallow (Debenedetti, 1982).

If the hypotheses of Hsu et al. (1977) are correct then the depositional setting during the deposition of the Messinian Evaporites was, to say the least, unusual, one for which no modern counterpart exists either in scale or diversity of depositional style. Probably the closest is the grotesque scenery of Djibouti in the Afar Triangle of Africa. Here the geology is that of an uplifted, dried out piece of the ocean floor. Prior to the "Messinian salinity crisis" the eastern end of the Mediterranean was supplied with seawater by a current of water flowing continually from the Atlantic. Seawater spilled from the Tethyus into the Mediterranean at the western end. The collision of Africa with Eurasia restricted the inflow, then some 5.5 million years ago a tectonic dam (including the hills of the Spanish Alps) cut off the surface inflow of seawater. Within 1000 to 10,000 years the Mediterranean dried out, creating a series of ephemeral and perennial lacustrine basins on what was previously the deep seafloor of the Mediterranean. These lakes were spattered across the abyssal plain and trenches of the seafloor at a level 2 km or more below the sea level in the nearby Atlantic. As the Mediterranean dried out, the French Rhône and the Egyptian Nile as well as other major rivers flowed out onto the desert plain (Barber, 1981). Their lowered base levels cut deeper thalwegs into the deposits of the former river deltas. Late Miocene erosion is seen seismically beneath many of the major rivers flowing into the Mediterranean.

Hsu envisioned the huge volumes of seawater required to form the Messinian Evaporites as being periodically supplied as a huge waterfall at the Atlantic end of the Mediterranean. A series of freshwater lakes mixed with seawater at the eastern end. Other authors give much more importance to a continual seepage mechanism supplying seawater and fresher waters to the desiccated Mediterranean. The final refilling of the Mediterranean some 5.3 million years ago must have been a spectacular event. When the Atlantic waters reentered the deep hole that was once again to become the Mediterranean, it did so as a series of gargantuan waterfalls, any one of which would have dwarfed Niagara or any other modern waterfall.

Facies models that include the hypothesis of a series of evaporite depositional settings ranging from continental to deep-basinal were first documented in the Messinian of Sicily (Fig. 4.7; Schreiber and Decima, 1978; Schreiber, 1986). At least eight facies are present in the various Mediterranean sub-basins (Fig. 3.9);

1. Nodular calcium sulfate within clastic sediments
2. Gypsum and calc-gypsum rudites, arenites, and arenitic marls, variably composed in part of fragmental gypsum (also some anhydrite) and in part other resedimented material, such as older carbonate and shale fragments, quartz, feldspars, glauconite, and somewhat altered volcanic glass. These are entirely clastic sediments and are subdivided on the basis of bedding type into
    a. Graded beds, with internal structures such as cross-bedding, ripples, and flames
    b. Graded beds, with well-developed Bouma sequences and ripples, flute casts on the soles of many beds

   c. Ungraded beds formed by mass flows, slumps
   d. Ungraded beds formed of dissolution residuum (collapse breccia)
3. Laminated calcium sulfate usually found as couplets with thin carbonate intercalations; these laminated beds may be
   a. Even and continuous (Fig. 3.9a)
   b. Wavy and irregular
4. Nodular calcium sulfate beds with discrete layers of nodules, coalescing beds of nodules, and/or chicken-wire nodular masses (Fig. 3.9b). These beds are found in association with laminated calcium sulfate
5. Bottom-nucleated crystalline gypsum, with an appearance of grass-like or cavoli (cauliflower) forms often found in association with the even, laminated calcium sulfate beds of facies 3a (Fig. 3.9c, d, and e)
6. Wavy, anastomosing flaser-like gypsum beds composed of tiny oriented crystals or broken, reworked cleavage fragments, marked by very thin irregular carbonate layers
7. Salt (halite) with chevron structures and hopper crystals (Fig. 3.9f)
8. Salt (halite, etc.) showing a continuous laminated structure and underlain by graded gypsum rudites and arenites which appear to be turbiditic deposits

These facies were deposited in environments ranging from continental for facies 1 to basinal in facies 2b and c and portions of facies 2a and 8 (Chapter 4). The remainder of the facies were deposited in shallow subaqueous shelf settings, where the most characteristic shallow-water fabrics occur in beds of selenitic gypsum.

Gypsum in the selenitic beds is composed mainly of orderly rows of vertically standing, elongate, and commonly swallowtail twinned crystals (Fig. 3.9c; Schreiber, 1986; Kendall, 1984) that range in height from a few centimeters to a few meters. The crystals are often euhedral and arranged either into a vertically aligned pallisade structure or into radiating-upward conical clusters (Fig. 3.9e, "cavoli"). Individual crystals are separated by layers and rinds of micritic carbonate, fine-grained gypsum, gypsum sands, organic matter, or overgrowths, which create an interlocking crystal mosaic. Crystals often contain faint laminations defined by carbonate and other impurities. Many impurities are pellets and algal filaments deposited in the zone of photosynthesis. The laminae either pass through the crystals as dissolution surfaces or outline the successive positions of the growing gypsum crystal, in much the same way that similar crystals grow today in the salinas of South Australia (Warren, 1982a). Other crystals show bizzare growth forms and twins as they split and grow into palmate and fan-shaped clusters of subparallel crystals (Fig. 3.9d; Orti-Cabo and Shearman, 1977). Splitting creates a variety of intercrystal angles, and it was this feature that first led Orti-Cabo and Shearman to conclude that the palmate crystals were not true twins. The splits are a result of the inclusion of organic material and the

(a)

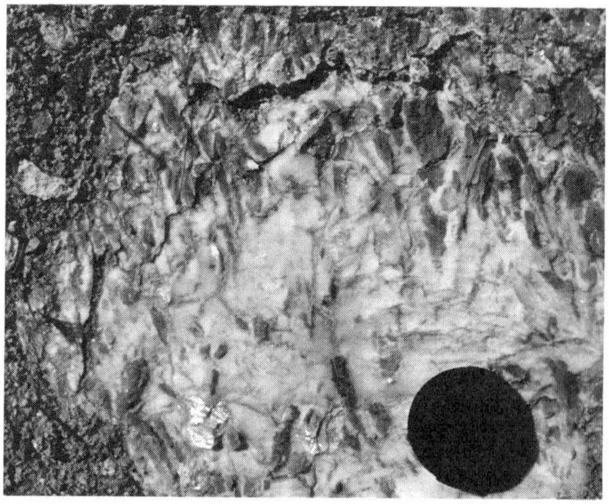

(b)

(c)

(d)

(e)

(f)

Figure 3.9 Shallow-water evaporites of the Late Miocene of the Mediterranean from Sicily, Italy; (a) laminated gypsum surrounding vertically aligned gypsum, Eraclea Minor. Lens cap scale; (b) nodular anhydrite due to tectonic alteration of laminated gypsum, Montallegro; (c) grass-like and spear-like gypsum, Passo Funnuto; (d) palmate and splayed gypsum, Passo Funnuto; (e) complete gypsum cycle showing laminated gypsum passing upward into domal layered gypsum-cavoli, Eraclea Minor; (f) molds of displacement halite, Favara. [part (f) is courtesy of B. C. Schreiber].

preferential growth of one side of a previously twinned crystal nucleus (see also Warren, 1982a).

Banks and mounds of clustered gypsum crystals outcrop in Spain and other areas of Messinian strata in Italy and elsewhere. The mounds grew up from the pond floor with a relief of 1 to 5 meters (Fig. 3.10). These banks grew rapidly

Figure 3.10 Gypsum mounds (''reefs'') in the Late Miocene gypsum of Spain; note the interfingering of the gypsum sand (photograph courtesy of B. C. Schreiber).

upward, perhaps aided by the binding activities of cyanobacteria. They accreted in much the same way that a coralgal patch reef grows up from the lagoon floor with the upper surfaces of these gypsum ''reefs'' standing proud on the floor of the brine pond. Debris aprons of gypsarenites and gypsum muds onlap the edges of the mound. This sediment was probably derived by the wave erosion of the ''reef,'' but the continued growth of the mound kept the crest of the structure above the surrounding sediment. The mounds probably grew up to the brine/air interface, as they sometimes have truncated tops created by dissolution. Similar Holocene mound structures grow upward from the floors of salinas in South Australia with a relief of 1 to 2 meters.

No single characteristic texture occurs in these different types of subaqueous gypsum–rather, a range of textures related to relative water depth and energy levels in the environment (Fig. 1.10). The package of sediments laid down in this unusual setting challenges many of the assumptions that geologists make as to the deep water origin of evaporite units overlying or sandwiched between deep water marine carbonates and siliciclastics. Sediments above and below the Messinian Evaporites contain a benthonic biota indicating marine abyssal depths, yet the bulk of the evaporites were laid down as shallow-water lakes. It underlines earlier statements that water levels in evaporite basins are subject to very rapid changes. Even so, because of evaporative drawdown and rapid sedimentation rates in these saline giants the evaporite depositional surface was rarely more than a few tens of meters below the brine surface (Table 1.4).

Ferry Lake Anhydrite, Fairway Field, East Texas Basin

The Fairway field is located in Henderson and Anderson counties in east Texas. The Ferry Lake Anhydrite is a Lower Cretaceous unit sitting above the James Limestone, which forms the major reservoir in the field. In the Ferry Lake Anhydrite, Loucks and Longman (1982) documented ancient subaqueous gypsum which occurs as vertically or near vertically aligned anhydrite nodules (''gypsum ghosts''). The outline was preserved as the original gypsum crystal was outlined by some sort of impurity—either carbonate, organic matter, or clay (Fig. 3.11).

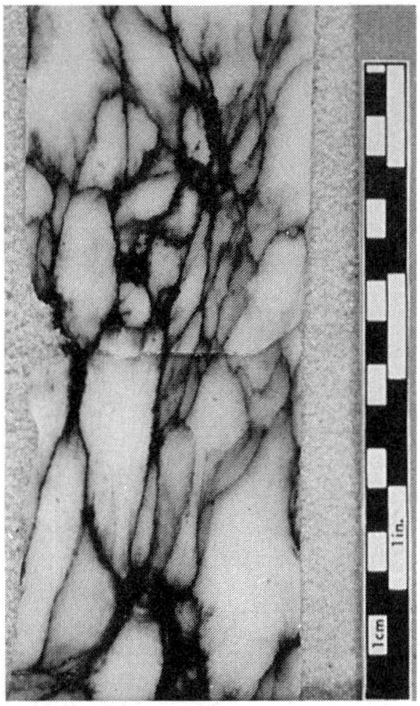

Figure 3.11 Vertically aligned anhydrite (''gypsum ghosts'') from the Ferry Lake Anhydrite, J. B. Kitchens No. 1 Well (photograph courtesy of Bob Loucks).

The outline is recognizable as the conversion from gypsum to anhydrite took place relatively slowly but not so slowly that the gypsum became overpressured and thixotropic.

The Ferry Lake Anhydrite is composed of alternating carbonate and evaporite units that were deposited in a broad, shallow subtidal hypersaline lagoon (saltern). The lagoon was up to 160 mi (260 km) wide and separated from the sea by shelf-margin rudist banks and possibly a tectonic sill (Fig. 3.12). Evaporites are vertically to randomly aligned (elongate) mosaic nodular anhydrite originally deposited as subaqueous gypsum. The anhydrite units are up to 22 ft thick and contain carbonate mudstone layers with anhydrite rip-up clasts which represent exposure or storm flooding of the lagoon. The carbonates, predominately mudstones and wackestones, were also deposited in a restricted lagoonal setting. The fauna is primarily ostrocodes and mollusks. Several units contain *Orbitulina* and/

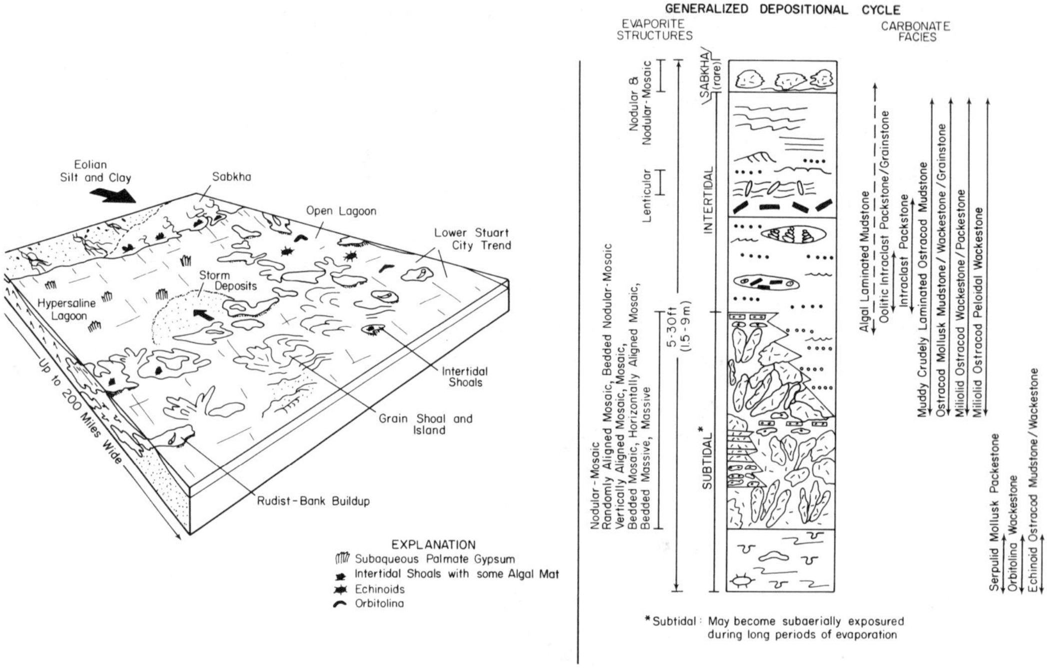

Figure 3.12 Depositional model for the Ferry Lake Anhydrite in east Texas and a general depositional cycle as interpreted from the J. B. Kitchens No. 1 Well (after Loucks and Longman, 1982).

or echinoids, indicating that normal marine waters were present at times in the depositional history. Sabkha evaporites are rare in Ferry Lake, most of the anhydrite formed in a subaqueous setting.

Ferry Lake Anhydrite carbonates are not highly productive, but their occurrence over deeper hydrocarbon horizons makes them a potential secondary objective. The Ferry Lake does produce in the Fairway Field and in other areas in the northern Gulf Coast Basin. Fox (1963) reported that production as of 1963 totaled 185,000 bbl of oil producing from thin oolitic stingers in the anhydrite section. Terriere (1963) reported that one well had produced 40,000 bbl from the Ferry Lake.

Jurassic Todilto Formation, New Mexico and Colorado

The Jurassic Todilto Formation is composed of a lower limestone member and an upper calcium sulfate member. It lies within and rims the San Juan Basin whose western edge is near the Arizona state line; the southern edge extends through Gallup, Grants, and Acoma, the eastern edge near the town of Los Vegas, New Mexico, while its northern edge is near Durango, Colorado (Fig. 3.13). The Todilto Basin was a craton-margin basin resulting from the upward bulge and

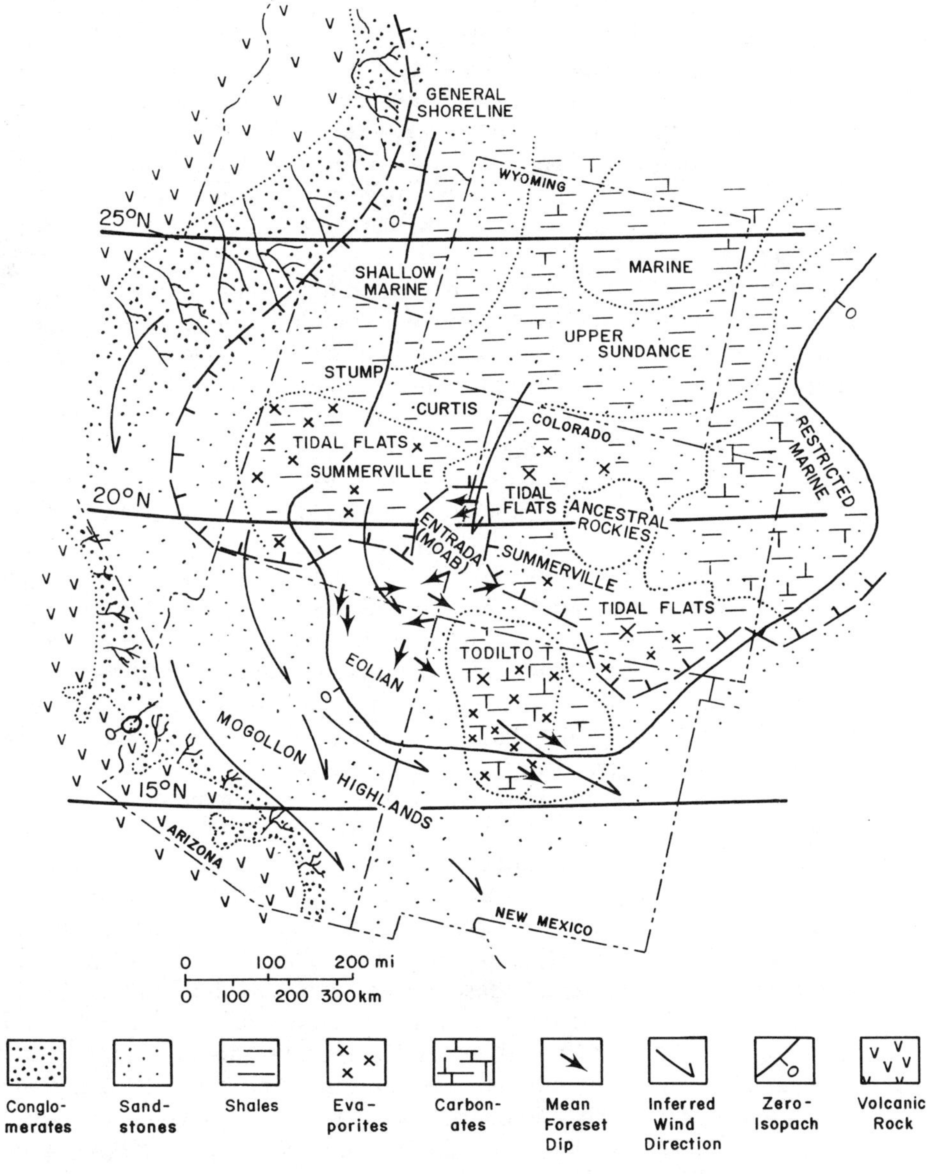

Figure 3.13 Location of the Todilto Basin, showing wind direction and paleolatitude (after Kocurek and Dott, 1983). Paleoclimate derived from a study of the Entrada Formation.

outward spread of an actively deforming mobile belt to the west. During Todilto time (Oxfordian) the San Juan Basin was located 15 to 35° north of the equator in a trade wind belt associated with desert conditions (Kocurek and Dott, 1983). The

Todilto is a well-studied unit and has been interpreted as a lacustrine or a marginal marine deposit (Tanner, 1970; Ridgley, 1984). McCrary (1985) looked at the Todilto on a regional scale and found that it is actually composed of a number of subfacies. The Todilto is not solely a marine or a lacustrine deposit, but contains elements of both settings. It was deposited under severely restricted marginal-marine conditions as a perennial water body in an arid environment. What is more, many features that have been used by previous workers to interpret the depositional setting have with more detailed study been shown to be largely diagenetic in origin.

The Todilto Formation was deposited in the Todilto Basin (Fig. 3.13) and is part of the San Rafael Group deposited in middle to late Jurassic time. The San Rafael is composed of the Todilto, the Entrada, the Carmel and where the Todilto is not present, the Curtis and the Summerville Formations. The Todilto lies conformably between the marine Summerville Formation and the marine and eolian Entrada sandstones throughout most of the San Juan Basin, but in the far southeastern edge of the basin it overlies the Entrada conformably but underlies the Morrison unconformably. In southwestern Colorado the Todilto is conformable between the Entrada and the Bilk Creek member of the Wanakah Formation.

The basal facies of the Todilto is a platy/planar laminated limestone composed of alternating mm-laminae of micrite and spar (Fig. 3.14a) It is a remarkably consistent and extensive facies which conformably overlies the sands of the Entrada Formation with a gradational contact. It grades laterally into a more sandy basal platy facies to the southwest and the southeast and into clean oo-sparites to the northwest (Fig. 3.13). The basal facies grade upward into a crinkly facies, a massive sparite facies, or in the northwest into a sandy, oolitic, ostrocod biomicrite facies. These three facies grade upward into laminated or massive breccias about the basin edge, and biomicrites containing sparite pseudomorphs after chicken-wire and nodular sulfates and layered gypsum in the basin center. In the central basin the layered gypsum grades up into a massive gypsum facies. The layered gypsum unit consists of beds of finely laminated micrite alternating with beds of nodular gypsum (Fig. 3.14b). The thickness and frequency of occurrence of the laminated micrites decrease upward, whereas the frequency and thickness of the gypsum beds increase. Gypsum nodule size also increases upward and seems to be related to the thickness of the gypsum bed. Finally, the bedded gypsum passes into the massive gypsum unit. The base of the layered gypsum unit is 62% gypsum and 28% micrite, while the top of the unit is 95% gypsum and 5% micrite. Gypsum makes up more than 95% of the massive gypsum unit. Pseudomorphs of palmate gypsum can be seen in the nodular gypsum in outcrop; this type of gypsum grows only in a subaqueous setting (McCrary, 1985; Warren and Kendall, 1985). The peripheral limestones of the southeastern, southwestern, and northwestern basin edges are laterally and time equivalent to the central gypsum facies. The lateral and vertical changes in the Todilto are due to varying salinities in the basin as the hydrology of the perennial basin waters changed with time.

(a)

(b)

Figure 3.14 Todilto Formation Facies: (a) basal Todilto Formation of mm-laminated mudstone deformed by soft-sediment flow near Grants, New Mexico; (b) regional view of Todilto Formation showing basal limestone (starts near person's head) overlain by layered gypsum which has passed from depositional gypsum to anhydrite with burial and back to gypsum with erosion near Laguna, New Mexico.

The lower limestone unit is a key unit in interpreting the Todilto. It is generally a fetid, sandy, allochem-free, very finely laminated micrite with abundant sulfate-related structures. It changes to the northwest in the area of a proposed opening to the sea. There the basal Todilto is a fetid but mud-free, plane- and cross-laminated and massive oo-sparite lacking any evidence of sulfates (Fig. 3.13). The contact between the two laterally equivalent units is gradational. The oo-sparite contains eccentric ooids in beds with sharp bases and graded tops. It was deposited during occasional storms under slightly higher-energy conditions than the more basinal finely laminated micrite. The lack of sulfate pseudomorphs in the facies of the northwest and their increasing abundance to the southeast suggests increasing salinity to the southeast of the basin. The presence of brackish-water(?) ostrocods in the northwest and the lack of ostrocods toward the southeast reinforces this conclusion.

At this time an ephemeral connection probably joined the basin to the ocean so that during storms a sheet of marine waters washed over a restricting sill into the basin. For the rest of the time the water input to the basin was probably from resurging marine and continental groundwaters flowing into the basin from the underlying aquifer of the Entrada dunes. Evaporation of this inflowing groundwater created the bull's-eye sediment pattern. The lake waters at this time were probably density stratified with the laminated organic-rich limestones being deposited beneath anoxic hypersaline bottom waters. The organics came from near-surface plankton blooms during times of freshening of the near-surface waters.

The deposition of fine-grained sediments restricted the subsurface inflow to the lake. At the same time the retreat of the sea from the former connection lowered the frequency of surface marine input to the basin and reduced the head between the sea and the basin, so lowering groundwater inflow rates. Conditions in the basin became more restricted and the micrite breccia facies were formed by the subaqueous, intrasediment growth of gypsum about the lake edge. Finally, the volume of inflow and the frequency of freshening events decreased to the point where relatively pure, subaqueous gypsum was deposited, first as bedded subaqueous gypsum with well-developed calcitic layers, and then as massive subaqueous gypsum when the brine pond shallowed and salinities increased.

San Andres Formation, Northwest Shelf, West Texas

The Permian San Andres Formation was deposited on a Late Leonardian to Guadalupian broad shelf which was the northwest boundary of the deeper Midland and Delaware Basins (Fig. 2.10). It preceded the Goat Seep and the Capitan reefs and was laid down on a ramp with no striking break in slope from the continental shoreline out to the open marine of the Delaware and Midland Basins. In the Northwest shelf, the San Andres is divided into two members, each containing several shoaling cycles previously interpreted as sabkha cycles.

The lower member contains two well-defined hydrocarbon reservoirs (P_1 and P_2) and forms the reservoir in the highly productive Slaughter-Levelland trend (Fig. 2.11). The producing style is stratigraphic with up-dip pinchouts from porous dolomite into nonporous dolomite and evaporites (Fig. 2.12). Production is slow but constant.

Cycles in the lower member can be divided into 5 lithofacies arranged in a shoaling upward succession; each cycle is 50 to 150 ft thick (see also Chapter 2). At the base of each cycle are shaley carbonate mudstones, the initial low-energy deepwater deposits formed after the initial transgression.

These are overlain by biomicrite units (≈30 ft) containing brachiopods, bryozoa, crinoids, mollusks, and cephalopods, sometimes forming what may be low relief bioherms (Fig. 3.15a). The unit was deposited on an open marine shelf, below wave base under low-energy conditions.

Next comes the dolomite, by far the most variable unit, but the most important, as it is the reservoir facies (Fig. 3.15b and c). It can be classified into three subtypes:

1. Oolitic peloid-foram grainstones forming as localized shoals (uncommon unit).
2. Wispy laminated to microstylolitized biomicrites, intramicrites, and peloidal micrites (abundant unit). It contains a poorly preserved open marine fauna now present as molds sometimes filled with anhydrite.
3. Inter- to supratidal mudstones with fenestra, pisolites, crypt-algalaminates, and possible desiccation cracks (uncommon unit).

Next comes a massive anhydrite unit previously interpreted as sabkha. It is meters thick and composed of a nodular anhydrite interbedded with laminated nonreplacive anhydrite (Fig. 2.11). In some cores there are vertically aligned anhydrite nodules which pass downward into laminated and bedded nodular mosaic anhydrite (Fig. 3.15d). The alignment, the lamination, the purity, the thickness of the units, and the fact that it is laterally equivalent to subaqueous gypsum to the north implies that this unit was laid down subaqueously in the Levelland–Slaughter area (Warren and Kendall, 1985; Elliott and Warren, in review).

Finally comes the capping halite unit, which is not present in the Roswell area but occurs to the north as subaqueous and recrystallized halite.

In outcrop to the west of Roswell the evaporites occur as dissolution breccias which can be used as mapping datum (Fig. 3.16). In fact, it is possible to correlate the five cycles of the lower member of the San Andres from outcrop down into the subsurface near Roswell and then in the subsurface all the way into the Palo Duro Basin. In outcrop the lateral facies changes can be seen in each cycle; the carbonate intervals were deposited as facies mosaics separated by the extensive evaporites. This has led Elliott (1985) to propose a revised model for San Andres deposition, a model that sees sabkha deposition as relatively minor in this area near Roswell, New Mexico. An area located a short distance along strike from the main Levelland-Slaughter trend (compare with the discussion of San Andres in chapter 2). Each complete cycle in the San Andres records a relative lowering of sea level (Fig. 3.17)

Time 1: high sea level stand—initially, quiet water deposition or non deposition, then the development of normal marine with open circulation, and the occasional bioherm.

Time 2: relative lowering of sea level—development of shoals, some oolitic, as well as the occasional island where supratidal carbonates were laid down but

(a)

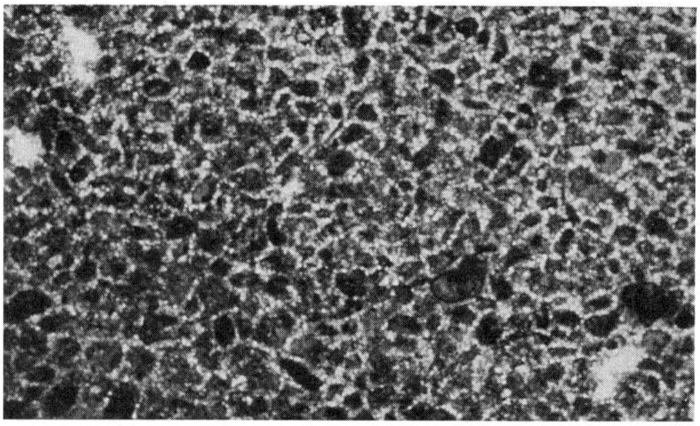

(b)

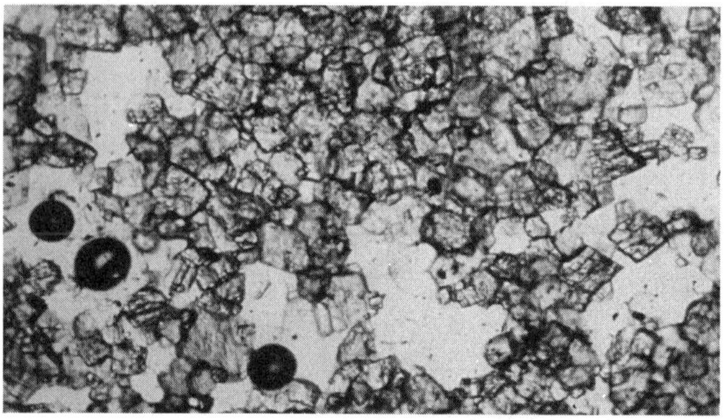

(c)

(d)

Figure 3.15 San Andres Formation, Northwestern Shelf: (a) open-marine facies of the basal limestone containing bryozoa, crinoids, mollusks, outcrop near Roswell, New Mexico; (b) dolomite facies, pelleted mudstone with intercrystalline porosity, subsurface near Roswell, New Mexico; (c) dolomite facies, oomoldic porosity, thin-section photomicrograph, subsurface near Roswell, New Mexico; (d) halite replacement of vertically aligned subaqueous gypsum, Palo Duro Basin (photographs courtesy of L. Elliott).

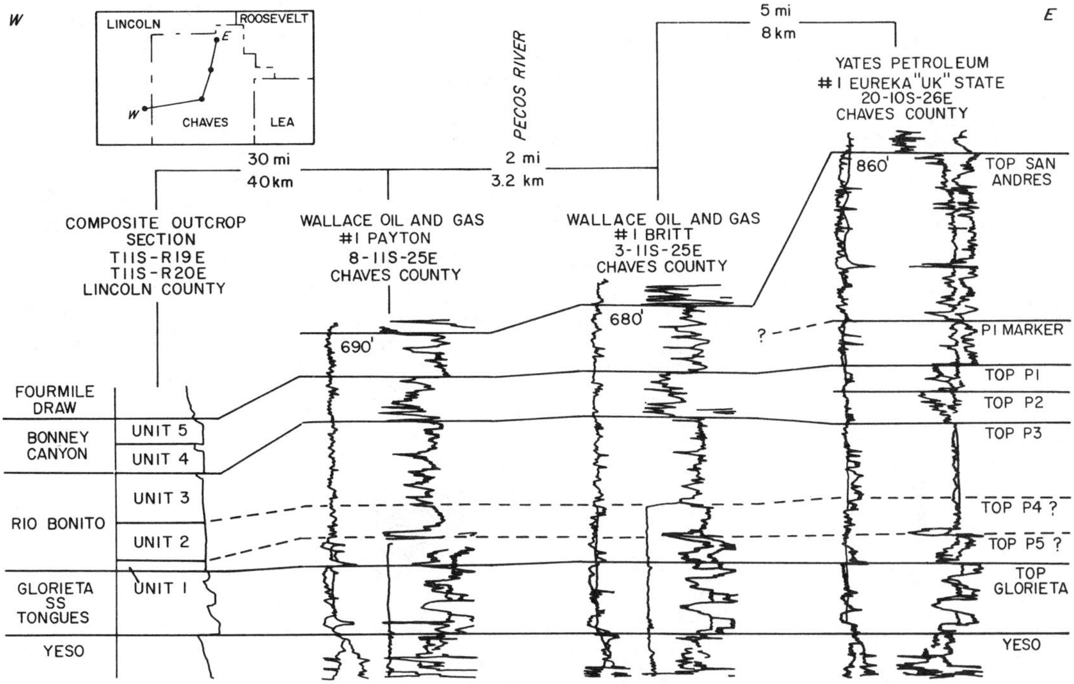

Figure 3.16 Surface to subsurface correlation of the San Andres Formation near Roswell, New Mexico (after Elliott and Warren *in review*).

the growth of sabkha anhydrite was limited. A facies mosaic developed, with wide channels allowing exchange of water between the more restricted ramp and the more open marine.

Time 3: a time of further relative lowering of sea level and climate may have become more evaporitic. An extensive gypsum-depositing lagoon formed still in partial contact with the more open marine. Few, if any sabkhas existed in this extensive shallow brine water body. At times of extremely high evaporation rates and very shallow water, the brine pond/ramp sometimes dried out. In the Palo Duro Basin and areas north of the Levelland-Slaughter trend, desiccation led to the deposition of subaqueous halite crusts crosscut by syndepositional karst (Hovorka et al., 1987).

The subaqueous model of San Andres deposition offers a new exploration strategy for the San Andres reservoirs on the Northwestern Shelf of the Permian Basin. The reservoirs did not form by the up-dip pinchout of marine grainstones and mudstones into stacked sabkhas. Rather, the reservoir trend marks the transition from an up-dip, evaporite-plugged, partially dolomitized subtidal—intertidal section into a dolomitized, subtidal-intertidal sequence characterized by intercrystalline and relict interparticle porosity not plugged with evaporites. This in turn passes down-dip into tightly cemented open marine subtidal limestones. Up-

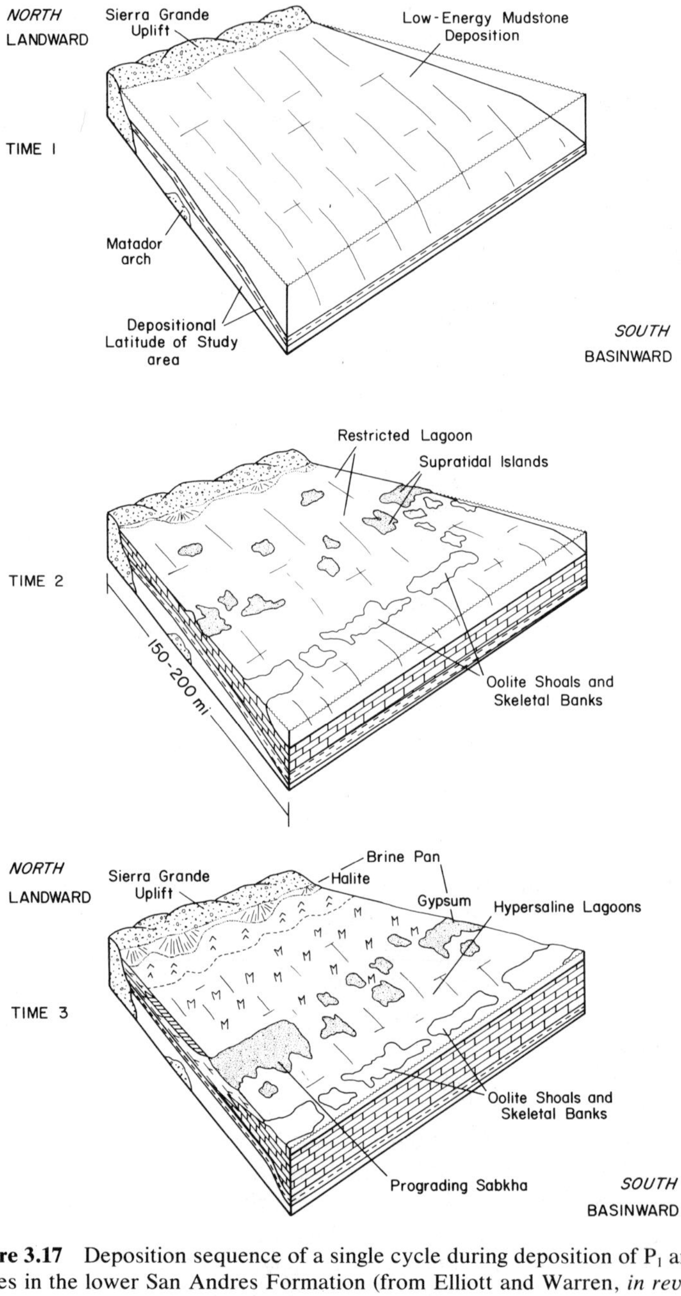

Figure 3.17 Deposition sequence of a single cycle during deposition of P_1 and P_2 cycles in the lower San Andres Formation (from Elliott and Warren, *in review*).

dip plugging was caused by long episodes of subaqueous evaporite deposition. Associated brine reflux filled the underlying carbonate porosity with sulfates. The down-dip subtidal—intertidal section saw some evaporitic episodes but did not suffer the long term massive influx of evaporite cements experienced by the up-dip section. Its marine margin position allowed the porosity to remain open and magnesium-rich solutions associated with brine reflux to enter the carbonates. The reservoir trend is not a paleoshoreline but a diagenetic front. Brine reflux dolomitization is discussed further in Chapter 5.

Arab Formation, Arabian Gulf

Another area of platform-ramp evaporite associated reserviors is the Arabian Gulf during the late Jurassic. The Upper Jurassic carbonates of the area include the Arab Formation which contains some of the most prolific oil fields in the world today including Ghawar and Qatif fields (Wilson, 1985). The Arab Formation is composed of 4 four cycles; all the members (A-D) are producers but the D is the most productive over much of the Arabian Gulf (Alsharhan and Kendall, 1986). Each cycle shoals up from marine to a restricted-marine to saltern-evaporites. The general cycle consists of a lower unit of bioclastic lime mudstone or wackestone with normal marine fossils, a middle grainstone unit of clean washed calcarenite composed of sorted and rounded skeletal particles and ooids, and an upper thin tidal flat dolomite that grades within a few feet into massive to nodular anhydrites that were deposited in a mud flat setting in sabkhas and brine pans (personal observations). Capping the whole sequence is a thick unit of subaqueous saltern anhydrite, the Hith Anhydrite which extends over much of the Arabian Gulf region.

The Late Jurassic paleogeography in this area is interesting as it demonstrates the ideal association of source, reservoir and seal. At that time the Arabian platform was a broad epeiric seaway with local organic-rich intrashelf basins depositing starved basin sequences of laminated organic-rich marls and mudstones (Diyab and Hanifa Formations: Ayres et al., 1982). High-energy grainstone shoals formed about the edges of these intrashelf basins (e.g., Diyab Fm. reservoirs in Abu Dhabi; Loutfi and El-Bishlawy, 1986). In the author's opinion, the intrashelf-basin onto carbonate-platform transition showed all the characteristics of a ramp profile; reefs or other boundstone facies were lacking about the rim of the intrashelf basins, although regionally the platform-margin may have been a rimmed-shelf. The regional platform-margin has been destroyed during later continental collision.

In the overlying Arab Formation, each reservoir of the various Arab cycles was capped by an evaporitic sabkha/salina complex. Wilson (1985) attributes the various evaporitic stages in the Arab Formation to prograding supratidal complexes. The present author feels some of the restriction may have been related to falls of relative sea level within an overall transgressive regime as much of the Arab Formation evaporites away from the structural culminations are dominated by subaqueous saltern evaporites. Personal observation of cores from producing intervals in the Arab Formation in some of the giant fields of the southeastern Arabian Gulf have shown sabkha signatures tend to predominate in Arab Formation evaporites over the crests of a field and subaqueous evaporites are more common off-structure. In some fields this syndepositional shoaling of the structure was perhaps generated by Jurassic flow of the underlying basinal Hormuz Formation evaporites (Mina et al., 1967).

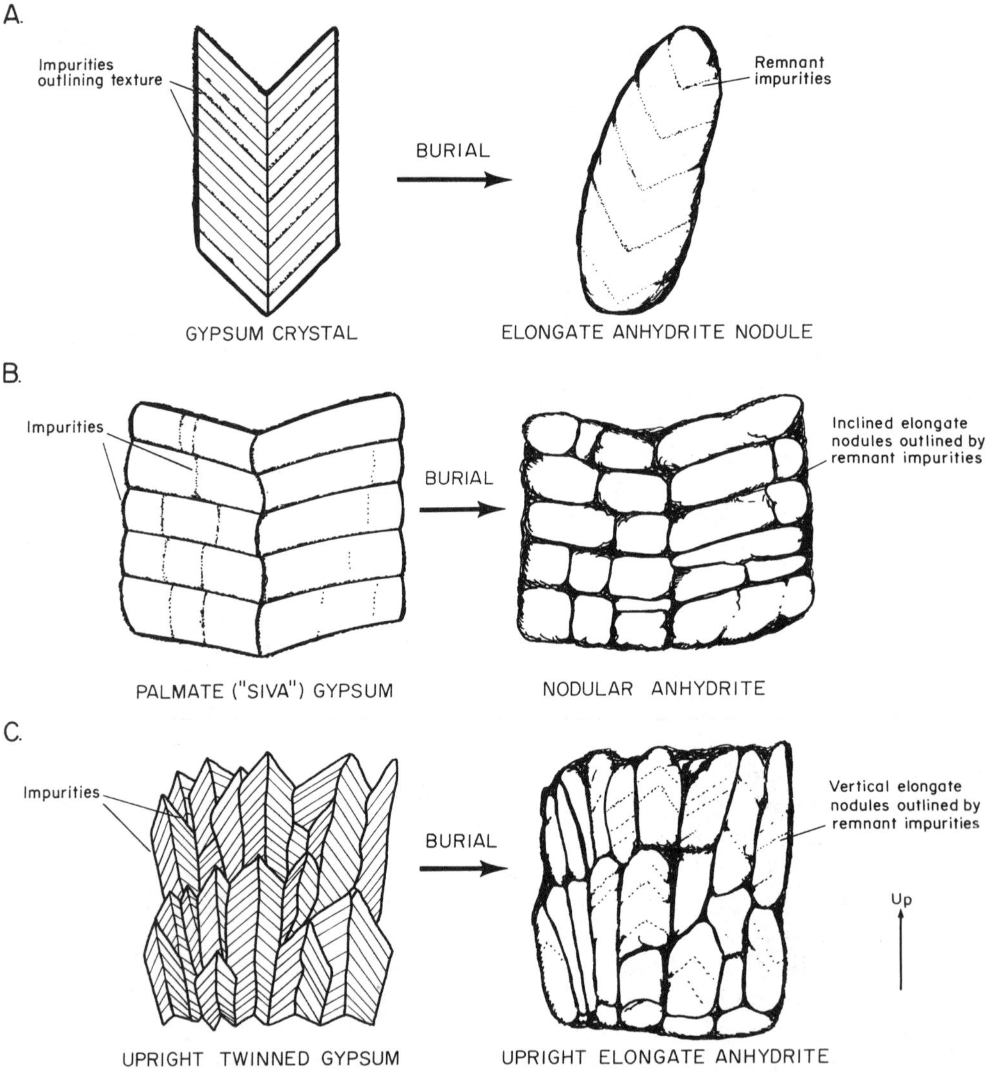

Figure 3.18 Schematic of the depositional significance of vertically aligned anhydrite nodules (after Warren and Kendall, 1985).

TABLE 3.1 DISTINGUISHING FEATURES OF SABKHA AND SUBAQUEOUS EVAPORITES FROM SOURCES QUOTED IN TEXT

Sabkha **Sulfate dominated facies**	Shallow-water **Sulfate dominated facies**
Evaporite units are supratidal, matrix dominated, usually <60% sulfate	Evaporite units are subaqueous, relatively pure, often >70% sulfate
Each evaporite (supratidal) depositional unit is thin, usually <1–2 m	Each evaporite (subaqueous) depositional unit is thick, 1–20 m
Displacive and replacive nodular and enterolithic textures	Bottom-nucleated evaporite crystal textures; often laminated, laminae can be laterally continuous, but do not extend across the whole basin
Evaporite crystals are diagenetic	Deposition can be mechanical; evaporites contain clastic textures; gypsolites, wave and current ripples, cross-beds, rip-up breccias, reverse and normal graded beds
Associated with sabkha tepees, flat-laminated and polygonal algal mats	Associated with groundwater tepees, laminar algal mats and domal subaqueous stromatolites
Facies units tend to be laterally extensive, parallel to shoreline; deposited as "peritidal" trilogy	Facies often symmetric or asymmetric bull's eye-patterns; sometimes as vertical aggrading facies
Carbonate matrix washed in from lagoon during storms; sands can be blown in from adjacent sand seas (ergs)	Carbonate facies outline areas of less saline water in basin; vadose pisolites can be common in such areas
Hydrology dominated by storm recharge and shallow brine reflux	Hydrology dominated by evaporative drawndown and deeper brine reflux
Halite dominated facies	**Halite dominated facies**
Tend to be matrix-rich units usually less than 2 m thick. Chaotic halite muds and halite karst are commonplace	Units are relatively pure and thick; composed of superimposed halite crusts separated by terrigenous or carbonate-sulfate laminae. Each depositional unit was as thick as the brine pond was deep, 2–10 m
Crystals tend to be displacive or replacive with common inclusions of matrix in the halite crystals; "pagoda" halite is a good indicator of mud flat deposition	Either bottom-nucleated chevrons or rippled and cross bedded accumulations; chevrons in crusts deposited in ponds with fluctuating salinities or on flats which may later be exposed. Chevron halite often composed of brine-inclusion-rich, upward aligned crystals occasionally crosscut with clear halite filling karst cavities

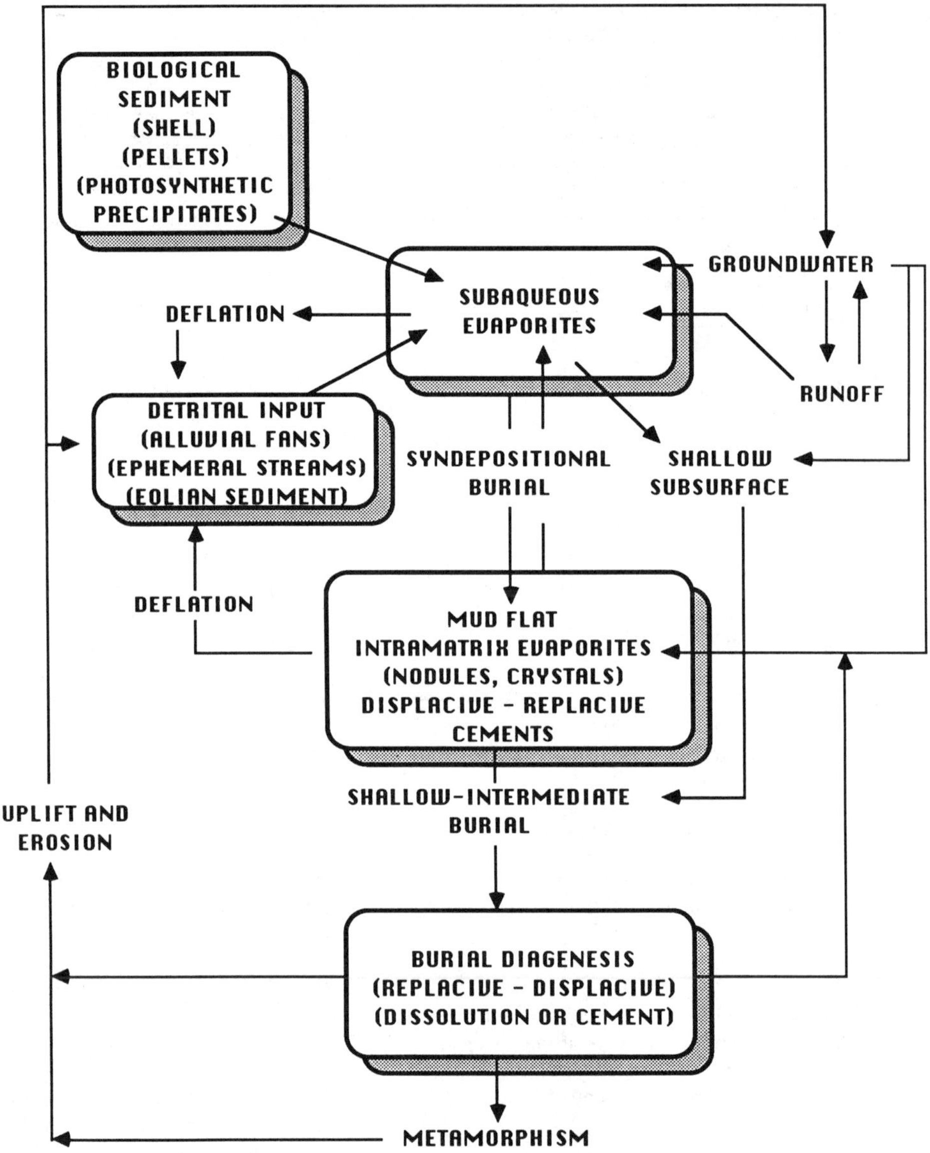

Figure 3.19 Flow chart of the complexities that an ancient evaporite experienced from the time it was deposited until it was examined by a geologist.

SUMMARY

Subaqueous interpretations and models for ancient evaporites have been given a footing in the Holocene in the last decade with studies of marine and continental lacustrine sediments in Africa, Australia, the Mediterranean, and the Middle East. Halite textures formed under shallow water conditions are now well understood (Figs. 1.8 and 3.3). Our understanding of ancient sulfate textures is growing year by year as the complexities of the gypsum-anhydrite transition are unravelled. A large step forward was made when vertically aligned and elongate anhydrite nodules were recognized as ancient buried equivalents of subaqueous gypsum crystals (Fig. 3.18). This knowledge, coupled with an understanding of the various vertical sequences laid down under different hydrological settings, is allowing more and more detailed environmental interpretations to be made in ancient evaporites. Stability of the water column and its salinity are paramount in controlling the type of succession. When conditions are stable, large bottom-nucleated crystals are deposited; under more changeable conditions of salinity and energy level, mechanically reworked accumulations of evaporites are deposited. The rapid depositional rate of evaporites compared to rates of change in the water level in the basin means that many evaporite sequences start off in relatively deep water depths (few to tens of meters) and quickly infill the basin to the water level. Any further evaporite deposition is in the capillary zone, and sabkha textures are the result. The resulting shoaling sabkha and salina evaporites can now be interpreted using textural criteria (Table 3.1). There is probably no such thing as a basin filled with a subaqueous section, nor a basin filled with sabkhas; these are interrelated systems subject to complex feedback interactions, related not just to what is happening at the surface but to subsequent interactions in the subsurface (Fig. 3.19). All ancient evaporites have felt these interactions and have textures indicating their effects!

CHAPTER 4

Deep-Water Evaporites

INTRODUCTION

Only a few sedimentological studies have been completed on deep-water evaporites, and of all the evaporitic depositional settings, deep-water is the least understood. Deep-water evaporites are difficult to model, as no modern marine deep-water evaporites are to be found anywhere on the earth's surface. The best approximation to a modern deep-water setting is the Dead Sea in the Middle East, where continental evaporitic sediments were laid down in brines over 300 m deep. However, as shown later, this transtensional lacustrine deposit is not necessarily an appropriate counterpart for ancient deep-water marine evaporite basins.

For a substantial thickness of salts to accumulate in deep water, the bottom waters must be saturated with gypsum and/or halite. In modern deep-water settings, most crystal growth occurs at the air-brine interface and crystals settle through the water column into deep-water as a "rain from heaven." The well-developed laminae so common in ancient deep-water settings were probably not due to changes in the salinity of the deep bottom waters which formed part of a well-buffered system. Rather, the bulk of the laminated sediment deposited in an ancient deep-water setting first precipitated in the near-surface waters (Fig. 1.8b). Laminae in ancient deep-water deposits were due to changes in the surface waters, often a response to changes in the rate of brine influx, temperature, or evaporation rates. These, in turn, affected mineralogy, sedimentation rate, and the amount of organic material settling through the water column. Brine mixing is another potential mechanism for the formation of evaporite crystals at various depths within the water column (Raup, 1970).

Small amounts of crystals may also grow on or in the bottom sediment, especially around deep-water brine springs associated with diapirs. Turbidites and mass flows are not uncommon in ancient deeper-water settings, especially about the basin edge. Such clastic evaporites were reworked from a shallow-water evaporite "factory" on the shelf. No modern counterpart exists to this evaporite platform–to–basin transition. Water depth estimates above reworked evaporites range from more than 175 m in the Mediterranean Miocene to 200 m in the Permian Zechstein Basin (Schreiber et al., 1976; Taylor, 1980). Water depths in deep-water evaporites are difficult to estimate, as the bottom waters are too saline for benthonic microfossils, so sequences lack good paleoecological control.

Geochemical evidence has been used to estimate water depth indirectly. The assumption is that the smaller the volume of brine, the greater the variation in the brine chemistry over an annual depositional cycle. A large volume of brine acts as a buffer and prevents significant changes in brine chemistry. Using such a model for strontium content of Pleistocene brines during deposition of the Lisan Formation in the Dead Sea region, Katz et al. (1977) estimated that water depths were 400 to 600 m at that time, much deeper than the 300+ m of the present Dead Sea.

There is a volume problem in attempting to explain mm-laminated deep-water evaporites as bottom-nucleated crystal growths. For example, a laminated calcite-gypsum sequence growing on a deep bottom requires bottom salinities to fluctuate between gypsum stability and carbonate stability. If the laminae are varves, this fluctuation must take place once a year. When the basin is a large deep-water evaporite basin, one cannot postulate an annual pulse of fresh water reaching the bottom. The fresher water is less dense than the gypsum-saturated bottom brine and will always float atop the more dense water. The only way to freshen the bottom waters back to carbonate precipitating salinities is to freshen and mix the entire brine body. With a basin hundreds to thousands of kilometers in area and hundreds of meters deep, this is a physically impossible feat on a short-term (annual) time scale. There is simply no regular source of fresher water (meteoric or marine storm flood) of sufficient volume to mix and completely freshen the brine column of a deep-water brine basin. Hence most ancient deep-water evaporites were either "rain from heaven" or reworked platform evaporites.

MODERN DEEP-WATER EVAPORITES

Dead Sea

The Dead Sea is the last episode in a graben-fill located on the transform strike-slip of the Jordan Rift separating Israel from Jordan. The Holocene–to–Miocene fill may be 8 to 10 km thick in places and the Holocene section at least 90 m thick. The water surface of the Dead Sea is 403 m below sea level, making it the lowest continental area on the earth's surface (Fig. 4.1). The Sea is 80 km long and 20 km wide, with an area of 810 km^2 and a volume of 144.5 km^3 (Neev and Hall, 1976). A Northern and a Southern Basin in the Dead Sea are separated by a shallow sill, the Lisan Straits. The Southern Basin is very shallow, and at the present time much of it is subaerially exposed; the Northern Basin is filled with brines up to 325 m deep. Feeding the Dead Sea are a number of rivers, wadis, and springs, by far the most important of which is the Jordan River, flowing into the Sea at the northern end of the basin. Outflow zones of these streams are defined by clastic aprons or fan deltas. The Dead Sea floor is asymmetric, with the western slopes averaging about 7° and the eastern slopes about 8°. The floor of the northern part of the Northern Basin is flat, and clayey deltaic sediments from the Jordan River are building a smooth planar slope into the northern end (Garber, 1980).

The Dead Sea Depression lies in a transtensional fault zone with sinistral displacement of up to 105 km along the fault (Garfunkel et al., 1981). Fault movement began in the Miocene with the opening of the Red Sea and is continuing at a rate of 7 to 10 mm/yr. The Dead Sea is a classic pull-apart basin. Zak and Freund (1981) showed how the depocenters in the Dead Sea region have moved 50 km northward along the shear zone. The floor of the basin appears to be more strongly coupled to the western margin (Levantine plate) and is being left behind by the northward moving Arabian plate. This has created an off-lapping style of sedimentation in the Dead Sea–Arava Valley reminiscent of the Ridge Basin in California. Continued seismic movement has triggered diapirism in the underlying Miocene evaporites. Diapirs subcrop along the major north-south-trending subaqueous fault defining the western border of the lake (Neev and Hall, 1979). These salt structures may be genetically related to the salt in the outcropping diapir of nearby Mt. Sedom.

Rainfall in the region is 47 to 88 mm, evaporation 1500 mm, and air temperatures between 11 and 21°C in winter and 18 to 40°C in summer. The average temperature of the upper 50 m of the water body is between 16 and 36°C, while at depths greater than 50 m, the temperature is 21.5°C. The density of the water is around 1.23 gm/cm^3 although historically the near surface waters were less dense than the deeper bottom waters. Until the mid-1970s the lake was density-stratified, with inflow, mainly from the Jordan, balancing evaporative loss. The lake was classified as "permanently stratified," with a well-developed pycnocline at a depth of 40 m (Neev and Emery, 1967). The Jordan River, flowing south from the Sea of Galilee, provided 80% of the water received by the Dead Sea (Eugster and

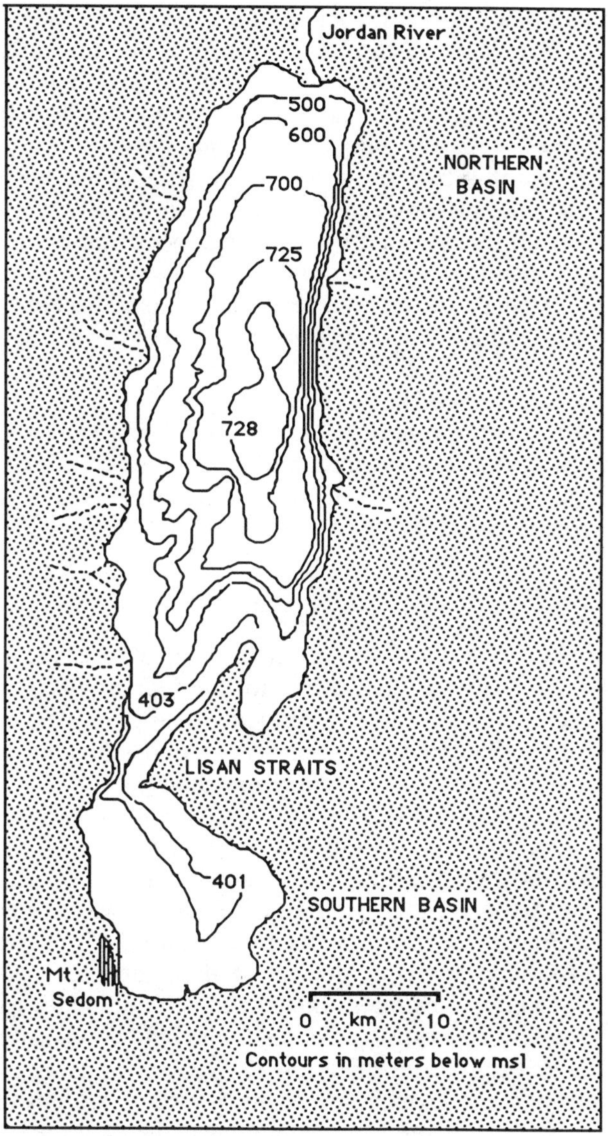

Figure 4.1 Bathymetry of the Dead Sea in the Middle East, contours in meters below mean sea level (after Garber, 1980). The present Dead Sea surface is 403 m below mean sea level. Major wadis are shown as dashed lines about the edge of the Sea.

Hardie, 1978). Storm runoff provided around 8% and spring inflow the remainder. Thus the bulk of the detrital sediment flowing into the Dead Sea was carried in by the Jordan River. However, the bulk of the solutes deposited as the chemical precipitates of the Dead Sea did not come from the Jordan but from the springs, which are up to 300 times more saline than the waters of the Jordan River.

The Jordan River inflow to the Dead Sea has been affected by irrigation projects along the banks of the Jordan. These projects have decreased the volume of water flowing into the Dead Sea while at the same time it increased the salinity of the inflowing river water. This lowered the lake water level and increased the salinity of near-surface waters. Density contrast between the upper and lower water masses lessened until in February 1979 the lake underwent overturn as the deep, lower water mass mixed with the upper water mass (Steinhorn, 1981, 1983). Today the lake is still a well-mixed brine body with an average density of 1.23 gm/cm^3. Before 1979 the lake last underwent overturn some 300 years ago (Stiller and Chung, 1984), and it appears that overturn has occurred naturally a number of times in the lake's history.

Lake water levels have changed throughout recorded history. According to Neev and Emery (1967), the paleolake level remained at −436 m for a long time; then around 1000 years b.p. the water level began to rise and waters spread out of the Northern Basin into the Southern Basin. Waters have occupied both Basins since around 500 years b.p. The highest recorded level of the Dead Sea (−392 m) was reached during the first three decades of this century; since then the lake level has been decreasing, reaching −403 m in 1979. Today the lowered water level has once again separated the Northern and Southern Basins (Fig. 4.1).

Basinal Sediments

Off from the present shoreline and away from the siliciclastic aprons, at water depths of more than a few meters and down to the deepest parts of the basin, the lake sediments are laminated muds. Sediments below the deeper waters of the Dead Sea can be subdivided into two major types, detrital and chemical precipitates (Garber, 1980). Detrital particles include limestone, quartz, and silicate clays, with minor feldspar and dolomite. Precipitates include aragonite, gypsum, and halite. The most common type of sediment is clayey silt, with terrigenous material by far the most important component. The proportion of terrigenous material increases toward the shoreline, especially toward river and stream mouths. Most of the deeper-water sediments are composed of fine-grained sediments (clayey silt or silty clay) showing alternating dark, detrital-rich, and light aragonite-rich laminae. Laminae thickness range from submillimeter to centimeters. Intercalations of sand layers, gypsarenite disks and halite cubes are also common near the sediment-surface interface (Fig. 4.2).

Aragonite precipitation occurs throughout the near-surface waters of the central Dead Sea and was first documented as "whitenings" in 1959 (Neev and Emery, 1967). Precipitation is enhanced by a periodic influx of bicarbonate-rich waters from streams and rivers, followed by periods of intense evaporation of the lake waters. The aragonite then settles to the bottom of the brine body to form a light-colored aragonite-rich lamina composed of microcrystalline ooze (Fig. 4.2a). Intrasediment gypsum today occurs as individual crystal disks making on average 6% of the deep basinal sediments. In fact, gypsum occurs today in the near-surface sediments of the Dead Sea in all subaqueous settings from shallow sea-marginal, through shelf to deep basinal settings (Garber, 1980; Garber et al., *in prep*). The gypsum rarely forms discrete laminae but is disseminated through the terrigenous (dark) layers, which contain mainly limestone fragments with lesser amounts of quartz, dolostone, and feldspar.

This distribution of evaporites is in marked contrast to the findings of Neev and Emery (1967), who found no gypsum in sediments laid down in waters more than 30 to 40 m deep. They explained the lack of gypsum as due to the effects of sulfate-reducing bacteria which flourished below the halocline in the deeper anoxic waters of the Dead Sea. Sulfate-reducing bacteria destroyed organic matter and gypsum, converting them to sulfides, energy, and H_2S. The presence of these anaerobic bacteria explained why gypsum observed in the surface waters never reached the bottom. It was destroyed as it settled through the water column. The fact that overturn of the water column occurred after the study of Neev and Emery could explain why Garber (1980) found gypsum in the deep bottom sediments, whereas Neev and Emery observed none. It may well be that overturn produced higher oxygen levels in the bottom waters and so reduced the biomass of sulfate-reducing bacteria.

Garber (1980; pers comm. Oct. 1985) believes that the gypsum in the deep-water sediments is not the result of overturn and that it was probably present even during the studies of Neev and Emery. The sand-sized gypsum is probably the settled-out remnant of gypsum precipitated in the surface waters. Any further gypsum crystal growth at depth may be a result of the upward diffusion through the sediments of brines supersaturated with respect to gypsum; such brines are perhaps related to the subsurface dissolution of nearby salt structures. Another reason for periodic sulfate saturation of the deep bottom waters may be the pumping of brine residue into the Dead Sea from the carnallite works in the Southern Basin. Garber attributed a detrital origin to the calcite fraction, whereas earlier workers have interpreted it as residue from the bacterial reduction of sulfate (Neev and Emery, 1967).

Deep-water accumulations of halite are found mainly in the southern part of the Northern Basin. The halite makes up less than 10% of the bottom sediment and is usually only a few percent. Most of the halite occurs as tiny clear cubes (40 to 500 μm in diameter) dispersed in the laminated bottom sediments (Fig. 4.2b). In the southern part of the Northern Basin, the halite also occurs as larger cubes of halite suspended in laminated sediment as well as rare accumulations of massive halite up to 30 cm across. The origin of the halite is not well understood, and several processes may have led to its formation:

1. Downwelling brines derived from the salt works in the Southern Basin
2. Upwelling brines derived by the near surface dissolution of shallow Miocene salt structures
3. Subaqueous brine-spring waters mixing with the bottom waters of the Dead Sea
4. The settling of salt precipitated in the near-surface waters. This is a recent development related to the increased salinities of the near-surface water. In

(a)

(b)

(c)

(d)

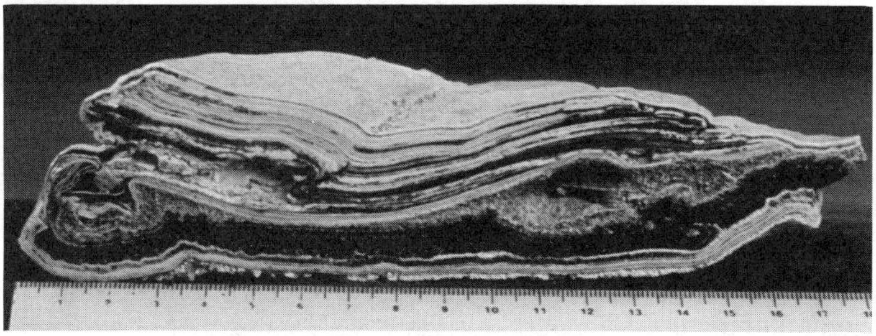

(e)

(f)

Figure 4.2 Dead Sea sediments: (a) laminated basinal sediments in core, greater than 200 m water depth; (b) clear halite cubes from the basinal sediments; note the lack of displacement shapes (centimeter scale); (c) elevated fan-delta at Ein Gedi; (d) coarse-grained sediments in the fan-delta, with an abundance of debris flow sediment; (e) inorganically formed laminated crusts from the margin of the Dead Sea, composed mainly of aragonite; (f) detail of laminated crust showing laminae due to varying degrees of alignment of aragonite crystals (photographs courtesy of Ray Garber).

February 1979, for the first time, halite crystallized in the surface waters above the deeper parts of the central Dead Sea (Steinhorn, 1983).

The most surprising thing about the deep-water deposits of the Dead Sea is the lack of evaporites. For the last few hundred years, until 1979, the Dead Sea was a stratified lake with a lower water mass or fossil water body which was at or near halite supersaturation. Except at the air-brine interface, the upper water mass was undersaturated with respect to aragonite, gypsum, and halite. During most of the year, but especially during the summer months, evaporite minerals (aragonite and to a lesser extent, gypsum and perhaps halite) precipitated at the surface. The minerals then sank through an undersaturated water mass above the halocline. Those that survived this, and the subsequent bacterial degradation below the halocline, sank to the bottom to be incorporated in the detrital bottom sediments. Irregularly spaced times of more intense evaporation, once every 5 to 20 years, resulted in a massive settling of aragonite to form the white layers in the bottom sediments. The laminae in the Dead Sea are not annual layers (varves); only 15 to 20 laminae have been deposited in the last 70 years. The bulk of the basinal sediment in the Dead Sea is detrital, carried in by the Jordan River or by storm floods over fan deltas which encircle the Dead Sea. The Jordan River has built a deltaic lobe which extends several kilometers into the Dead Sea. Detrital clay minerals make up a significant proportion of the Dead Sea sediments. Illite averages about 40% of the carbonate-free clay fraction, smectite about 32%, and kaolinite about 28% (Garber, 1980).

At the present time the Dead Sea is in a "humid phase" with relatively elevated levels of meteoric input. Although evaporites have not accumulated in the basin center for the last few hundred years, it is interesting to speculate on the effects of overturn. Today, the water column has been homogenized by overturn and evaporites are precipitating from the surface waters. After the overturn, halite crystal dust was collected at water depths of 60 and 80 m (Steinhorn, 1983). If overturn has occurred naturally during the Holocene, as it probably has during the more arid phases of the Dead Sea, the conditions would have been ripe for the accumulation of deep-water evaporites (Stiller and Chung, 1984). The whole water column would have been supersaturated with respect to halite and gypsum. It would have been well mixed throughout and so may have lacked sulfate-reducing bacteria. At such times evaporites could have accumulated as laminar or bedded deposits on the deep-water bottom. To date, no one has drilled the Northern Basin to depths that would test this hypothesis.

Yet in the southern end of the Northern Basin, cores taken below water depths of 40 m bottomed on a hard halite after passing through 10 to 80 cm of laminated carbonate sediment (Fig. 4.3; Neev and Emery, 1967). What was the origin of this halite? Was it deposited some 300 or so years ago during a previous overturn event? If the crystals forming in the surface waters of the Dead Sea today are any indication of what ancient deep-water accumulations should look like, then the bulk of ancient deep-water accumulations form by a "rain from heaven"

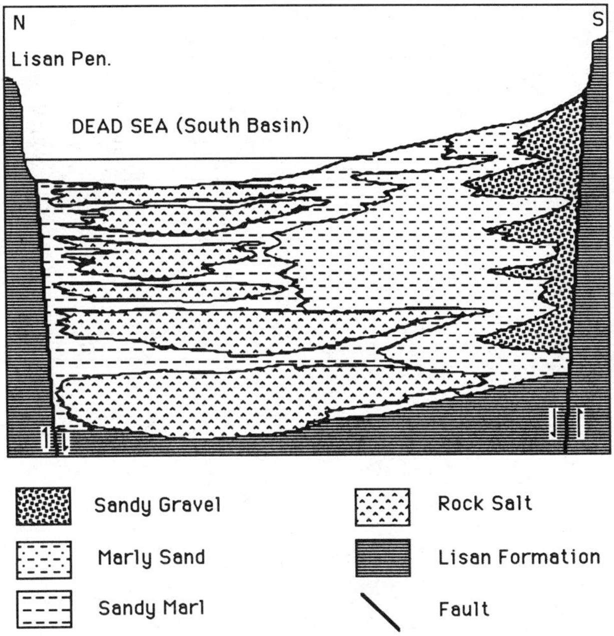

Figure 4.3 Schematic cross section of the Southern Basin of the Dead Sea, showing the massive bedded halite units and alluvial fans on the southern side. The recent sediments of the Dead Sea are enclosed by the Pleistocene Lisan Formation (after Neev and Emery, 1967).

of tiny halite cubes and small gypsum disks. These fine-grained crystals are the types that dominate ancient deep-water accumulations such as the Mediterranean Messinian basinal sequences, the Castile Formation of west Texas, and the Zechstein of Europe.

Basin-margin Sediments

The fine-grained laminated sediments discussed so far are not going to make a potential hydrocarbon reservoir. Chemical and detrital muds of the Dead Sea are surrounded by an apron of permeable siliciclastic sediments brought into the basin as debris flows and fan deltas. Fan deltas, especially in those areas of shoreline reworking, are likely future reservoirs. Sediment is dispersed to this apron by a series of fan deltas spilling out into the edge of the Dead Sea. Thick gravelly fan deltas lie at the mouths of all the larger canyons on both sides of the Dead Sea graben (Fig. 4.1). A few deltas are active and reach down to the present shores of the Dead Sea. Most are inactive and hang some 200 m above the present shoreline, reflecting earlier, high water stages in the graben during Lake Lisan and the earlier Lake Samra time. The asymmetric tectonic setting of the basin is also reflected in the distribution of fluvial sediments. Along the eastern side of the Dead Sea is the active transform escarpment. On this side of the Dead Sea, slopes are steep and the fans are dominated by poorly sorted debris flows which often reach all the way to the water. On the western side the slopes are not as steep and there is a greater potential to deposit a sand fairway. Today this is not happening because of the dominating effect of Late Pleistocene mudstones. In terms of its ability to generate potential sand reservoirs the Dead Sea suffers from the same problems as Death Valley (Chapter 2). The sides are simply too steep, and the hole too deep, to allow well-sorted sand bodies to accumulate about the lake edge. In addition, a Late Pleistocene highwater episode in the Dead Sea has covered much of the basin strandplain with fine grained mudstones.

Modern shoreline sediments range from conglomerates to mudstones. The grain size, shape, and lithology of the shoreline sediments depend on the following factors:

1. Distance from the main border escarpment
2. The nature of nearby outcrops
3. Direction and energy of nearshore waves
4. The nature of recent changes in lake level

When the present shoreline is a long distance from the border escarpment, the sediments tend to be muddy, derived from the reworking of Lisan Marls. Beaches and bars of sand and gravel are common on moderately steep shores, especially where shoreline waves are reworking older Lisan fan delta sediments. Where the shoreline is a cliff at the base of the border escarpment, the sediments tend to be reworked slumps. Grain size analysis of gravels and sands in active shorelines show that both types are well-sorted mature sediments (Neev and Emery, 1967, p. 99). Between and on the deltas, and along both the east and west border escarpments of the Dead Sea, more than 30 near-horizontal sandy shore terraces and gravel bars line the current strandline (Fig. 4.3c, d). Many of these linear shoreline-parallel features are relict strandlines, reflecting former higher levels of the lake ranging from the present −403-m level to the −180-m shore of the Pleistocene Lake Lisan.

Gravels along the lake shoreline are coated by a mm-laminated aragonite crust stained with manganese (Fig. 4.2e and f; Garber, 1980; Garber and Friedman, 1983). Some of these precipitates are botryoidal, others are encrusting, but both are interpreted as inorganic. Although inorganic precipitates, they are reminiscent of the stromatolitic crusts of the South Australian salinas (Warren, 1982b) and like them are strandline indicators.

Lake Lisan (A Dead Sea Precursor)

Below and surrounding the Holocene sediments of the Dead Sea is the Quaternary Lisan Formation (Neev and Emery, 1967; Begin et al., 1974). The Lisan is the typical cover of the rift floor and can be found outcropping in the valley walls at elevations as high as −180 m below m.s.l. Like the present Dead Sea sediments, the Lisan Formation contains substantial thicknesses of laminar dark-light carbonate couplets and lesser amounts of gypsum-carbonate couplets. Lake Lisan was much more widespread than the present lake of the Dead Sea, but like the

Dead Sea it occupied the Dead Sea Rift Valley and was a tectonic closed lake (Fig. 4.4). Salinities during Lisan time were hypersaline but probably less than those of the present Dead Sea. Deposition of the Lisan Formation began some 60,000 years b.p. and ended 18,000 years b.p. The uppermost part of the Lisan Formation forms an almost uninterrupted blanket over the entire Jordan Valley and in places covers the fan deltas of the even earlier Samra Formation (Fig. 4.5a).

The Lisan Formation is composed of two members, a lower Laminated Member and an upper White Cliff Member (Fig. 4.6). The Laminated Member is the typical Lisan Formation composed of very fine, varve-like white and dark laminae (Fig. 4.5b). White laminae are usually monomineralic, composed mainly of aragonite needles with diatom frustules. In some areas near the top of the Laminated Member, the white laminae are composed almost entirely of gypsum

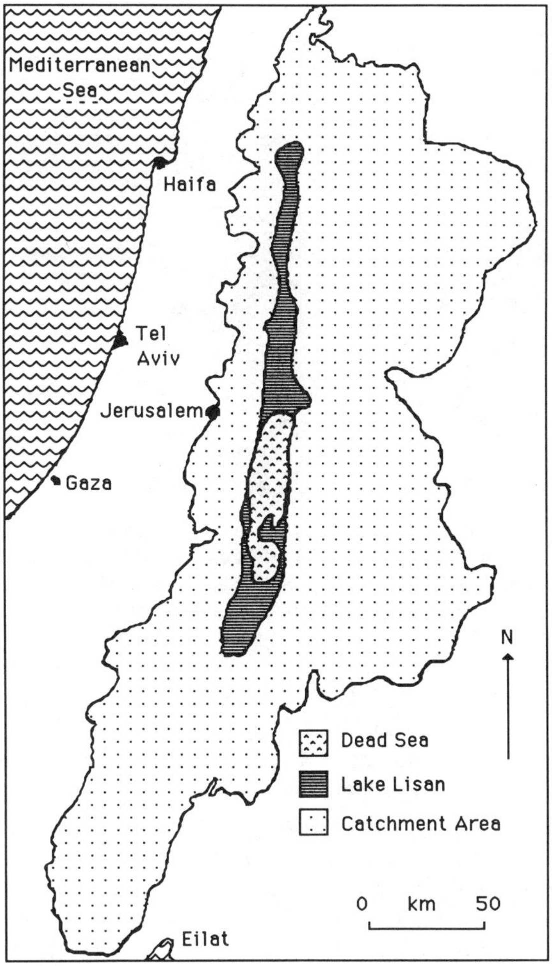

Figure 4.4 Locality map of the Dead Sea region showing the far great extent of Pleistocene Lake Lisan compared to the present Dead Sea (after Begin et al., 1974).

(a)

(b)

Figure 4.5 Lake Lisan sediments; (a) general view of the Dead Sea margin showing the distribution of the lighter-colored Lisan Formation sediments up to elevations of 180 m below sea level; (b) Lisan Formation showing its bedded and laminated nature with obvious soft-sediment deformation (photographs courtesy of Ray Garber).

or mixtures of aragonite and gypsum. Gypsum laminae are composed of fine lenticular gypsum crystals (0.25 to 0.50 mm) arranged parallel to the bedding. Dark laminae are predominantly detrital mud-sized calcite, the reworked remnants of microfossils, nannofossils and rock fragments. Quartz, dolomite, clay minerals (kaolinite, smectite, illite, and palygorskite), halite, and gypsum also occur in these calcitic layers. In the basin center the Laminated Member contains intercalated gypsum beds with laminae which are sometimes crenulated and contain sulfur concretions. About the basin margin the Laminated Member lacks gypsum and is termed the "aragonite facies" by Begin et al., (1974). To the north

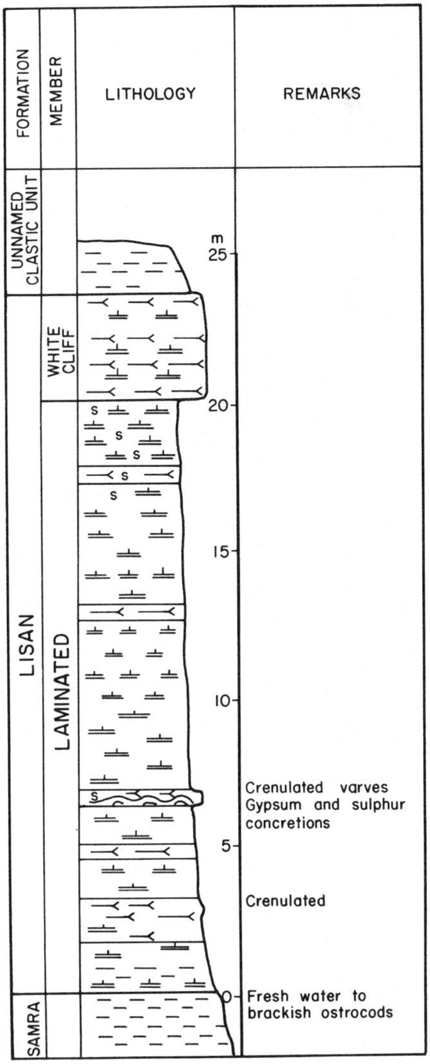

Figure 4.6 Vertical section of the Lisan Formation after Begin et al., 1974. Symbols: S sulfur nodules, —< gypsum, ⊥ laminated aragonite, ≡— clay and silt.

the Laminated Member is almost entirely detrital with almost no aragonite laminae. Even so, the sediment there is laminated, almost a pure diatomite. Overlying the Laminated Member is the White Cliffs Member; it contains two or more relatively thick beds of gypsum or aragonite. The White Cliffs is still well bedded and contains laminated sediment. Overall, the white laminae are more gypsiferous in the White Cliffs Member than in the Laminated Member. The sediment distribution in the Lisan constitutes an asymmetric bull's-eye.

Many layers in the Lisan contain sediment slumps and overfolds, water escape structures, growth faults, and other evidence of soft sediment deformation

(Fig. 4.5b). The Lisan Formation was laid down as a water-saturated succession; earthquakes in the rift kicked off multiple episodes of sediment movement both syndepositionally and in the shallow subsurface. Some of the couplets in the Lisan are interpreted as deposited in lakewaters some 400–600 meters deep (Katz et al., 1977). The sedimentary structures resemble many of the structures to be seen in the basal limestone of the Todilto Formation (Fig 3.14a). However, some gypsiferous sediments in the Lisan Formation show evidence of much shallower deposition in the form of large vertically aligned gypsum crystals poikilitically enclosing carbonate laminae. These giant crystals fit better an earlier model of Picard (1943) who described the Lisan Formation as being deposited in a combination of settings, a shallow lake, a large playa, and/or a giant sabkha. When the Lisan is studied from a sedimentological perspective, it will probably become the best modern counterpart for deep-water evaporite deposition.

If the Quaternary sediments of the Dead Sea graben are predominantly nonevaporitic, then can ancient marine evaporites composed of thick laminated sequences realistically be interpreted as deep-water? If a strictly uniformitarian model is used, taking the Dead Sea Graben as the counterpart, the answer is probably no. But every sedimentary basin is different, and the Dead Sea is no exception. The Dead Sea is much smaller than many ancient evaporite basins discussed later in this chapter. It is continental, not marine, and is the lowest point on the earth's land surface. The basin is enclosed by steep walled cliffs hundreds of meters high, as well as fan deltas, and alluvial fans, all supplying large volumes of detrital material to the basin depocenter—hence the dominance of detrital material even in the evaporite depositing areas of the lower parts of the Dead Sea Basin. It is in a tectonically active area underlain by active salt structures and fed by groundwaters depleted in sulfate and bicarbonate (compared to seawater). For these reasons the Dead Sea cannot be used to model the textures and sediments that were deposited from the deep-waters of many ancient evaporite basins. No known Holocene example exists which can explain ancient deep marine evaporites.

ANCIENT DEEP-WATER EVAPORITES

A wealth of information can be found in the literature on deep-water marine-derived evaporites in pre-Quaternary evaporite basins. Textures in deep-water settings range from laminated units which are still gypsum in the Late Miocene of the Mediterranean, to the more deeply buried laminar nodular anhydrite units of the Permian Delaware Basin of Texas, to the anhydritic mass flows and turbidites of the Permian Zechstein Basin of Germany, to the laminated halites of the Salina salts in the Silurian Michigan Basin. The deep-water sulfate and halite units of the Mediterranean Upper Miocene are relatively unaffected by the burial dewatering of gypsum and contain more easily recognized depositional structures. They will be considered first, followed by a discussion of their ancient anhydritic equivalents.

Messinian Evaporites, Sicily

The Sicilian Basin is filled with almost a kilometer of evaporites deposited during the Late Miocene (Messinian) Mediterranean "salinity crisis" (Fig. 4.7). The basin was formed during a complex series of crustal movements which climaxed

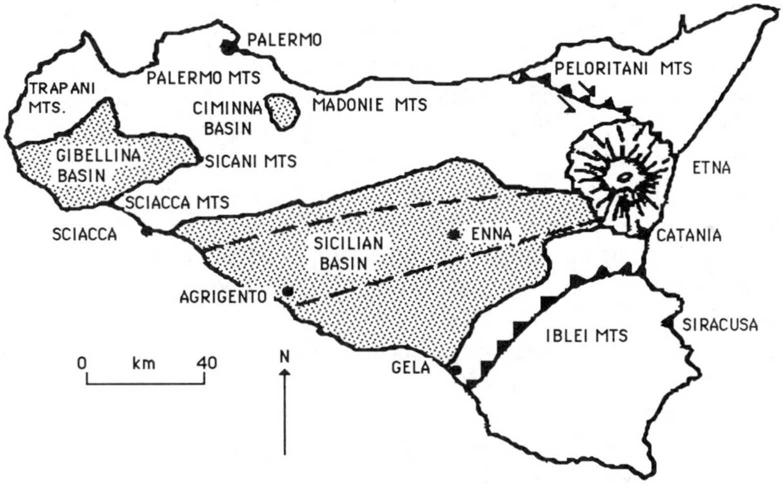

Figure 4.7 Extent of thick Messinian evaporite deposits in Sicily (shaded basins, after Catalano and D'Argenio, 1982). Also shown are major thrust faults and the area of either a former subduction or collapse zone beneath the Sicilian Basin (dashed lines).

in the Tortonian as a series of massive nappes. This formed the depressions of the present Sicilian Basin, a basin measuring 60 × 100 km and of some considerable depth. It was bounded on three sides by mountains or plateaus and open to the sea on the fourth. The Basin is floored by deep-water marls of Late Tortonian age and rimmed by reefs of the same age. In the Late Miocene the whole basin was some 1 to 2 km below its present position with respect to sea level but that is not to say that it was below 1 to 2 km of seawater. Most of the evaporites deposited at the end of the Miocene were deposited as shallow water evaporites (Chapter 3). Occasionally, interbedded among saltern evaporites are deeper water units deposited in structural lows such as the Gibellina and Cimina sub-basins (Fig. 4.7).

Heralding the onset of evaporite deposition in the Sicilian basin and the rest of the Mediterranean was a period of rapid pinnacle reef growth (Catalano and

D'Argenio, 1982). In Sicily the position of the pinnacles was controlled by bottom highs often located along the edges of faulted and thrusted blocks. These pinnacles were subsequently dolomitized and plugged with evaporites. The *Porites* reefs were buried by Messinian evaporites and evaporite-cemented fluviodeltaic clastics derived locally from rising thrust blocks. In terms of depositional and diagenetic facies distribution the reefs are excellent counterparts for many ancient pinnacle reefs, such as those of the Alberta and Michigan Basins (author, personal observation). The Messinian pinnacles of Sicily do not contain any oil as the basin is not yet mature. However as noted by Warren (1986a) with further burial plus a few more million years for maturation, the Messinian carbonates and evaporites above the pinnacles could become excellent source rocks and feed the associated pinnacles. This is not to say that the Miocene pinnacles are not prospective elsewhere in the Mediterranean in other Messinian sub-basins, areas where the thermal gradient is higher or the hydrocarbons come from another more mature source. Selli (1973) stated that most of the gas in the Po Basin is probably sourced in Messinian strata. The basin-wide evaporites that cover the reefs were laid down in water depths ranging from less than a few centimeters to hundreds of meters, but it is the latter that are relevant in this chapter.

Such basinal evaporites are composed of widespread laminar beds, even-bedded massive gypsum, halite, and potash beds, and extensive turbidites (Catalano et al., 1976; Schreiber and Decima, 1976). The sub-basins were deep enough to establish a stratified water body and slopes sufficiently steep and extensive as to permit turbidite deposition. Yet the basin was large enough to allow undisturbed quiet water "rain from heaven" deposition in the central areas. Much of the basin center is composed of laminar organic-rich beds of gypsum associated with mass flows and turbidites (Fig. 4.8a). Deep-water gypsum in the Gibellina basin is a potential source rock, as it contains up to 7% organic matter (Schreiber and Warren, 1984). Gypsum laminae show occasional current features such as cross-beds, as well as some soft-sediment deformation features (Fig. 4.8b). The small gypsum crystals and flakes in these laminated beds formed at or near the water surface and sank to the bottom or were eroded from shallow-water crystals and then moved downslope to be deposited in an environmental setting far different from where they formed. Turbidite beds are underlain and in a few places intercalated with mass flow deposits. The turbidites contain Bouma cycles and associated flame structures (Fig. 4.8). They are overlain by laminated gypsum which passes up into a thick sequence of shallow-water selenite. This suggests that the basin was relatively deep during turbidite deposition but had shallowed to tens of meters by the time the selenite was deposited. Such rapid water level changes characterize many ancient evaporites and may be repeated again and again through an evaporite unit. Many of the Messinian sequences show that during evaporitic drawdown a basin often starts off deep and rapidly shoals. In such settings it is only the exceptionally high depositional rates that allow the short-lived but widespread deep-water succession to be preserved.

(a)

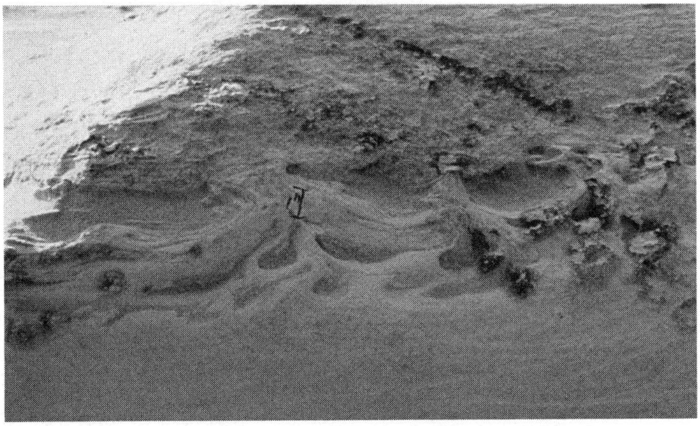

(b)

Figure 4.8 Deep-water evaporites, Gibellina Basin: (a) layered and laminated gypsum turbidites. Beds are graded and many contain complete "Bouma" sequences; also contains laminated organic-rich interbeds. Figure used for scale is Roy Park (photograph courtesy of B. C. Schreiber); (b) soft-sediment deformation structures—flame structures in gypsum sand overlain by mechanically deposited bed of coarser-grained gypsrudite.

The Permian Delaware Basin, Texas

The Delaware Basin began as part of the Ordovician Tobosa Basin. By the Pennsylvanian it had developed into two deep-water basins, the Delaware and the Midland Basins separated by the Central Basin Platform (Fig. 2.9). In the Permian a series of platforms and ramps rimmed the basin. By the end of Guadalupian time a subtidal reef mound complex (Capitan Formation) rimmed the Delaware Basin, which then covered an area measuring some 200 × 300 km (Figs. 2.10 and 4.9). Behind the rim was an extensive complex of restricted lagoons and

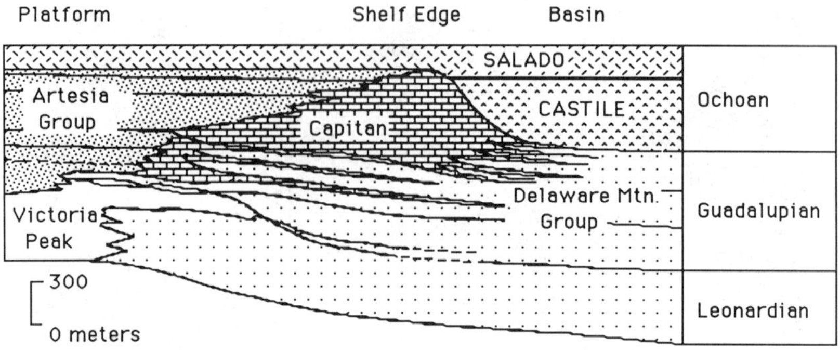

Figure 4.9 Schematic cross section of the Delaware Basin margin, showing the basinal position of the Castile Formation within the Delaware Basin and the more widespread deposition of the Salado Formation. The Bell Canyon Formation is the last episode of deposition of the Delaware Mountain Group. (*see also* Fig. 1.14)

sabkhas discussed in Chapters 2 and 3. In front of the reef province, up to 300 m of deep-water clastics were being deposited as a series of limestone and siliciclastic turbidites, with associated laminated and mass flow carbonates. The Bell Canyon Formation (uppermost Delaware Mountain Group) was the last carbonate turbidite unit deposited in the Delaware Basin before the onset of Ochoan time. Toward the end of Bell Canyon time the bottom waters had become more restricted and marine fossils disappeared except for redeposited shallow water forms. The upper part of the Bell Canyon contains syndepositional evaporites, tiny crystals, and rosettes of displacive celestite and gypsum (Fig. 4.10a). The Bell Canyon Formation passes up into the Castile (Ochoan) with no break in deposition (Williamson, 1978).

Ochoan time (Upper Permian) signaled the beginning of widespread evaporites in the Delaware basin. They were deposited on the floor of a deep-water, steep-sided basin. These evaporites form the world-renowned Castile Formation, a thick sequence of mm-laminated calcite-sulfate couplets, one of the best-documented ancient deep-water evaporites. In hand specimen, the Castile is composed of dark-colored organic-rich calcite laminae alternating with lighter-colored gypsum/anhydrite laminae. Dolomite is absent in the Castile Formation. Four halite beds 10 to 30 m thick break up the Castile into its widely distributed members. A number of complete cores (400 to 800 m long) have been taken through the Castile and the overlying Salado Formation, they have been the subject of intense study by Anderson and Kirkland (1966), Anderson et al., (1972), Dean and Anderson (1974), and Dean (1978a, b). Their studies indicated that the Castile is a deep-water evaporite deposited as sand- to mud-size gypsum crystals interlaminated with micritic calcite. The Castile is a classic "rain from heaven" evaporite with carbonate laminae deposited during times of freshening of the near-surface waters. Carbonate laminae are organic rich due to the settling of organic detritus from the associated planktonic blooms. The original sulfate was gypsum (now

(a)

(b)

(c)

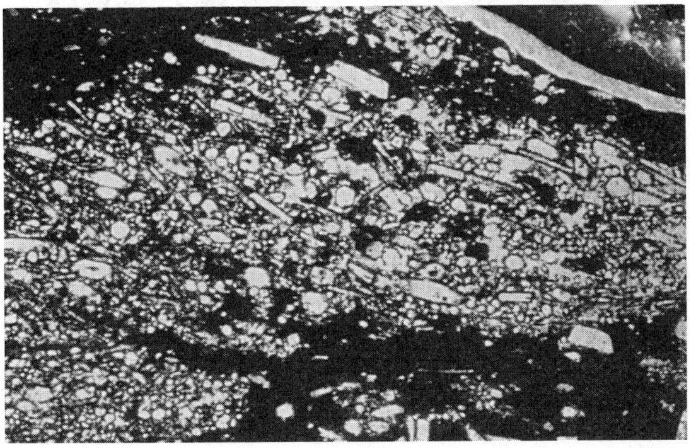

(d)

Figure 4.10 (a) Uppermost Bell Canyon Formation showing displacive gypsum concentrated along bedding surfaces; (b) portions of three cores in the Castile Formation, composed of alternating layers of light-colored anhydrite and darker organic-rich calcite; laminae can be correlated yet two of these wells are 114 km apart (photograph courtesy of W. Dean); (c) anhydrite pseudomorphs of bottom-nucleated shallow-water gypsum in the Salado Formation. Crystals are up to 8 mm long; (d) cross section of a ripple composed of mechanically reworked gypsum, now pseudomorphed as halite, Salado Formation. Photo approximately 15 mm wide [photographs in parts (c) and (d) courtesy of Tim Lowenstein].

anhydrite in the subsurface) deposited as small lenticular gypsum crystals with their long axes parallel to bedding. Laminae are interpreted as varves, suggesting that the Castile was deposited in 260,000 years. Groups of laminae can be correlated over horizontal distances greater than 110 km, implying a below-wave-base depth of deposition (Fig.4.10b). Milankovitch climatic cycles (related to the earth's orbit) are recorded in the laminar spacings of the Castile, in much the same way that Milankovitch cyclicity is forming in the deep ocean sediments of the present Ice Age (Anderson, 1982). The last major Ice Age before the present was in Permo-Carboniferous time, although not during Castile deposition; widespread glacial deposition ceased millions of years before Castile time. Not only are the Castile laminae correlated over large distances, but the trace-element chemistry is also relatively unchanged across the basin during Castile time (Dean, 1978b). This too implies a relatively deep-water body buffered by the large volume of brine in the basin.

Toward the top of the Castile the laminae thicken and display vertically aligned pseudomorphs indicative of shallow-water gypsum (Schreiber, 1986). At the same time the beds become less and less regular and the amount of halite increases. Finally, the Castile passes up into the bedded halite of the Salado Formation (700 m thick), which in shelf settings contains extensive deposits of potash. In contrast to the calcite mineralogy in the carbonates of the Castile Formation abundant dolomite is found in the Salado Formation. During Salado time, a strong continental component influenced the chemistry of the waters flowing into the Delaware Basin. Unlike the Castile, which occurs only in the Delaware Basin, the Salado Formation extends up out of the Basin to the north, where it covers the Northern Shelf and to the east across the Central Basin Platform and the Midland Basin. Detailed studies by Lowenstein (1982) demonstrated that the Salado Formation was deposited in a shallow-water playa-lake setting, in environments similar to those seen today in perennial lagoons and salt pans (Figs. 4.10c and d).

In Ochoan time the Delaware Basin began as a relatively deep-water basin which shoaled as it was rapidly filled with evaporites. The Castile is the world's best documented example of laterally extensive mm-laminated evaporites, yet even here no objective indication of absolute water depth can be found. The water depth need not have been "basin full" (1000 m) but it was deep enough to have been below the wave base and of sufficient volume to have acted as a chemical buffer. This implies that the water depth was greater than 50 m and could, at least in the early stages of Castile deposition, have been hundreds of meters deep. The outcropping Castile evaporites were deposited in a basinal setting well away from the basin margin, where slope evaporites and turbidites would have been deposited. On the eastern margin the sequence is buried and intact and may well be dominated by reworked evaporites, but it has not been cored for study. Deep-water basin margin–to–basin center relationships in an evaporite basin have not been documented in the Delaware Basin but have been studied in the Permian anhydrites of Europe.

The Permian Zechstein of Europe

The Zechstein deposits of Europe were laid down in a broad, well-developed basin/graben extending from the British Isles across the North Sea, Denmark, and the Netherlands through Germany to eastern Poland and Lithuania. The basin measured some 900 by 1500 km, much larger than any modern evaporite area and was divided into a Northern and Southern Permian basin. The basin-wide evaporites in the basin are divided into five depositional cycles (Z_{1-5}), four major and one final minor one; each cycle changes up-section from evaporitic carbonates at the base to evaporite salts, including potash salts, at the top. The depositional setting during early Zechstein time (Zechsteinkalk-Z1C) was an extensive evaporitic carbonate platform surrounding a deep-water basin where water depths were greater than 100 meters (Fuchtbauer and Peryt, 1980; Taylor, 1980, 1984). This was overlain by an extensive basin-wide evaporite unit, the Werraanhydrit (Z1A). Like carbonate deposition, the evaporite episodes of basin infill were characterized by well developed platform into basin profiles with the shoal water evaporitic platforms supplying much of the sediment to the slope and rise as well as the more proximal parts of the basin (Clark, 1986). After the "open marine" conditions at the base of the first cycle, the basin never returned to open marine conditions, only to less saline conditions at the base of each new cycle.

When bottom conditions were suitable, stromatolitic reefs formed in association with grainstone belts in the shoal-water areas of the basin. The patch reefs and shoals of the Eichsfeld high in the Zechstein Basin (a deep-basin shoal-water rift basin) are outcropping examples of this association (Paul, 1980). A number of oil and gas discoveries have been made in the Zechstein carbonates in and around the North Sea, and there is production from numerous onshore wells in the Netherlands, Germany and Poland (Fig. 4.11a; Taylor, 1984; Peryt, 1985). Most of the finds are in the oncolitic and oolitic rocks of the platform margin-barrier facies with the most prolific finds concentrated near the basinward margin of the shelf in the Southern Basin. Such shoal-water carbonates are similar to the pisolitic island-crest facies of the Guadalupian mid-shelf highs in the Permian Basin of west Texas (Ward et al., 1986). Gas finds predominate, and the Hauptdolomite (Z_2 carbonate) is the most common reservoir (Fig. 4.11b). Gas has migrated from the humic-rich Carboniferous to the Z_1 reservoirs and to those reservoirs in Z_2 and Z_2 when the seat seals provided by the underlying evaporites have been breached. Onshore small but economic accumulations of oil and minor condensate, rather than gas, are found in the Z_2 and Z_3 cycles. This oil is thought to be sourced in the basinal sediments of these cycles not in the Carboniferous. These internal oil-prone sources are adequate for onshore exploration but have yet to be demonstrated to be sufficiently prolific to justify a basin-margin play in the offshore Zechstein (Fig. 4.11b; Taylor, 1984).

The only giant fields in Zechstein strata are Auk and Argyll fields; they are the first Zechstein discoveries known in the Northern Zechstein Basin. Their occurrence is unusual for Zechstein reservoirs as both fields occur on the rifted

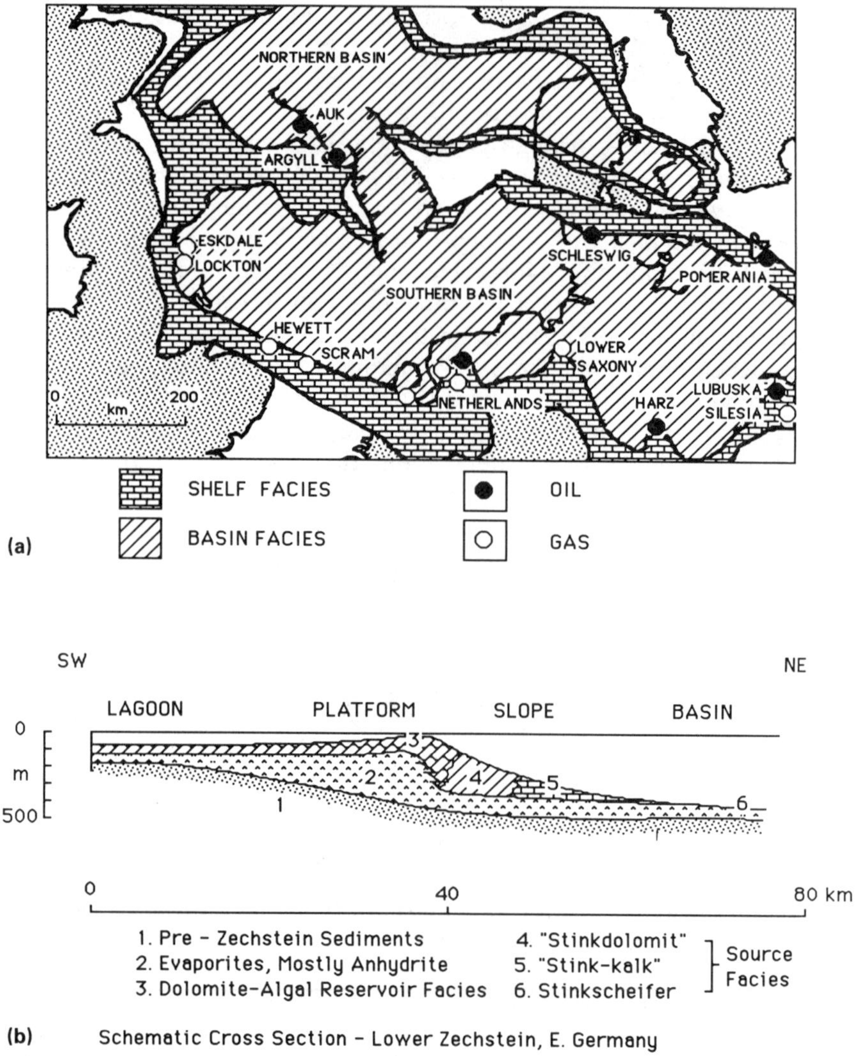

Figure 4.11 Zechstein sedimentation in Northwest Europe (after Taylor, 1984): (a) distribution of producing fields in Zechstein sediments; they occur in shoal-water carbonates of the platform margin; (b) schematic cross section of platform margin during lower Zechstein time.

edge of the Central Graben. Late Jurassic, followed by Late Cretaceous faulting, first put the Zechstein in a structurally elevated region allowing secondary porosity enhancement to occur and then allowed a mantling by Kimmeridgian Clays—the source and seal to the brecciated and vuggy reservoir facies in the Zechstein. Its unusual setting led Taylor (1984) to conclude similar reservoirs may be relatively uncommon in the Zechstein Basins away from the central graben.

Regionally, Zechstein strata form the seal for the underlying gas sand reservoirs in the Rotliegend Formation of the Southern Basin.

The Southern Permian Basin extends 1500 km from eastern England to the Polish-Russian border with a maximum width of 400 km. Subsidence was greatest at the southern end of the Danish Peninsula where over 1500 m of Rotliegend sediments accumulated in a desert lake mainly as evaporitic lacustrine halites and sabkhas. Wadi and dune sands, later the main gas reservoir sands, piled up along the southern margin of the basin, a consequence of fluvial transport out of the Variscan highlands to the south and N.E. trade winds. These highly porous sands were subsequently sealed by the thick basinal evaporites of the Zechstein series. The surface of the Rotliegend rift lake in the Southern Permian Basin probably lay some 200–300 m below global sea level at the time of the mid-Permian Zechstein marine transgression. This transgression initiated the major phase of thick evaporite fill across both basins (Smith, 1979; Ziegler, 1980, 1982). Glennie and Buller (1983) have shown that the influx of the first Zechstein waters into the isolated desert basin was so rapid (rises in water level probably of decimeters/day) that the transgression did not have enough time to extensively rework the dune sands of the Rotliengend. Thus the rapid transgression of the Zechstein Sea preserved draa morphology beneath the Zechstein evaporites. Locally some reservoir sands in the Rotliegend owe their structural closure to this partially preserved draa shape (Glennie, 1986). Regionally the Rotliegend play is controlled by block faulting, basin inversion and source maturity of the underlying Westphalian Coal Measures. It is beyond the deep-water scope of this chapter to discuss much more of the shallow-water Zechstein fill, but the interested reader is referred to the excellent papers in Fuchtbauer and Peryt (1980).

The Zechstein series is famous for its "varved evaporites," of the basin center documented in great detail by Richter-Bernburg and co-workers. The rhythmically banded halite and clay–dolomite–anhydrite laminites can be correlated over considerable distances, arguing for below wave-base deposition (Richter-Bernburg, 1955). The laminar spacing has been related to sunspot cycles; not to the Milankovitch cycles interpreted in the Castile Formation of west Texas. Based on a deep-water origin and a total thickness of 1200 m for the laminated unit in the basin, Richter-Bernburg (1955) argued for a deep "water-full" basin with water depths on the order of 1000 m in the basin center. Surrounding and laterally equivalent to these basinal laminates are a series of mass flows and turbidites (Fig. 4.12).

Some of the best documented reworked deep-water sulfates are in the Z_1 and Z_2 anhydrites of Germany and Denmark (Herrmann and Richter-Bernburg, 1955; Schlager and Bolz, 1977; Clark, 1980). In both units the structures of interest occur in a wedge of anhydritic sediment that thins away from the marginal sulfate platform into "deep-water" varves (Fig. 4.12). Millimeter–to–centimeter bedded couplets dominate in the thicker part of the wedge. The anhydrite in the centimeter-scale layers is usually graded; some layers are ripple cross-laminated or structureless. Millimeter bedded units often show convolutions, kink bands, mi-

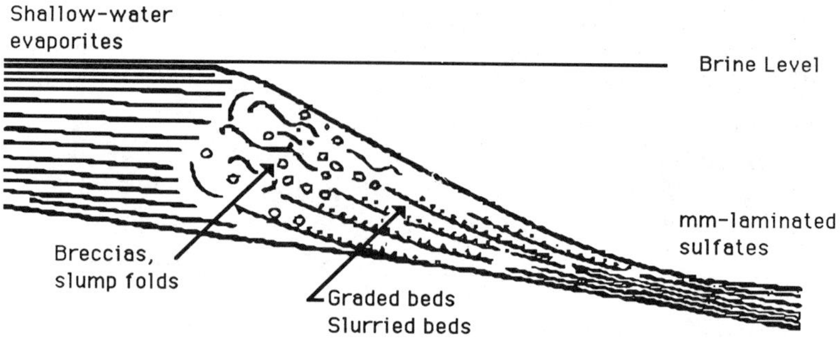

Figure 4.12 Schematic cross section of the shelf-to-basin transition in the Zechstein Basin (after Schlager and Bolz, 1977). The saltern shelf acted as an evaporite factory supplying sediment to the slope where it was deposited as slumps and turbidites.

crofaults, and rare ripple-laminated beds. Toward the basin the wedge thins, and the millimeter laminated facies increases in importance as folded and brecciated beds appear. The folded beds are intercalated with undeformed sediment and may represent syndepositional slumps (Herrmann and Richter-Bernburg, 1955). The breccias contain features that are typical of submarine debris flows; they are very poorly sorted beds with normal or reverse grading and contain imbricate clasts (now nodular anhydrite) and Bouma sequences. Still farther basinward, these disrupted beds pass into a thin "starved basin" sequence of relatively undisturbed millimeter laminated anhydrite.

Schlager and Bolz (1977) concluded that the sulfates comprising the wedge were reworked into deeper water from the adjacent shallow-water sulfate platform (saltern) by gravity processes. Clark (1980) agreed the sediments were reworked into deep water from the sulfate platform but postulated that storm reworking of the sulfate platform was the most important factor in the formation of the wedge. The latter poses important questions in terms of the origin of the laminated basinal sulfates. Traditionally, these laminates have been interpreted as "varves" produced by the annual precipitation and settling of sulfate within the basin center (Richter-Bernburg, 1960). Clark (1980) argued that this is not the case; rather, the laminae were produced by deposition below the storm wave-base, from a layer of turbid water that covered the basin floor after storms. Such a mechanism would also account for the lateral persistence of laminae across the basin, for the turbid layer would have covered a large area in the deeper part of the basin. What is more, if the sulfate laminae were not produced by direct precipitation, the salinity of the basin waters may have been much less than was imagined previously.

Just how deep was the water in the Zechstein Basin? Was it 1000 m as suggested by Richter-Bernburg, or was it much shallower, as the adjacent shallow water carbonates and evaporites suggest? The presence of sulfate turbidites implies significant basinal relief at the time of sulfate deposition. Schreiber (1986) postulated a minimum relief of 100 to 200 m for the shelf–to–slope transition in the Zechstein. Regardless of water depth, the most important observation to be made from the Zechstein is that even when mechanically deposited gypsum has been converted to anhydrite, it is still possible to infer an environment of deposition from the resultant textures.

The Upper Silurian of the Michigan Basin

The Michigan Basin is a classic area for the study of evaporite deposition and the interpretations of depositional setting are as varied as the geologists who have worked there. The pinnacle reefs around the margin of the basin contain hydrocarbons and their growth is related intimately to evaporite deposition. The inner part of the Michigan Basin is surrounded by a massive stromatoporoid barrier reef some 150 to 180 m (500 to 600 ft) thick (Fig. 4.13). Pinnacle reefs occur in a belt 16

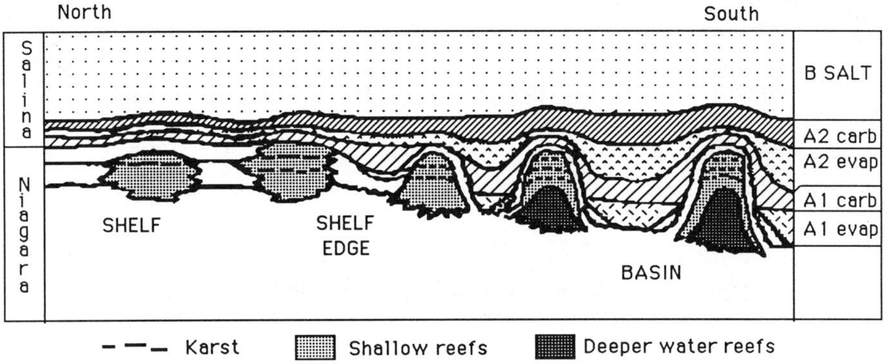

Figure 4.13 Cross section of the Michigan Basin showing the shelf-to-basin transition. Complied from sources mentioned in text.

to 24 km (10 to 15 mi) wide extending from the steep inner face of the barrier reef toward the basin center. The basinal equivalent of the reefs is a thin carbonate some 18 to 24 m (60 to 80 ft) thick. The barrier reef is dolomite and most of the pinnacles are dolomitized, although a few of the more basinal pinnacles are limestone (Sears and Lucia, 1979, 1980). Some argue that reef growth preceded evaporite deposition in the basin; others argue it occurred simultaneously with evaporite deposition. In the latter case the reefs would be associated with shallow-water evaporites and the intervening areas with predominantly deeper-water evaporites. A determination of the depositional setting of the salina evaporites encasing the reefs will resolve some of the problems concerning time of reef growth.

The basin center evaporites of the Lower Salina Group are of the most interest in solving this problem and in a discussion of deep-water evaporites (Fig. 4.13). Classically, these halites have been interpreted as a deep-water deposit (Dellwig, 1955; Dellwig and Evans, 1969). The Salina Group salts exhibit a clear salt–cloudy salt banding (as couplets 2 to 9 cm thick) in addition to dolomite-anhydrite laminae. The banding is absent or poorly developed in basin margin locations, where there is evidence in the anhydrite and other facies for shallow water deposition. The cloudiness in the banded salt comes from the presence of brine inclusions, which according to Dellwig, outline numerous pyramidal-shaped hopper crystals (rafts) that grew on the brine surface and subsequently sank to the basin floor. On reaching the basin floor they developed syntaxial inclusion-free overgrowths which resulted in a cubic habit. Dellwig (1955) inferred that the sinking hopper crystals created halite-saturated conditions in the bottom waters. Any temperature rise in the bottom waters caused undersaturation and some dissolution of previously accumulated bottom halite. Subsequent cooling of the brine once again allows the brine to become supersaturated. This promoted the growth of inclusion-free halite as overgrowths on the surviving cloudy hoppers, so creating a new clear halite band above the cloudy band. The early diagenetic origin of the clear salt layers and overgrowths can be seen in core when carbonate-anhydrite laminae overlie flat dissolution surfaces that cut across both the hoppers and their overgrowths. If Dellwig is correct, then the same explanation can explain the recrystallized nature of other deep-water halites. However, the mechanism for varying the bottom water temperature in a deep basin is not well understood nor is the level or intensity of temperature change needed to dissolve halite.

Nurmi and Friedman (1977) have challenged many of Dellwig's conclusions. They interpreted the cloudy halite as having formed as a crust covered by shallow waters subject to periodic freshening. Such crusts are forming today in Salina Ometepec in Mexico (Chapters 2 and 3; Shearman, 1970). They also interpret the clear halite as primary, the clear nature of the crystals a reflection of slow precipitation from a brine that did not suffer rapid compositional changes. The large inclusion-free halite crystals occur only in the basin center and only in the basal A1 evaporite, not in the more marginward A1 nor in any of the overlying A2 or B evaporites. Nurmi and Friedman suggested that large clear halite crystals are the only deep-water halite; the remainder of the halite above the basal A1 was deposited as crusts in shallow or ephemeral waters (Fig. 4.13).

The arguments of Nurmi and Friedman are rooted strongly in processes observed in the Holocene and in my opinion are a more likely interpretation. Thus the evaporites argue for periods of evaporative drawdown which would have subjected the shelf carbonates and pinnacles to at least periodic subaerial exposure. This is seen in the karst horizons crosscutting many of the pinnacle reefs (Kahle, 1978). This interpretation leads to another dilemma in reef growth. In a detailed biostratigraphic study, Shaver (1977) showed that paleontologically there were no time gaps in reef growth. Yet the reefs contain soils and karst, indicators of subaerial exposure and no deposition. The two apparently disparate observations can be reconciled if rates of deposition for evaporites are taken into account (Table 1.4). Extremely rapid basin infill during evaporative episodes meant that times of drawdown and exposure were so short that not enough time existed to develop paleontological (evolutionary) breaks in the reefs (Schreiber pers. comm.).

SUMMARY

Deep-water evaporites are not being laid down in any modern sedimentary basins, with the possible exception of the Dead Sea in the Middle East. Yet extensive deposits of deep-water evaporites are known from many large ancient evaporite basins. They are characterized by laminar sequences with lamina that can be correlated over long distances. Turbidites and mass flows can be found toward the basin edge. Vertical profiles through these reworked portions about the basin margin can show fining-upward profiles reminiscent of Bouma sequences. Most of the clastic evaporites in these basin margin deposits are from reworked shelf evaporites, an evaporite "factory" in shallower water. Deep-water evaporites can be interpreted in much the same way that deep-water deposits are interpreted in carbonate and siliciclastic depositional systems. An important difference is that the bottom waters in evaporite basins are density-stratified, so are usually anoxic or at best dysaerobic, and so deep-water evaporite basins lack a bottom-dwelling biota (Kirkland and Evans, 1981). For the same reasons these sediments can contain significant volumes of organic matter (e.g. the Castile Formation of West Texas and the Messinian gypsum of Sicily). Burrows and other forms of bioturbation do not occur; laminae are almost always preserved until early burial. Provided that the laminae survive burial diagenesis they are also good indicators of periodic changes in the chemistry of the basin waters.

CHAPTER 5

Dolomites and Dolomitization

INTRODUCTION

Dolomite is probably one of the best studied and yet most poorly understood rock types. About half of the oil and gas reservoirs in North American carbonate rocks are in dolomite and a quarter to half of the world's carbonate reservoirs are dolomites (Wilson, 1979). The reasons for studying dolomite are obvious, but why then is its origin so poorly understood? Dolomite is a common carbonate in many evaporitic settings, hence its inclusion is this book, but the origin of dolomite is not simply as an evaporite mineral. Dolomite is a complex diagenetic mineral and forms in many depositional and diagenetic settings. Some argue that dolomite forms from hypersaline brines, others argue that it forms from schizohaline waters (Friedman, 1980; Folk and Land, 1975). Dolomite is an unusual mineral in the Holocene, yet is common in ancient carbonates. The lack of abundant modern dolomite is one of the main reasons that its genesis is so poorly understood.

Dolomite, the Mineral

The original name *dolomie* was given by N. T. Saussare, in 1792, in honor of the French geologist Deodat Guy de Dolomieu (1750–1801) and was first applied to rocks in the Tyrolean Alps. Dolomite, the rock, contains a large proportion of dolomite the mineral. Ideal dolomite has a chemical composition of $CaMg(CO_3)_2$ with a crystal lattice containing layers of Ca and Mg, each layer separated by CO_3 layers (Fig. 5.1). The unit cell is a rhombohedron with a symmetry structure $\bar{3}$ in the trigonal subsystem of the hexagonal system. Dolomite in natural systems is

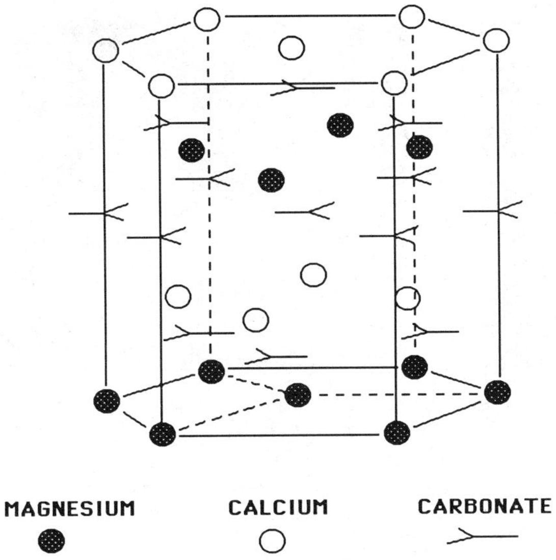

Figure 5.1 Ideal structure of stoichiometric dolomite consisting of layers of carbonate separated by alternating layers of calcium and magnesium ions.

better represented by the formula $Ca_{(1+x)}Mg_{(1-x)}(CO_3)_2$ as no sedimentary dolomite is truly stoichiometric. Naturally occurring dolomite ranges from $Ca_{1.16}Mg_{0.84}(CO_3)_2$ to $Ca_{0.96}Mg_{1.04}(CO_3)_2$ (Land, 1985). Most sedimentary dolomites are calcian-rich; they form slowly and once formed, an ancient well-ordered dolomitized limestone is less soluble than its undolomitized equivalent. Compared to calcite, dolomite is relatively brittle and fractures easily at depths greater than 3000 m.

Studies of modern and ancient dolomites have shown that the bulk of ancient dolomite is diagenetic. The time of diagenesis varies from penecontemporaneous with deposition to late-stage, deep-burial dolomite. The use of the name *dolomite* will probably become more circumfluent as the chemistry and atomic structure are better understood. Dolomite should be used to describe a group of rocks with similar chemical and lattice structures. The rock name *dolomite* should be used in much the same way that feldspar is used. Just as there are many kinds of feldspar, so dolomite exhibits variation not only in chemical composition but in subtle atomic arrangements (Land, 1985; Reeder, 1981).

Holocene dolomite is a metastable phase, but nevertheless, a true dolomite in that it possesses at least partial ordering of the lattice structure. Its lattice is characterized by high densities of defects (Reeder,1980) as well as variation in the degree of ordering and/or the degree of calcium enrichment existing in domains tens to hundreds of angstroms wide. Ancient post-Paleozoic sedimentary dolomite is also calcian-rich but exhibits a much more regular lamellar structure when compared to Holocene dolomite (Fig. 5.2). Reeder (1980) interprets this lamellar structure to be a structural or compositional modulation on the scale of a few

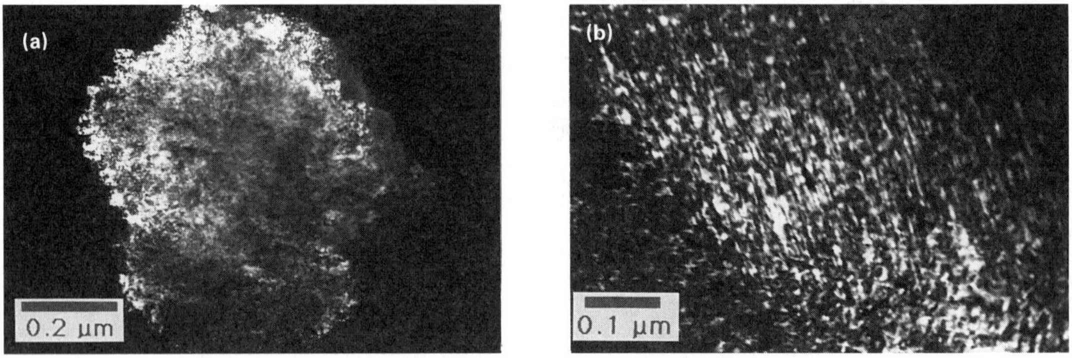

Figure 5.2 Transmission electron photomicrograph (TEM) of a sample of Holocene magnesian-rich Coorong dolomite. The irregular light and dark areas of the crystal are due to lattice strain localized about irregular and ill-defined lattice defects that formed during the growth of the crystal. The extremely high density of lattice defects is typical of Coorong and many other non-marine Holocene dolomites. (b) TEM dark field photomicrograph of Cretaceous calcian-rich dolomite from the Edwards Formation of central Texas. The oscillatory light and dark regions represent slight, somewhat oscillatory changes in lattice orientation and form a modulated structure. Modulated microstructure is common to ancient calcian-rich dolomites, some stoichiometric dolomites, and some marine Holocene dolomites (photomicrographs and interpretation courtesy of Don Miser; see also Rosen et al., 1987 and Miser, 1987).

hundred angstroms. This structure, like the disordered structure of Holocene dolomite, is metastable. Even "nearly stoichiometric" ancient Paleozoic dolomite is composed of numerous defects, mainly faultlike phenomena. Its defect density is less than the other types of sedimentary dolomite and the crystals are well ordered, but even this type of dolomite is less stable than structurally and compositionally ideal dolomite (Land, 1985). Truly ideal dolomite is common in metamorphic but not in sedimentary rocks.

All sedimentary dolomites are metastable, although older dolomites often are better ordered and "more stoichiometric" than younger dolomites. Dolomite is modified extensively during diagenesis, perhaps in a "stepwise" fashion, as earlier dolomite phases are replaced by later, less-soluble phases. Extensive dolomitization can be accomplished only when a long-term supply of magnesium is available. Such a situation requires active, long-term fluid circulation. Dolomitization is probably a dissolution-reprecipitation process where more stable dolomite phases replace earlier, less stable phases. On a larger scale the metastable phases can be dissolved in one place and reprecipitated in another, causing porosity enhancement in one area and porosity occlusion in another. If one can model the geometry of the plumbing of the dolomitization process, perhaps one can model porosity trends in ancient dolomite reservoirs.

DOLOMITIZATION MODELS

The terms ''*primary dolomite*'' and ''*secondary dolomite*'' need to be defined before this discussion of models of dolomite formation can proceed. ''Primary dolomite'' is strictly defined as dolomite crystals precipitated in areas where nothing was there before. This definition covers cements as well as crystals deposited at the sediment–water interface. In this sense few, if any, extensive beds of thick primary dolomite are forming today on the present surface of the earth. A few areas of widespread but patchy, modern ''penecontemporaneous dolomite'' can be found in the Caribbean, the Arabian Gulf, and South Australia. Small, local areas of true primary dolomite may exist in thin dolomite beds in Baffin Bay, Texas and the Coorong in South Australia (Land, 1983; von der Borch and Jones, 1976; Rosen et al., 1987). Geologists working in ancient sequences often use the prefix ''primary'' in a slightly different sense to imply an early dolomite, one that preserves fine-scale depositional textures and forms dolomite beds with well-defined stratiform upper and lower contacts. Whichever definition is used for ''primary dolomite,'' the bulk of ancient dolomite is a ''secondary'' or diagenetic phase.

This then is the essence of the ''dolomite problem'': Most modern dolomite is penecontemporaneous, patchy, and forming units at most 1 m thick; most ancient dolomite is secondary, extensive, and has affected whole shelf sequences hundreds of kilometers wide and occurs as units hundreds of meters thick. The reasons for this disparity are many, the most important one is probably kinetics. There just is not enough time in the Holocene to form extensive areas of secondary dolomite. Modern dolomitizing solutions include marine brines (Behrens and Land, 1972; McKenzie, 1981), continental brines (Clayton et al., 1968; von der Borch et al., 1975), essentially normal seawater (Saller, 1984; Carballo and Land, 1984), seawater modified by extensive sulfate reduction (Baker and Kastner, 1981; Kelts and McKenzie, 1982), and seawater mixed with meteoric water (Land, 1973; Magaritz et al., 1980). Seawater, albeit modified by evaporation, sulfate reduction, or mixing with meteoric water, is the only natural solution that contains sufficient magnesium to cause massive dolomitization (Land, 1985). Therefore, most ancient dolomite probably forms relatively early in the depositional and burial history of the sediments, when seawater can be actively pumped through the rock. Under what conditions this pumping actually forms dolomite is not well known at the present time. Hardie (1987) extends this idea in a thought provoking essay that questions many of our current models of dolomitization, and shows how they all contain geochemical inconsistencies.

To dolomitize a typical carbonate sediment containing 6% $MgCO_3$ with 40% porosity requires that 807 pore volumes of seawater flow through the rock. With seawater diluted 10 times with meteoric water (mixing zone) something like 8100 pore volumes are needed. With a halite-saturated, seawater-derived brine, only 44 pore volumes are needed. If the brine has reached the bittern stage, only 10 to 12 pore volumes are required. However, even with these large volumes of water flowing through the rock, no pore volume reduction has taken place. To reduce the porosity to values common in ancient dolomites requires even greater volumes of solution to flow through the rock. The fact that much smaller numbers of pore volumes of hypersaline water are required to dolomitize a given volume of limestone and the observation that most modern dolomite is associated with evaporitic settings has led Friedman to state "dolomite is an evaporite mineral" (Friedman, 1980).

Geochemists have attempted to mimic the surface conditions of modern dolomite formation in the laboratory and have failed. Dolomite formation is probably controlled by reaction kinetics, as dolomite can be formed in the lab when experiments are run at temperatures in excess of 150 to 200°C. Where low temperature geochemists have failed, a dog has succeeded! At a temperature of 38°C and approximately 1 atm pressure, a Dalmatian managed to precipitate dolomite uroliths in its urinary tract (Mansfield, 1980). The physical and chemical conditions in this dog's urinary system may more closely approximate modern and ancient inferred carbonate depositional systems than most previously successful dolomite-forming laboratory runs, especially in terms of temperature, pressure, total salinity, and possibly biota. The presence of urease-producing and/or uric acid—fermenting bacteria in urea- and/or acid-containing sediments, such as fecal pellets and algal mats, could promote the formation of authigenic dolomite or other carbonate minerals.

Using Models

Four major facts must concern a geologist invoking a model to explain an ancient dolomite:

1. What is the source of the magnesium in the dolomite? Is the model invoking a source of magnesium sufficient to explain the volume of magnesium in the rock?
2. Is there a pump mechanism to explain the large volume of solution that must have passed through the rock?
3. Is the composition of the proposed dolomitizing solution conducive to dolomitizing the rock?
4. Does the model allow sufficient time for the proposed mechanism to dolomitize the measured volume/thickness of limestone?

Over the years, many models of dolomitization have been proposed but only seven seemed to have stood the test of time. They are:

1. Sabkha model
2. Brine-reflux model
3. Coorong model
4. Mixing-zone model
5. Sulfate-reduction model
6. Burial-stage model
7. Solution-cannibalization model

Of the seven, only three, or perhaps four, are invoked to explain thick sequences of ancient shelf dolomites (Fig. 5.3). They are models 2, 4, 6 and perhaps 5.

Sabkha Model

As described in Chapter 2, the Arabian (Persian) Gulf sabkhas, centered on Abu Dhabi, are composed of subtidal, intertidal, and supratidal sediments. The intertidal zone contains laminated algal mats; the supratidal zone is characterized by nodular and enterolithic anhydrite, while the whole sequence is capped by an erosion surface. Dolomite occurs most commonly in the supra/intertidal sabkha sediments but is also found in the subtidal sediments (Illing et al., 1965). The dolomite is not a depositional mineral, as it is not found in the present lagoon nor in the growing algal mats. It occurs below the surface of the prograding sabkha, usually as units less than 1 m thick. On the sabkhas bordering the Qatar Peninsula, the dolomite is most abundant beneath the high intertidal zone, where it forms 1 to 5 μm euhedral crystals in aragonitic muds (Fig. 5.4). An inverse relationship exists between aragonite and dolomite in the sabkha, as the proportion of dolomite increases the proportion of aragonite drops. In other words, the dolomite is replacing an earlier aragonite phase.

This penecontemporaneous "secondary" dolomite of the Arabian Gulf forms in isotopic equilibrium with brines at temperatures between 34° and 49°C (McKenzie, 1981). Aragonite and perhaps high–magnesium calcite serve as intermediates in the formation of dolomite, which proceeds via a dissolution-reprecipitation process. With time, the dolomite appears to "age"; increased ordering is observed in the crystal lattice, as is an increase in the crystal size and a continued isotopic equilibration with brines at the lower temperatures found deeper under the sabkha surface. Diagenesis and isotopic reequilibration also occurs in the coexisting $CaCO_3$ phases.

The dolomite is forming in areas of maximum flux of seawater through the sabkha (Patterson and Kinsman, 1982). Areas of dolomite growth follow the strandline, heading inland in the vicinity of tidal channels and moving seaward

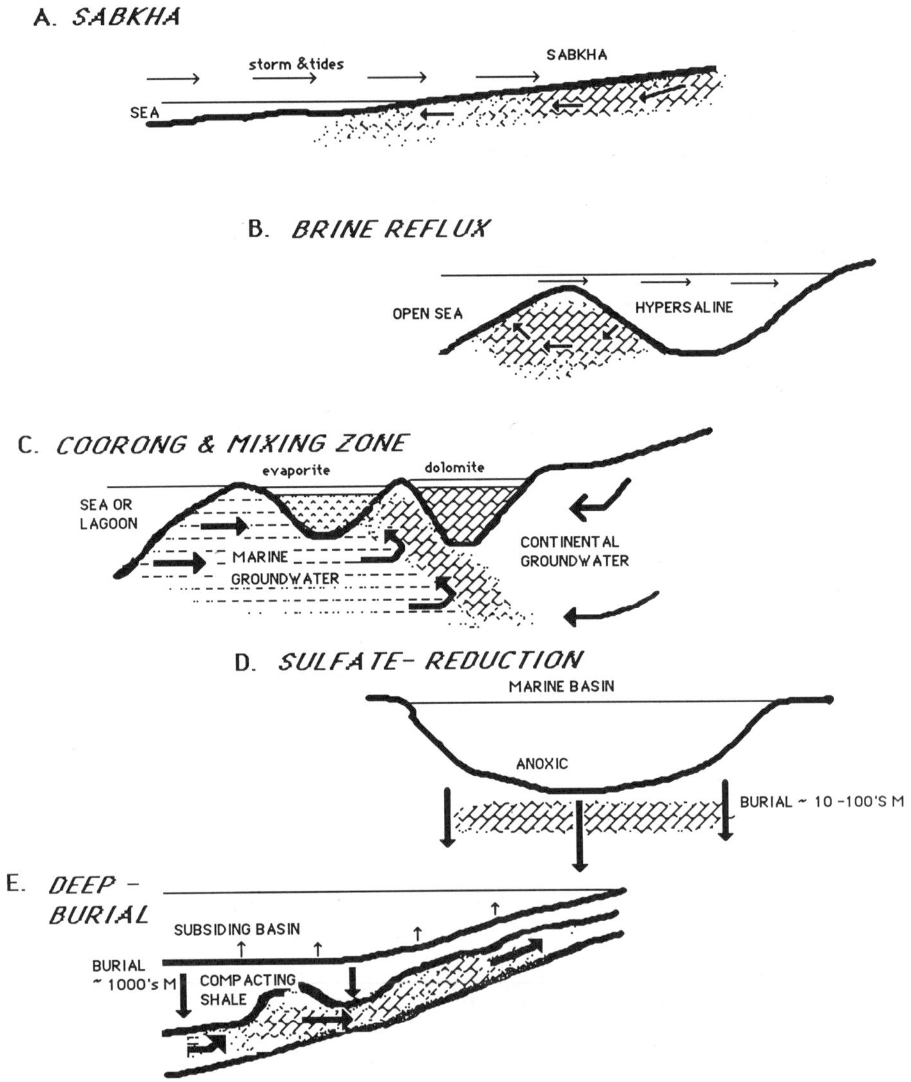

Figure 5.3 Major models of dolomite formation applicable to the interpretation of ancient dolomites.

near tombolo islands (Fig. 5.5). Storms and high spring tides drive seawater from the lagoon up onto the sabkha flat. Some of this seawater sinks into the sabkha and only later returns to the atmosphere by evaporation of waters in the capillary zone. As it evaporates the remaining water concentrates to where it precipitates aragonite and then a $CaSO_4$ phase. Precipitation of these diagenetic minerals removes calcium from the solution, raises the proportion of magnesium, and so raises the Mg/Ca ratio. The dense Mg-bearing hypersaline pore fluids now sink

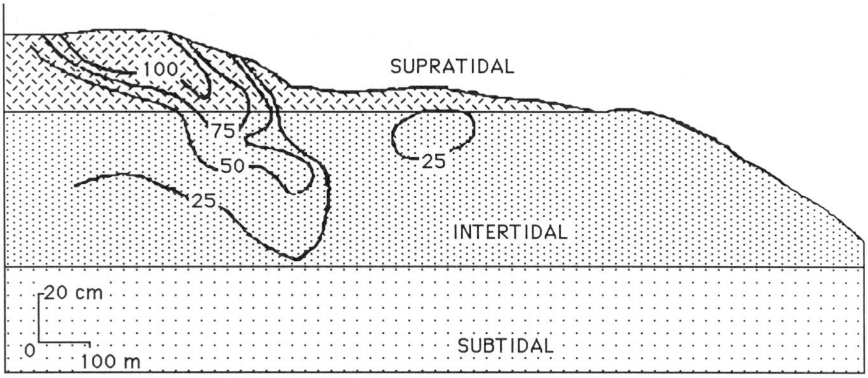

Figure 5.4 Cross section of the peritidal zone on the Qatar Peninsula sabkha, Arabian Gulf. Contours are the percentage of dolomite in the fine fraction (after Bush, 1973).

downward and flow seaward through the sabkha sediments, dolomitizing the aragonite as they go. Dolomite occurs in intertidal and subtidal sediments to depths of 2 to 3 m usually in areas landward of the algal mat. The most intense dolomitization occurs in buried algally-bound and/or lagoonal muds (Patterson, 1972; Bush, 1973).

In their attempt to explain the distribution of sabkha dolomite Hsu and Schneider (1973) proposed an alternative hydrological model which they termed evaporative pumping. This invokes a continuous subsurface flow of lagoonal seawater into the sabkha, as it replaces groundwater lost by evaporation at or near the sabkha surface. Under evaporative pumping the seawater flows landward in the subsurface; the flow pattern is the reverse of the seaward flow described in the preceding paragraph. The field evidence for evaporative pumping is contradictory, many authors feel that evaporative pumping is not a feasible mechanism to explain the distribution of dolomite in the sabkha (Morrow, 1982; Land, 1983).

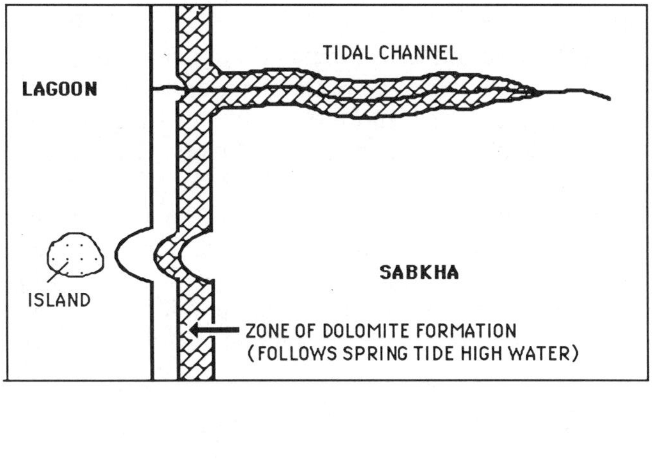

Figure 5.5 Schematic of the zone of dolomite formation on the Abu Dhabi sabkha, Arabian Gulf (after Patterson and Kinsman, 1982).

Evaporative pumping of seawater is unlikely to create large volumes of sabkha dolomite because the rapid deposition of marine evaporites from landward-flowing seawater would quickly plug the pores, reduce the permeability of the sediments, cancel the head difference, and so choke off flow before sufficient volumes of magnesium have flushed through the system.

Perhaps the mechanisms supplying seawater to the sabkha are not as simple as the preceding either/or presentation implies. Work by McKenzie et al., (1980) has shown that the sabkha hydrology may be characterized by three sequential stages: flood recharge, capillary evaporation and evaporative pumping. All three stages occur as the dolomitizing brine evolves. They also demonstrated that the sabkha hydrology is not operating within a simple homogeneous aquifer; rather the sabkha plumbing is composed of three aquifers separated by two crusts which perform as aquitards. The question of sabkha hydrology and the plumbing of the associated dolomitization will not progress much beyond our current understanding until more detailed facies and hydrogeochemical studies are conducted in the Arabian Gulf sabkhas (Warren and Kendall, 1985).

Brine Reflux Model

This model is very similar in mechanism to the sabkha model, but the brine reflux model is usually invoked as an explanation of thicker, much larger-scale ancient dolomites associated with evaporites (Fig. 5.3b). It was first proposed by Adams and Rhodes (1960), who used brine reflux as an explanation of the extensive lagoonal dolomites of Guadalupian age in the Permian Basin of west Texas. Reflux dolomite forms when "hypersaline brines eventually become heavy enough to displace the connate waters and seep slowly downward through the slightly permeable carbonates at the lagoon floor." The dense sinking brines are derived from the evaporite-depositing waters of the lagoon, waters whose Mg/Ca ratios have been increased by the precipitation of a $CaCO_3$ and a $CaSO_4$ phase. Unlike the sabkha model which explains thin supratidal-associated "primary" dolomites, the brine reflux model is used to explain thick, laterally extensive, ancient "secondary" dolomites associated with shelf evaporites.

The model is usually set as a restricted lagoon behind some sort of sill or reef acting as a physical barrier to the free inflow of unevaporated seawater. In ancient epeiric shelves, such as the Northwest Shelf of the Permian Basin, no actual physical barrier existed. Large areas of shelf more than 600 km wide were covered by very shallow water. The shelf width may have sufficiently restricted flow until evaporites were precipitated on the platform floor. Once $CaCO_3$ and gypsum began to form, the Mg/Ca ratios of the remaining brine began to rise, the solution density began to increase, and these magnesium-enriched solutions sank down through the underlying limestones, dolomitizing as they seeped seaward. In this way large volumes of magnesium-rich water passed through previously deposited shelf carbonates, so converting them to dolomite. In ancient carbonates the process was enhanced by changes in the basin water level, generating a groundwater head (see later).

In the 1960s seepage reflux gained a large number of proponents. Deffeyes, Lucia, and Weyl (1965) reported modern magnesium-rich brines in a marine-fed gypsum lake (Pekelmeer Lagoon) on Bonaire in the Southern Caribbean. From the water chemistry they predicted that the brines should be sinking and forming dolomite beneath the lake. They did observe Holocene dolomite in the lagoon and dolomite in a Pleistocene limestone on the opposite end of the island, which they inferred formed during an earlier seepage reflux episode. Lucia (1968) drilled beneath the Pekelmeer Lagoon and found no widespread dolomites in the underlying Pleistocene carbonates. He also found the sublake pore waters were of normal marine salinity in areas where a brine was predicted by a seepage reflux model. Pekelmeer's claim to forming extensive seepage reflux dolomites was laid to rest when Sibley (1980) concluded that the solutions responsible for extensive dolomitization on Bonaire were probably fresh to brackish waters.

Other tropical islands have been found where seepage reflux associated with hypersaline conditions is forming Holocene dolomites. Muller and Teitz (1971) documented seepage dolomite replacing an earlier cement in skeletal grainstones from the shoreline in Fuerteventura in the Canary Islands. Kocurko (1979) found seepage reflux dolomite a few meters above the high tide line in the spray-zone pools of the shoreline of San Andres. However, all modern examples of seepage reflux are small in scale and operate along an island shoreline. They never approach the scale of similar processes which dolomitized shelf carbonates in ancient successions.

From a study of modern seepage dolomites, one would conclude that reflux probably cannot operate on the regional scale for which it was first proposed (Land, 1983). In modern settings the small fluid potentials caused solely by density differences cannot move brine very far through low permeability sediments. Elevation head is even required to generate a sabkha dolomite which in itself is a form of reflux dolomite. In ancient evaporite settings the lack of a short-term groundwater head is probably not a problem. Modern examples are not at the scale of ancient evaporitic settings; the lagoonal area of many ancient evaporitic shelves and platforms were huge compared to modern evaporite depositing areas. Ancient epeiric shelves often measured hundreds or thousands of kilometers across. Such areas would have generated huge quantities of refluxing brines seeping in a seaward direction. In addition, when an evaporite depositing basin was closed or isolated from a direct marine connection it was subject to rapid changes in water level. This could also have generated a substantial gravitational head to drive subsurface brines into the lower areas of the basin. Changing brine levels were probably inherent in all large closed evaporite basins due to fluctuations in the delicate climatic balance required to deposit and preserve a thick evaporite unit (Chapter 1). For example, the Zechstein evaporites formed in less than 1/40 of the time that the basin was in existence. For the rest of the time, there were marked water-level changes in the brine level in the basin (Schreiber and Hsu, 1980). Such brine-level changes would easily generate the elevation head to drive the groundwater flow and so form seepage dolomites. Outside of a single evaporite-forming basin, more regional sea level changes could have been generated by eustatic/tectonic events. They would have changed sea level, and so with enough time generated the required groundwater heads beneath open marine epeiric seas.

Coorong Model

The Coorong Region of South Australia was first documented as an area of "primary" dolomite deposition by Mawson (1929) but was not studied in any detail until the work of Alderman and Skinner (1957), Skinner (1963), von der Borch (1965, 1976), and von der Borch and Lock (1979). The bulk of the studies in the area have been concerned with the mineralogical and hydrological setting of the Coorong dolomites. The most recent work by Warren (1985, 1986a, in press) is the first to put the Coorong sediments into a model capable of defining similar ancient sequences in drill core. The dolomite forming lakes of the Coorong are situated on the prograding coastal dune plain of southeastern South Australia (Fig. 5.6). The famous dolomite is usually the last phase of a shallowing-upward Holocene lacustrine cycle 3 to 4 m thick whose exact vertical sequence is determined by the geological setting of the lake.

Lakes in the Coorong region are filled with a diverse suite of Holocene carbonates; mineralogies including aragonite, calcite, Mg-calcite, magnesite, hydromagnesite, and dolomite. Dolomite is the main mineral of concern in this study yet it makes up no more than 5% of the carbonate minerals forming surficial deposits in the coastal plain of southeast South Australia. The hydrological setting of the Coorong lakes does not allow this area to be used to generate a model to explain the occurrence of widespread ancient supratidal and shelf dolomites. Coorong counterparts in the rock record are lacustrine carbonates underlain by a paleoaquifer and formed as a facies mosaic in an evaporitic setting.

Coorong dolomites are one of the few examples of true "primary dolomite" forming at the earth's surface. That is dolomite precipitated as dolomite and not replacing an earlier carbonate mineral. Coorong dolomite occurs as three stratigraphically and mineralogically distinct forms; a basal, a marginal, and an evaporative dolomite (Rosen et al., 1987; Warren, in press). Of these the evaporative dolomite is the most common in most Coorong lakes and is the only dolomite type described by previous workers in the Coorong lakes. *Evaporative dolomite* was laid down as the last episode of sedimentation in those Coorong lakes which contain dolomite. In most lakes it is a capstone unit no more than tens of centimeters thick although in Milne Lake it forms dolomitic units up to 4–5 meters thick. Evaporative dolomite is often magnesian-rich. In some lakes there is a second type of surficial dolomite, a more calcium-rich *margin dolomite* about the edges of the lake. Like the evaporative upper dolomite this dolomite is not intergrown with other carbonate phases and defines areas where resurging continental

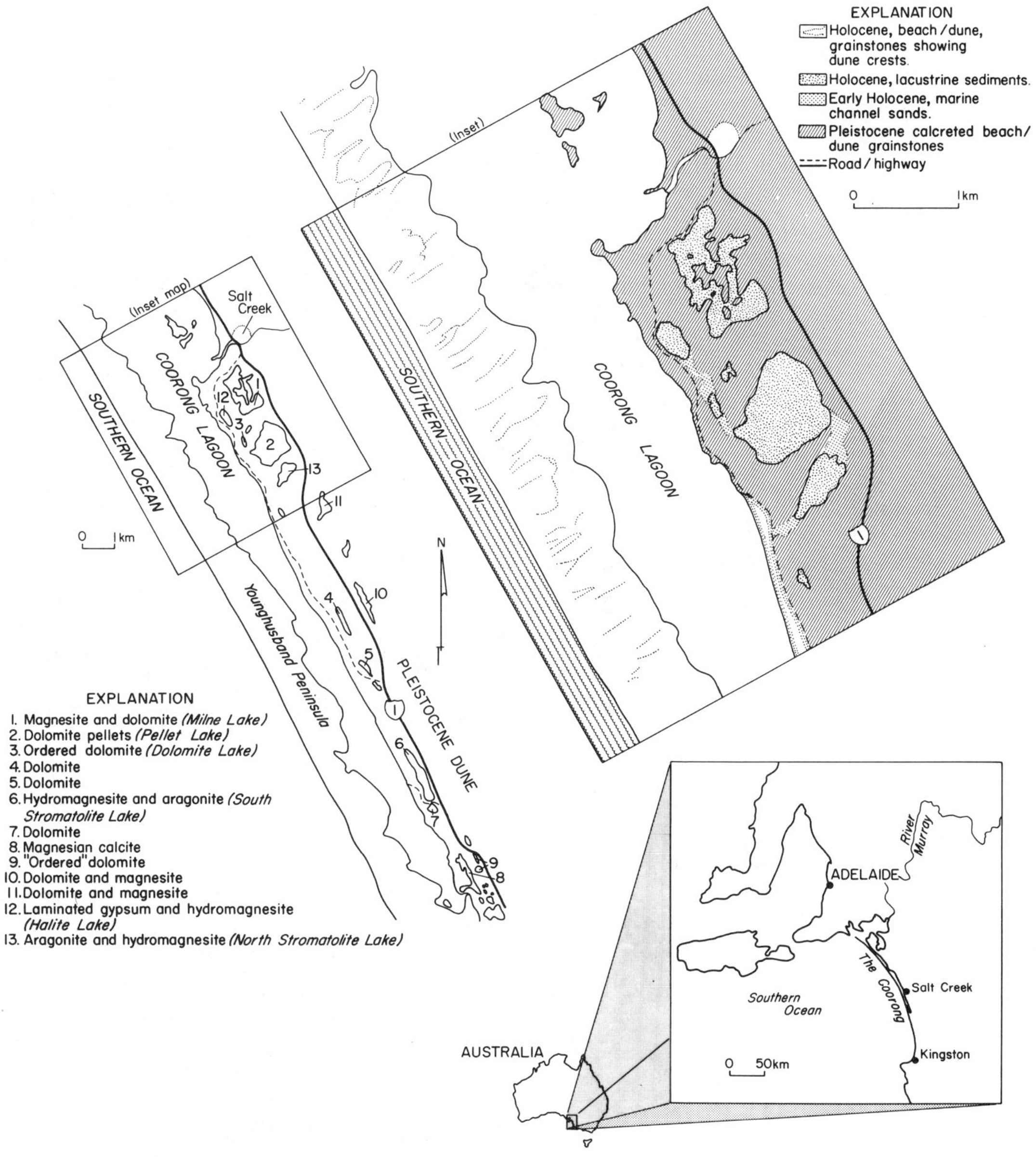

Figure 5.6 Regional setting and surficial geology of Coorong Lakes in the Salt Creek area. Shows the locality and major mineralogy of surficial sediments. Surficial mineralogies in part after von der Borch (1985).

groundwaters first enter the lake margin. There is a third type of dolomite in some Coorong lakes. This is a *basal dolomite* which is more crystalline than the other two forms of dolomite. It formed some 6000 years ago when the rising Pleistocene water table (driven by a transgressing sea) first caused continental groundwaters to outcrop and evaporate at the surface.

Textures found in vertical sequences from Coorong lakes are largely independent of mineralogy, similar vertical sequences, in terms of texture, occur in Coorong lakes filled with either lacustrine carbonate or gypsum. Coorong lake sediments were deposited in two settings (Fig. 5.7a). Facies changes up section and across the various lakes are controlled by two factors (Fig. 5.7b) here

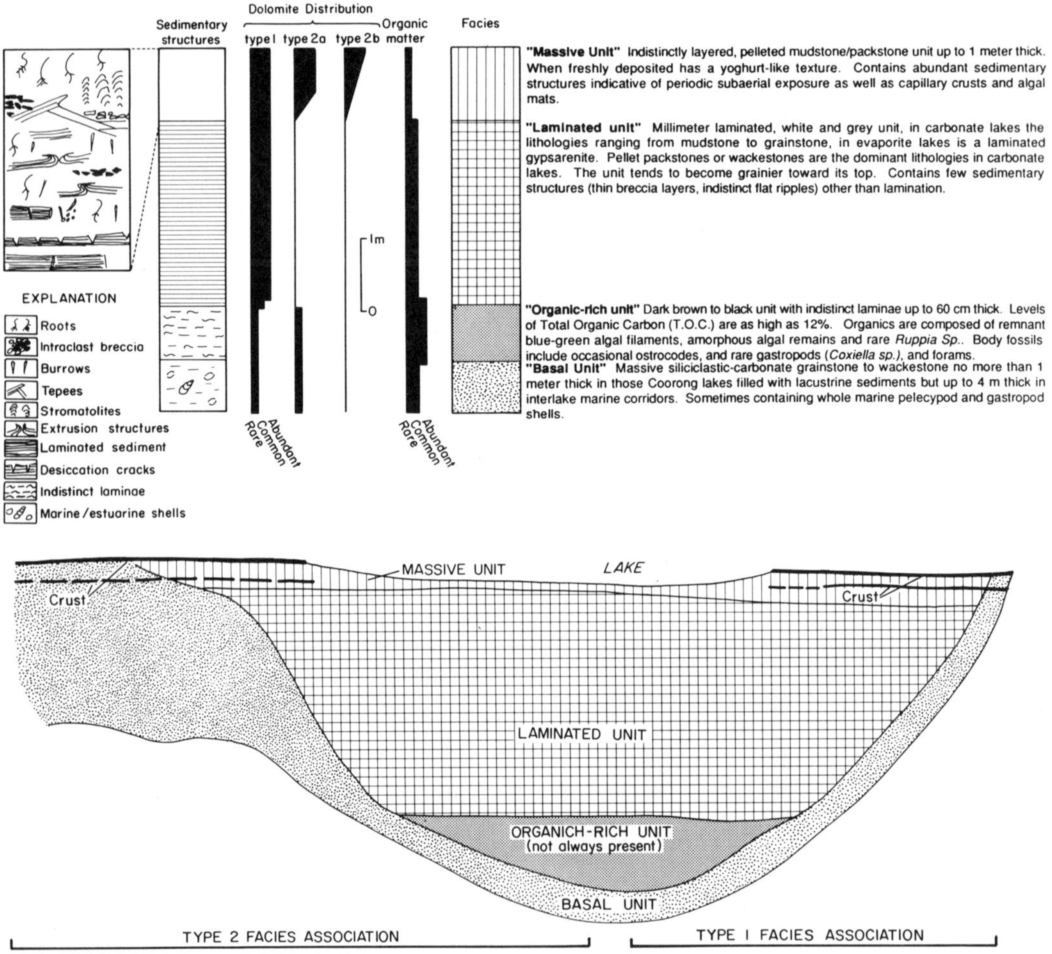

Figure 5.7 (a) Schematic of lake stratigraphy showing sedimentary structures, dolomite distribution, and unit descriptions. (b) Schematic cross section of a Coorong Lake showing facies associations.

1. Did the lake ever possess a Holocene connection to the marine/estuarine waters of the Coorong Lagoon?
2. What were the energy levels and degree of bioturbation at the depositional site?

If the lake possessed an early Holocene connection to the marine waters of the lagoon then the *basal unit* is a marine/estuarine skeletal grainstone/packstone. If there was no marine connection the basal unit is often a quartzose packstone to wackestone. Basal facies can contain small percentages of basal dolomite. Above the basal facies in some lakes is a massive to faintly laminated *organic-rich unit*. In lakes with an early marine connection the levels of T.O.C. in this unit can be as high as 12%, usually as an oil-prone proto-kerogen. Overlying this unit is a *mm-laminated unit* of pelletal packstone to mudstone deposited once the estuarine connection to the open Coorong Lagoon was completely cut off by beach-ridge accretion. The proportion of grains in the laminated unit is a direct reflection of energy level and organic binding when the lake sediments were deposited. In areas of higher wave energy/bottom currents and little or no algal binding, the sediment is a packstone, such areas were more common about the lake edge. In areas of lower wave energy/bottom currents in relatively deeper water and algal binding, the sediment is a wackestone or at times a mudstone. These muddier sediments are more common in the central parts of the lakes. In a few lakes such as Milne Lake this laminated unit composed of evaporative dolomite up to 5 meters thick, more often it is composed of varying proportions of aragonite, hydromagnesite, magnesian calcite and rarely of laminated gypsum/aragonite couplets. Capping the lake sediments is a *"massive" unit* of poorly layered packstone/mudstone unit usually less then 60 cm thick and showing a varying degree of induration. This unit usually holds the bulk of the evaporative dolomite found in the Coorong but can also be composed of aragonite, hydromagnesite, magnesian calcite and magnesite. The cap contains bioturbation structures, mud cracks, extrusion tepees, breccia fragments, all features indicative of at least occasional desiccation.

The thickest evaporative dolomites form in sea-marginal lakes which are now fed by resurging continental groundwaters and never possessed a surface connection with the marine waters of the Coorong Lagoon. Such dolomite-filled lakes are rare in the Coorong lakes and are herein defined as type 1 dolomite forming lakes. Sediments in these lakes, like all the Coorong lakes containing dolomite, are dominated by a lower laminated subaqueous carbonate and are capped by a massive micrite unit. Only in the type 1 lakes are these laminated sediments composed predominantly of dolomite (Fig. 5.7). In other lakes of the Coorong Region which contain dolomite only the upper massive micrite unit contains large quantities of dolomite. Milne Lake in the Coorong National Park is a type 1 lake, it contains a laminated magnesian-rich dolomite/magnesian-calcite unit up to 4 meters thick, capped by a massive, bioturbated dolomite/magnesite unit, half a meter thick.

The other two lake types are dolomite lake types 2a and 2b; they make up the great majority of all those Coorong lakes which contain dolomite sampled during this study of the Salt Creek region. In these lakes the evaporative upper dolomite does not fill the lake, but forms a massive capping unit up to 80 cm thick above other carbonates and magnesian carbonates (Fig. 5.7). Type 2 dolomite lakes possessed an early Holocene connection with the marine realm, sediments in this type of lacustrine depression have a basal marine unit which is lacking in those lakes with no early connection to the marine realm. Pellet Lake, Halite Lake, North and South Stromatolite Lakes, Dolomite Lake and numerous other unnamed lakes sitting in interdunal areas behind the landward margin of the Coorong Lagoon are type 2 lakes.

In type 2a lake sequences the deposited sediment was relatively fine-grained and the majority of the section composed of a laminated mudstone/wackestone. Type 2b vertical sequences are coarser grained with the bulk of the laminated unit composed of skeletal packstones/grainstones. Type 2a sequences often have higher T.O.C. than type 2b lakes due to the finer grained nature of the organic rich unit (Warren, in review). Lakes dominated by type 2a sequences often contain a basal dolomite within basal sands which were deposited immediately prior and during to the Holocene incursion of estuarine/marine waters into these lakes. The calcian-rich marginward dolomite is also common in the lakes dominated by type 2a sequences. Areas of marginward dolomite are often found beneath capillary crusts about the lake margin.

Type 2b sequences pass directly up from the laminated to bedded packstone/grainstones of the open marine Coorong Lagoon into a "massive" unit which in some lakes contains a few ten's of centimeter of evaporative dolomite. Lakes dominated by type 2b sequences do not seem to contain the basal dolomite unit nor is the marginal dolomite well developed in these lakes. Type 2b dominated lakes only form dolomite when the lake depression was isolated from the open waters of the Coorong Lagoon. This often happened when arms of the Coorong Lagoon were cut off by spit accretion and coastal-dune migration, type 2b sequences are usually found in the southern reaches of the Coorong Lagoon.

In winter, when the ephemeral lakes are filled with water, the more basinward dolomitic mudstones possess the consistency of yoghurt and are partially bound by algae and "duckweed" rootlets. In late summer and early fall, when dried, the same evaporative dolomite mud often has a leathery consistency and it is cross cut by dessication cracks (Fig. 5.8a). About the edges of most of the lakes, the same mud forms a hard indurated crust where almost year-round evaporation has converted the mud to a well cemented capillary crust (Fig. 5.8b). The crusts contain extrusion tepees and intraclast breccias (Fig. 5.8c and d; Kendall and Warren, 1987). Marginward dolomites are often associated with this crust along with local cherty cements.

Fresh evaporative dolomite mud in the massive capping unit consists of amorphous spherular aggregates (0.5 to 1.0 μm diameter) which von der Borch and Jones (1976) argued are a result of true "primary" dolomite precipitation from

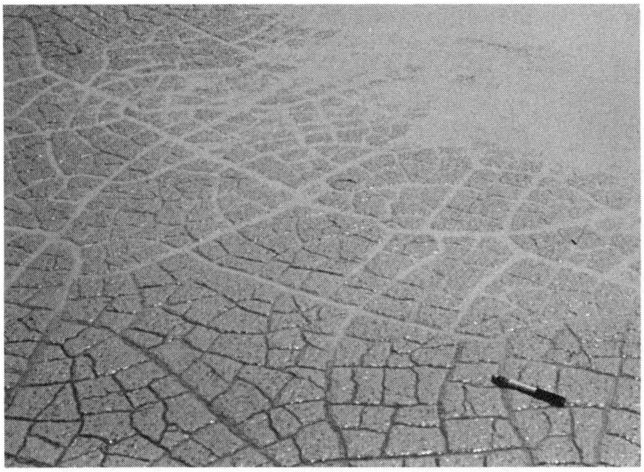

(a)

(b)

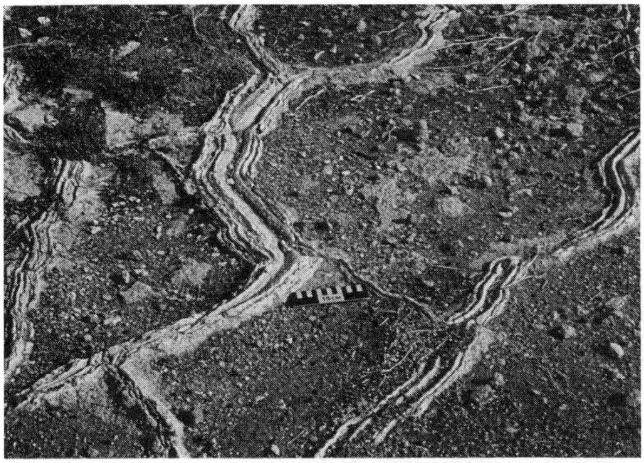

(c)

(d)

Figure 5.8 Sediments and structures in the Coorong Lakes, South Australia: (a) desiccation cracks forming along lake strandline; (b) mm-laminated domal stromatolites exposed during lowered water levels of a drought year; (c) extrusion tepee in capillary crust adjacent to lake strandline; (d) magnesite breccia in a dolomitic matrix located on the strandline.

the "yoghurt" gel. In contrast, the dolomite in the capillary crusts and some of the buried laminated dolomite muds are composed of small dolomite rhombs.

Volumetrically, most Coorong dolomites form by the evaporation of magnesium-rich continental groundwaters driven to the surface as they float up over a more dense seawater wedge (Fig. 5.3c). That most dolomite forms from resurging and evaporating continental waters explains why the thickest Coorong dolomites form adjacent to the present coast, yet similar Coorong-style dolomites are currently forming thin capping units in ephemeral lacustrine settings 20 to 40 km inland on the coastal plain, well away from the seawater wedge but still in areas where continental groundwaters are returning to the surface. Some Coorong lakes are filling with aragonite and gypsum, not calcitic and magnesian carbonates. Such lakes are always in a near-coastal position and are fed by resurging seawater (Warren et al., 1984). The occurrence of one such gypsum-filled lake adjacent to the shoreline and less than 1 km from the best dolomite-forming lake in the whole Coorong region poses important implications for general models of Coorong Dolomites.

In fact, many misconceptions can be found in the literature as to the features important in recognizing an ancient Coorong equivalent, the most basic of which is the statement that Coorong dolomites are penecontemporaneous carbonates never associated with preserved evaporites. The best type 1 lake in the Coorong region is situated less than 1 km from a lake filled with a laminated gypsum-aragonite sequence. The dolomite-forming lake is fed by continental groundwaters; the gypsum-filled lake is fed by marine groundwaters. Another misconception is that because dolomite is forming over an area measuring 90 × 120 km, the Coorong model can be used to explain laterally extensive ancient dolomites. If the present distribution of dolomite-forming lakes is measured against the aerial extent of the present 90 × 120 km coastal plain, then Holocene dolomite-forming areas constitute less than 5% of the total area. The inflow hydrology in the Coorong Lakes is such that the dolomite-forming lakes must be surrounded and underlain by an extensive aquifer. Coorong dolomites form by the evaporation of meteoric continental waters, not seawater; the volumes of groundwater required to evaporate to form a given volume of dolomite are substantial. Thus ancient Coorong equivalents are most likely to be thin lacustrine units which show facies mosaics in plan view with dolomite most likely in local topographic lows. A Coorong model cannot be used to interpret extensive, shelf dolomites tens of meters thick (Warren et al., 1984).

Probably the best ancient equivalent to Coorong Dolomite occurs not in a marine marginal setting, but in a lacustrine realm. The carbonate mud flat setting of the Green River Formation, Utah, contains all the structures and textures present in the Coorong as well as many of the same vertical transitions from calcium carbonate sediments to dolomite. The Green River was deposited in a large lake, part of an Eocene foreland basin that contained no marine waters. Green River carbonates consist of a series of stacked shoaling lacustrine cycles composed of laminated dolomites and limestones (Wolfbauer and Surdam, 1974).

Individual units shoal upward and contain stromatolites, ripples, flat pebble breccias, oolites, mudcracks and extrusion tepees (personal observation and Ron Surdam personal communication, Nov. 1985).

Mixing-zone Model

The model involves forming a dolomitizing solution by the subsurface mixing of seawater and meteoric water (Fig. 5.3). It was first used by Hanshaw et al., (1971) and by Land (1973) to explain dolomite in Quaternary limestones in Florida and Bermuda. It was used subsequently by many workers to explain extensive shelf dolomites not associated with evaporites. An analogous term, "Dorag dolomitization," was coined by Badiozamani (1973) in his excellent paper on Middle Ordovician shelf dolomites in Wisconsin. *Dorag* is Persian for mixed blood. In the chemical sense the mixing zone process is quite simple. When CO_2-saturated meteoric groundwaters mix with seawater, the resulting solution may be undersaturated with respect to $CaCO_3$ and supersaturated with respect to dolomite. Thus mixing-zone dolomites are replacement dolomites. The ideal proportion of seawater in the mix is between 10 and 35% (Fig. 5.9). The magnesium ions are

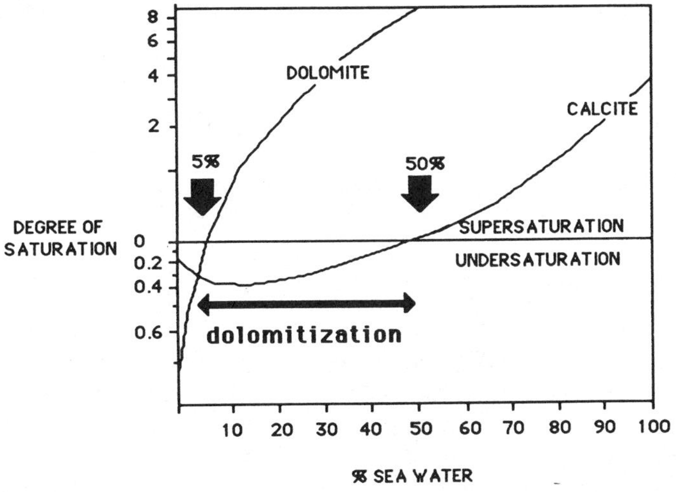

Figure 5.9 Percent saturation of various carbonates for various mixtures of seawater and a typical groundwater with a P_{CO_2} of 10^{-2} atm (after Land, 1983).

derived mainly from seawater, but some can also come from the carbonates dissolved in the mixing-zone.

Very few examples are known of dolomite forming in a modern mixing zone. The only undoubtable case comes from the coastal zone of Israel, where Magaritz et al., (1980) reported dolomite cements composed of subhedral rhombs up to 50 μm across. The dolomite occurs in two distinct horizons in the Quaternary calcareous sands of the coastal aquifer. The shallower horizon is in the modern mixing zone, the deeper horizon is interpreted as an earlier mixing zone.

However, the dolomites are found only as void-filling cements, the dolomite never appears to replace or develop pseudomorphs of a $CaCO_3$ phase as the ideal mixing zone model requires.

To form the extensive shelf dolomites of the past probably required a more stable hydrological setting than has prevailed throughout the Quaternary, a time of rapid, glacially induced sea level changes. Few, if any, stable hydrological settings have formed along the present coastlines of the world. Most present coasts, their depositional systems, their dynamics, and their subsurface hydrologies are all still adjusting to a sea level rise of 120 to 140 m during the last 18,000 years. Such rapid changes in sea level and the associated climatic changes have led to rapid changes in the position of the shelf mixing zone over the last 1 to 2 million years. The mixing-zone model shows definite application in ancient sequences, yet cannot be fully tested in processes active in the present carbonate shelves of the world.

Ancient mixing-zone dolomites were originally porous, as most of their Neogene equivalents possess porosities which commonly exceed 20%. To get sufficient volumes of meteoric water carrying magnesium through a limestone and so create a dolomite means that the precursor limestone must have acted as an aquifer. To form a reservoir, the dolomite must retain some of this porosity during burial. Predicting reservoir occurrence and geometry with a mixing-zone model is difficult, as the model is affected by many factors including:

1. *Sea level fluctuations,* which move the mixing-zone back and forth across the shelf
2. *Climatic variations*, which affect the amount of meteoric water in the system so changing flow rates and the thickness of the mixing zone
3. *Permeability of the sediments,* which controls the ease of passage of pore solutions through the sediments

Mixing-zone models can be used only when sediments were dolomitized in humid settings. Arid-zone settings lack sufficient fresh water to form an extensive mixing zone and their meteoric water flow rates are too sluggish to supply the required volume of magnesium. Thus a mixing-zone dolomite is rarely associated with evaporites or their pseudomorphs. The most reliable tool for predicting the occurrence of mixing-zone reservoirs is to map either paleohighs or shoreline trends, but paleo-sea level changes during and soon after deposition can greatly complicate the dolomite trend. Ancient mixing-zone dolomites often occur directly below regional unconformities, which fortunately are recognizable in seismic sections.

Ericksson and Warren (1983) used a mixing zone model to interpret secondary dolomites in the 2.25-billion-year-old Malmani dolomite of South Africa. Coarse secondary dolomite only alters grainstone units and subadjacent tepee-affected sediments of the Malmani. Tepees are capped by stromatolites and are groundwater tepees, a type of tepee that forms only in areas of groundwater resurgence. It is likely that the same solutions that were generating the dolomite in the subsurface were also creating the tepees at the surface. Mixing-zone cherts precipitated in conjunction with these secondary dolomites.

Sulfate-Reduction Model

Baker and Kastner (1981) demonstrated experimentally that the rate of dolomitization is controlled by the level of sulfate in the dolomitizing solution and may not be controlled by the Mg/Ca ratio. High levels of sulfate such as are found in modern seawater inhibit dolomite formation. Low levels such as are found in marine pore fluids undergoing microbial reduction in organic-rich sediments are suitable sites of dolomite formation. This same argument can be used to explain mixing-zone dolomites, forming in areas where the sulfate levels of seawater were reduced by dilution with seawater. Sabkha dolomites form in areas of the supratidal flat where the levels of sulfate were reduced by the precipitation of a sulfate phase. Baker and Kastner also argued that with the evolution of carbonates into the pelagic realm (coccoliths and forams) the fertility of calcareous nannoplankton increased using more and more $CaCO_3$. Thus since the late Cretaceous, the oceans have gradually become more undersaturated with respect to the carbonate minerals and the volume of aragonite and magnesian calcite has decreased as the proportion of calcite has increased. Of all the carbonate minerals, calcite is the most resistive to dolomitization, sulfate levels which are 5 to 7% of that present in normal seawater are enough to inhibit the dolomitization of calcite. This may be one reason that dolomite is an uncommon mineral in Holocene sediments.

Deep murine sulfate-reduction dolomites are present in the Miocene Monterey Formation of the Santa Barbara Basin, California (Fig. 5.10a). There the dolomites form thin stratiform, pinch and swell beds of kilometer-scale lateral continuity, with local swells up to 2 to 3 m thick. They are sandwiched in between dark organic-rich laminated diatomaceous opaline cherts. In total, the dolomitized horizons make up no more than 10% of the chert section. Internally, the dolomites are massive micrites and contain remnant foraminifera and diatom frustules. Evidence for the early origin of the dolomite occurs in faults that fracture the dolomite beds when the surrounding organic rich cherts are folded. The dolomites must have formed prior to the induration of the surrounding rocks. The beds directly above a local swell in the dolomite also show evidence of compactional drape as well as erosion above the crest of the drape, implying that the dolomitization of the precursor bed was sometimes penecontemporaneous with deposition.

Kelts and McKenzie (1982) documented extensive Quaternary equivalents forming in the subsurface beneath the Gulf of California in areas covered with water more than 700 m deep (Fig. 5.11a). These Quaternary examples occur as shallow dolomites located 90 m below the sediment-water interface. The dolomites are fine grained; sometimes they replace carbonate shells (e.g., forams) and

(a)

(b)

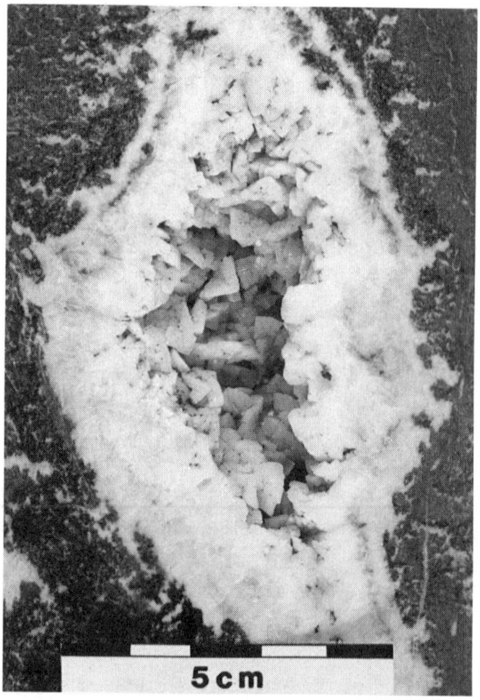

Figure 5.10 (a) Deep-water dolomite layer in the Miocene Monterey Formation exposed on the Santa Barbara coast of California; the carbonate layer has responded in a brittle fashion to compression, while the surrounding diatomaceous ooze has flowed; (b) coarse-grained saddle dolomite in the Cisco Formation of the Permian Basin, west Texas.

sometimes form cements. Invariably, they retain the fine structure of the original sediment (Fig. 5.11b). Dolomitic horizons make up no more than 3% of the sediment column. Kelts and McKenzie concluded that the dolomite formed by diagenetic precipitation at 40 m subbottom in zones of high alkalinity and low sulfate. Not all of the magnesium is derived from the dissolution of microfossils; some comes from the interstitial waters.

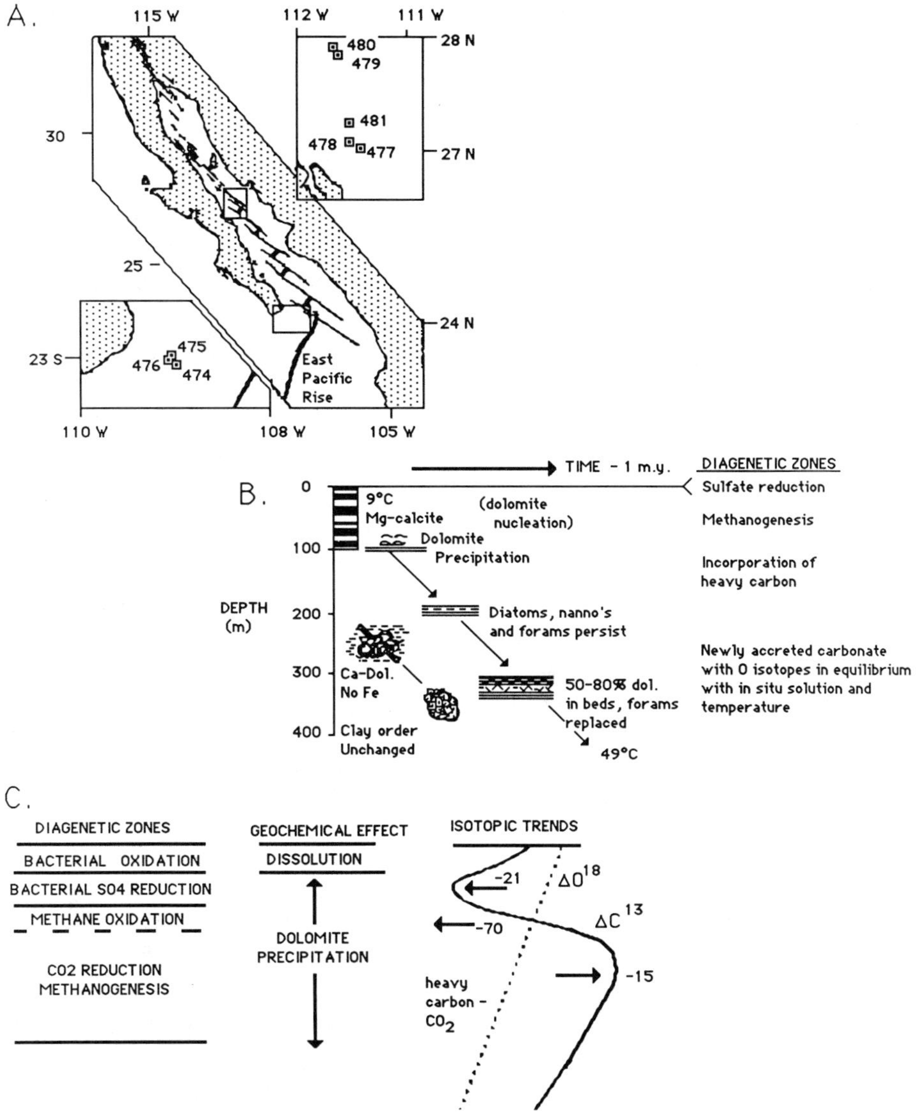

Figure 5.11 Quaternary deep-sea dolomite, Gulf of California (after Kelts and McKenzie, 1982): (a) locality map of DSDP sites 474 to 481, rectangles indicate areas with documented diagenetic dolomite; (b) summary model for depth of nucleation and slow accretion of dolomite layers with burial, based on site 479; (c) schematic model of diagenetic zones in anoxic sediments and hypothetical trends in isotopic compositions of diagenetic carbonates formed at various subbottom depths.

Few would doubt that sulfate reduction is a common process in anoxic organic-enriched environments, but the amount of dolomite that forms is small, as the diffusive resupply of magnesium is an ineffective pump (Land, 1985; Kelts and McKenzie, 1982). The effects of sulfate-reducing bacteria can be modelled by the equation:

$$2CH_2O\ \text{(organic material)} + SO_4^{2-} \rightarrow H_2S + HCO_3^-.$$

The release of bicarbonate depleted in $\partial^{13}C$ from the bacterial reduction of the organic matter is an inevitable part of this process. Dolomite formed in such sulfate-depleted environments is typically depleted in $\partial^{13}C$ (Fig. 5.11c). The lack of widespread $\partial^{13}C$ depletion in ancient dolomitized shelf sequences argues that this mechanism may not be volumetrically important, although it may play a minor role in sabkhas and mixing zones.

In a summary paper on the occurrence of Quaternary deep marine dolomites sampled by the Deep Sea Drilling Program, Baker and Burns (1985) found that all deep marine dolomites share the following features. Overall, sedimentation rates range from 45 to 500 m/m.y. The calcium carbonate content ranges from 8 to 45%, with much of the carbonate consisting of coccolith and foraminiferal tests. Carbonate accumulation rates are around 20 m/m.y. An inverse correlation exists between total sediment accumulation rate and percentage of carbonate. The average organic content ranges from 0.5 to 3.4%. At most sites the organic material came from a marine source. Noncarbonate sediments are dominated by terrigenous detritus and biogenic silica. By weight the dolomite content is only 1 to 5%; dolomitization is not pervasive or complete. The $\partial^{13}C$ of the dissolved CO_2 decreases downward from seawater values to around -20 per mille in the zone of sulfate reduction but becomes more positive at greater depths in the zone of methanogenesis (Fig. 5.11c).

In an earlier variation on the same sulfate reduction theme, Dow (1971) suggested a model for some of the Red River dolomites in the Williston Basin which involved bacterial degradation. By acting on disseminated organic material (which had accumulated in burrows), bacteria preferentially dolomitized the burrows and so caused the mottling found in many Red River dolomites. Other workers have explained the same texture as a result of permeability differences between the burrow fill and the surrounding matrix.

Burial-Stage Model

Burial-stage dolomites are a late-stage dolomitization event found in deep subsurface carbonates and formed by the migration of hot magnesium-rich basinal fluids out of the basin (Fig. 5.3e; Zenger, 1983). Prior to the 1960s this was considered an important mechanism of dolomite formation, as it explained the paucity of modern dolomites and the abundance of Paleozoic and Precambrian dolomites. The idea lost its appeal in the 1960s as carbonate workers moved into detailed studies of Quaternary depositional systems and formulated the mixing zone and brine reflux models. Today, as geologists realize the limitations inherent in the

Quaternary, the model has regained a number of proponents especially among geologists working in the Middle and Upper Devonian reefs of western Canada (Illing, 1959; Mattes and Mountjoy, 1980; Machel and Mountjoy, 1986).

At its simplest the model explains the occurrence of dolomites by the up-dip migration of hot, organic- and magnesium-rich basinal fluids. The magnesium in the solution forming the dolomite probably comes from a number of sources: from connate sources, from absorbed Mg^{2+} derived during clay mineral transformations, from structural Mg^{2+} expelled during the montmorillonite-to-illite transformation, and from the deep subsurface dissolution of bittern salts (carnallite–$K_2MgCl_3 \cdot 6H_2O$, polyhalite–$K_2MgCa_2(SO_4)_4 \cdot 2H_2O$, kieserite–$MgSO_4 \cdot H_2O$). Work in the Alberta Basin assumes that the fluids carrying this magnesium are driven by dewatering deeply buried basinal shales. Clay mineral transformations in the shales generate magnesium ions according to the formula

$$\text{Ca \& Na montmorillonites} + Al^{3+} + K^{+} \rightarrow \text{illite} + Si^{4+} + Mg^{2+} + Fe^{2+} + Na^{+} + Ca^{2+}$$

Problems abound with all the burial mechanisms when the volumes of magnesium required to dolomitize shelf limestones are considered. At best, burial-stage dolomites can introduce only a small amount of new dolomite into the basin. Clay mineral transformations in shales are the most commonly cited source of magnesium for the dolomitizing solutions, yet the formation of chlorite commonly consumes magnesium. In the compacting shales of the Gulf of Mexico, the formation of chlorite and illite from smectite provides an efficient internal sink for magnesium. The massive release of magnesium by compacting shales is undocumented (Land, 1985). The utility of connate waters is limited, as most subsurface brines are much richer in calcium than magnesium. Subsurface fluids are commonly derived from the original marine pore waters and may even attain their Ca/Mg ratios via shallow dolomitization. Most deep connate waters possess Ca/Mg ratios that are initially in equilibrium with calcite and dolomite. If this water moves up-dip and cools, it becomes undersaturated with respect to dolomite and will either dissolve or dedolomitize the surrounding dolomites. The huge volume of connate marine-derived waters that must flow through a limestone to dolomitize the succession at depths where compaction has probably reduced the porosity to less than 5 to 10%, means that little dolomite can be generated merely by compacting away seawater. A magnesium-rich brine derived from the dissolution of bittern salts is a more suitable means of dolomitization then dewatering shales, but the volume of bittern salts found in most sedimentary basins rarely, if ever, approaches the quantity needed to dolomitize an ancient shelf sequence. This is not to say that dolomitization does not occur during deep burial. Both Wanless (1979) and Moore and Druckman (1981) have documented deep dolomitization, but in neither case was there an extensive introduction of new dolomite into the basin.

The geometry of deep-burial dolomite reservoirs is difficult to predict as no consensus exists among proponents of deep-burial dolomitization as to how extensive this process is in shelf limestones. Illing (1959) explained the dolomitization of the Upper Devonian shelf in western Canada by dewatering basinal shales. His evidence was rather circumstantial; he argued that those interconnected reef chains which possessed substantial early-burial porosities and were part of an interconnected basin plumbing system were dolomitized by the upward lateral migration of basinal fluids. Isolated reefs and offshoots which were not part of the main plumbing system were left relatively unaltered. Mattes and Mountjoy (1980) studied a single Upper Devonian buildup (Miette Buildup) and found that the dolomite was fairly well confined to the margin of this buildup. They listed five different types of dolomite about the edge of the buildup, with four of the five related to late-stage burial, stylolitization or the migration of hot fluids forming breccias cemented by saddle dolomites (see later). Given the problems in generating the large volumes of magnesium required to dolomitize limestones, most burial dolomites are probably restricted to the margins of carbonate bodies in direct contact with the dewatering basinal sediments or to aquifers carrying these fluids.

Solution-Cannibalization Model

This is the rather simple process of dissolving magnesium-bearing limestones and then reprecipitating the magnesium along with some of the calcium to form dolomite. Pressure solution along solution seams and stylolites has been cited as a major cause of dolomitization (Logan and Semeniuk, 1976; Wanless, 1979). The model suffers from the deep-burial setting of stylolites and the common shallow removal or conversion of high-magnesium calcite to low-magnesium calcite. For a large influx of magnesium another source must be found, and to date, proponents of solution cannibalization have not found it. On a local scale, the process undoubtedly works, but on the scale of dolomite reservoirs in a basin it probably is not a useful model. As Land et al. (1975) pointed out, it is ‘‘selfconsuming and unable to supply large quantities of magnesium necessary to dolomitize thick sections of carbonate rocks without dissolving adjacent continents away.’’

HOW TO RECOGNIZE THE MODELS

No single criterion permits a geologist to place an ancient dolomite into a single dolomitization model, nor may it be wise to do so (Table 5.1). Many ancient dolomites show evidence of having passed through at least two dolomitization stages, probably associated with at least two dolomitization models. For example, the Arab Cycles in a single field in Abu Dhabi show evidence of an early sabkha dolomite, a later mixing-zone dolomite, and a still-later deep-burial saddle dolomite (pers. obs.). This multistage dolomite occurrence is also seen in the Smackover Formation of the Gulf of Mexico and in many other productive dolomites (Worrall and Warren, 1986). The occurrence of more than one type of

TABLE 5.1 CHARACTERISTICS OF DOLOMITES FORMED IN VARIOUS DEPOSITIONAL AND DIAGENETIC SETTINGS

Criteria	Sabkha	Brine Reflux	Coorong	Mixing Zone	Deep burial	Sulfate reduction
Crystal Size	Fine; <15 μm	Medium; 10–100 μm	Fine; <15 μm	Medium; 20–100 μm	Coarse; >60–100 μm	Fine; <20 μm
Primary sedimentary structures	Yes	No	Yes	No	No	Yes
Replacement textures	Rare	Common	Rare	Common	Common	Common
Important feature(s)	Occur in intertidal and subtidal muds	Replace open-marine grainstones or paleoaquifers	Laminated and massive lacustrine muds	Replace open-marine grainstones or paleoaquifers	Replacement in porous units	Replacement in organic-rich diatomaceous muds
	Arid association	Arid association	Semi-arid/Humid association	Humid association association	Late diagenesis association	Deep-marine association
	Sabkha associated	Evaporite associated	Coorong associated	Platform associated	Hydrocarbon and mineralization associated	Bacteria associated
	Stratiform; part of predictable vertical sequence	Often crosscuts depositional contacts	Stratiform; part of predictable vertical sequence	Often crosscuts depositional contacts	Often crosscuts depositional contacts	Stratiform part of predictable vertical sequence
Depth of formation	Near-surface	Shallow-burial	Near-surface	Shallow-burial	Deep-burial	Shallow-burial
Pore fluids	Modified marine brine	Descending brine	Continental schizohaline brine	Mixed marine and meteoric groundwater	Hot basinal brines	Modified marine

dolomite associated with more than one dolomitization model is not surprising, as dolomite is a mineral that evolves over time with evolving chemical conditions. Often, the passage of one dolomitizing event into the next does not completely obscure earlier stages. One type of dolomite certainly dominates in a particular reservoir unit in a particular field, but other types can be seen, sometimes in a single thin section.

Sabkha- and Coorong-style dolomites are usually recognized by their fine-grained crystal sizes (<10 to 20 μm), their stratiform occurrence, and the characteristic sedimentary features associated with sabkhas or Coorong-style settings (Table 5.1). Sulfate-reduction, deep-water dolomites are also fine grained and show a characteristic sedimentary association of deep-water sediments and perhaps chert deposition and diagenesis. The fine-grained crystal sizes of these three settings is due to relatively rapid crystal growth as dolomite replaces aragonite within meters of the sediment water interface.

Mixing-zone dolomites and seepage-reflux dolomites tend to occur on basin-scale paleotopographic highs. Mixing-zone dolomites occur on highs because these were the areas that could most often hold a freshwater lens. Reflux dolomites occur on highs because such areas would most often restrict the platform and allow sabkhas and/or salinas to form behind the restriction. Mixing-zone and seepage-reflux dolomites tend to be coarser-grained than penecontemporaneous dolomites; crystal sizes are often in excess of 20 μm. Coarser crystal sizes are probably the result of more stable pore fluid chemistries. Such ancient coarser-grained dolomites are often termed "sucrosic dolomites" and are usually composed of euhedral dolomite rhombs with crystal sizes often between 20 and 120 μm. Many dolomite crystals from the Plio-Pleistocene mixing zones of the Caribbean region display a distinctive cloudy center surrounded by a clear limpid rim. Sibley (1980) suggested this was a result of pore fluid evolution from a state of near-calcite saturation in the early stages of dolomite crystal growth to a state of calcite undersaturation as grain growth continues in the mixing zone.

Burial-stage dolomites are most often characterized by coarsely crystalline sparry dolomite, often in the form of saddle dolomites. Saddle dolomite is a variety of dolomite that shows a warped crystal lattice and often contains significant levels of iron (Radke and Mathis, 1980). Such dolomite is turbid, coarsely crystalline with curved crystal faces (the saddles), and shows sweeping extinction in cross-polarized light. It occurs as a void cement in all porosity types as well as a replacement mineral. Other names for saddle dolomite include baroque dolomite, pearl spar, late ferroan dolomite, white sparry dolomite, hydrothermal dolomite, and gangue dolomite (Fig. 5.10b). The origin of saddle dolomite is the least understood of the various dolomite types. It has not been observed in Quaternary sediments, and the deep-burial environments where it forms are not easily accessible. Saddle dolomite shares an association with epigenetic sulfide mineralization, sulfate-bearing carbonates, and hydrocarbon accumulations (Radke and Mathis, 1980). A growing consensus is developing that saddle dolomite is precipitated at temperatures between 60 and 150°C from hydrothermal solutions carrying sulfides and hydrocarbons. In other words, saddle dolomite is a burial indicator which forms in the oil window perhaps from solutions associated with oil migration. In some oil fields in Arab cycles of the Middle East, an intimate relationship exists between saddle dolomites and stylolites; in one field saddle dolomites up to 1 cm across grew in thick bitumen infills in stylolites (personal observation). The intimate relationships between stylolites and saddle dolomites may well be the key to the timing of oil migration in some of these major reservoirs.

Gregg and Sibley (1984) argued on the basis of field and laboratory studies that aside from saddle dolomite, another petrographic indicator of the temperature of dolomite crystal growth exists, namely the presence of what they define as xenotopic dolomite. Xenotopic dolomite is composed of tightly packed anhedral dolomite crystals. When a solid mosaic, the crystal boundaries are curved, lobate, serrated, indistinct, or irregular. When a cement, the crystals are saddle dolomites. The irregular noneuhedral boundaries are a result of dolomite growth above a critical roughening temperature (CRT). They concluded that the petrographic recognition of xenotopic dolomite provides a quick and easy way to determine if a dolomite formed under epigenetic conditions above 50°C, either by neomorphism of an earlier dolomite or replacement of a limestone.

Timing of Dolomitization

The relative timing of the dolomitizing events is a relatively simple matter in Quaternary examples but in ancient examples timing events are much more difficult to define. Some of the rules of thumb are:

1. Fine-grained dolomites with well preserved sedimentary structures are early (Sabkha, Coorong, or Sulfate Reduction).
2. The occurrence of limestone blocks in porous debris flows in basinal settings adjacent to now-dolomitized shelf edges implies that the shelf dolomite is not a burial-stage dolomite.
3. Saddle dolomites are burial-stage dolomites forming in a diagenetic environment some hundreds of meters below the surface.
4. Xenotopic dolomite formed above 50°C (usually, medium to deep burial).

Summary of Models

Predicting dolomite reservoir occurrence is a difficult task for the exploration geologist. On the scale of an individual unit(s) in a productive field, most dolomite occurrence is controlled by the permeability of the sequence at the time of dolomitization. If a grainstone was tightly cemented early in its diagenetic history, either in the marine or in the meteoric-phreatic zone, it will not be extensively dolomitized. In such sequences it is the associated mudstones that will be dolomitized especially if deposited in an evaporitic lagoonal or sabkha setting. If the grainstone retained a high porosity into the shallow burial stage, it was often dolomitized in preference to the mudstones as it was an excellent conduit for dolomitizing solutions. The understanding of dolomitization controls and timing within individual fields is often a postmortem study more useful to the development or reservoir geologist. More often it is on the larger scale of basin occurrence of dolomite that dolomitization models can be predictive tools. At the basinal scale dolomite formation is controlled by:

1. *Paleogeographic position of the originally permeable limestone* units relative to a source of syndepositional dolomitizing fluids. Where are the shorelines and sabkhas? Do not ignore extensive mudflats!
   Sabkha, Coorong, and sulfate-reduction models
2. *Paleogeographic distribution relative to syndepositional highs* within the basin and on the shelf. Basin margin positions are often extensively dolomitized, especially in evaporitic settings.
   Seepage-reflux and mixing-zone models
3. *Exposure at unconformities* and zones of evaporite dissolution.
   Mixing-zone and seepage-reflux models
4. *Timing of eustatic and tectonic changes* which control the relative position of the regional water table and the evolution of connate waters. Being able to discern on-lap versus off-lap in this setting is very useful in determining potential reservoir occurrence.
   Seepage-reflux, mixing-zone and deep-burial models

These environments are not mutually exclusive and are often interrelated. However, on the basinal scale such features are recognizable in seismic even in a frontier area. It is now time to move to actual case examples of dolomitized sequences.

ANCIENT DOLOMITES AND RESERVOIRS

Dolomitic reservoirs are created by diagenesis, either deep or shallow. It is often difficult, if not impossible, to describe a single dolomitic unit and say that this reservoir dolomite formed by a single dolomitization process. Most often the reservoir dolomite contains two or more styles of dolomite, often with one type, for example a saddle dolomite, dominating in a particular oil field. This same dolomite unit may, for example, also contain subordinate mixing-zone dolomite and even some sabkha-style dolomite. In another area in the same formation, but perhaps further up-dip, the mixing-zone dolomite may dominate and control reservoir trends while the saddle dolomite is less important. The Red River Dolomite described in Chapter 2 demonstrated this multi-stage origin of dolomite. For this reason this section on ancient dolomites does not present a particular dolomitic model as an explanation of a particular formation. Rather, it attempts to show the variation in dolomitization models in particular units, as it describes productive dolomite formations and the types of dolomites within particular fields.

San Andres Formation, Palo Duro Basin, Texas

The San Andres Formation in the Palo Duro Basin is thought by many to differ from its equivalents farther to the south and southwest as the predominant depositional setting in the Palo Duro Basin was a subaqueous evaporite lagoon rather than the peritidal sabkha so common in the productive intervals to the south (Fig. 2.9; Chapter 2). Recent work by the author has shown that the San Andres to the south of the Palo Duro Basin may, in fact, not be a sabkha (chapter 3). The San Andres in the Palo Duro Basin comprises extensive, thick evaporite units separated by several thick carbonate units in its lower part and many thin units in its upper part (Fig. 5.12; Bein and Land, 1982). The San Andres in the Palo Duro

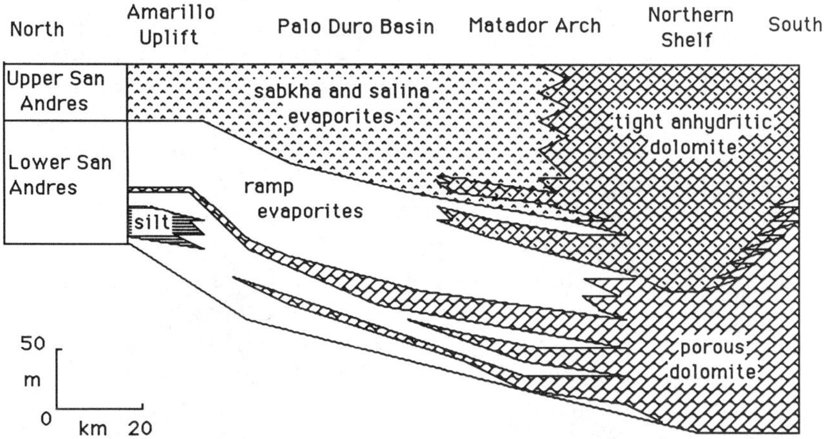

Figure 5.12 Idealized north-south section of the northern margin of the Delaware Basin—Northern Shelf area, west Texas, showing the lower laterally correlatable facies of the Lower San Andres passing up into the facies mosaic of the Upper San Andres. The productive intervals of Lower San Andres on the Northern Shelf mark the transition from open-marine to restricted evaporitic conditions. For locality of section see Fig. 2.10.

Basin is interpreted as a regressive unit with five major cycles, with incomplete subcycles within these cycles. Recent work by Havorka et al., (1985) and Bein and Land (1982) have shown that the halite-rich San Andres of this area was not laid down as the sabkha complex but was deposited subaqueously, as aggrading shallowing saltern units on a very broad, shallow shelf perhaps accumulating behind the carbonate shoals and channels of the Northern Shelf (Fig. 3.17). No modern counterpart exists for the scale of subaqueous evaporite deposition in the

Palo Duro, a basin measuring at least 100 by 2500 km and perhaps extending all the way up into Kansas. Havorka argues that it may have been possible to reach halite precipitation in the Palo Duro Basin even in the absence of any restrictive barrier, given that the shelf/ramp was shallow and broad enough. To the south, across the Northern Shelf of the Midland and Delaware Basins, the evaporites pinch out into carbonates.

Six lithofacies are present in the carbonates of the Palo Duro Basin and the Northern Shelf: dolomudstone, pellet-oolite packstone/grainstone, filamentous (*Girvanella*-like) grainstone, sponge-spicule packstone, wispy-laminated crinoid packstone, and skeletal packstone/grainstone. Facies distribution in the San Andres was controlled by the salinity of the brine water body, which showed increasing salinities in a northward direction. In the Palo Duro Basin, as in the northwestern shelf discussed in Chapter 3, the carbonates in the upper part of the San Andres Formation differ from the lower part. They lack the skeletal lithofacies of the lower units and display higher concentrations of manganese, iron, and terrestial organic matter (Bein and Land, 1982). The lower part of the San Andres was deposited within a broad-shelf evaporitic basin or lagoon which was deep enough to sustain extended periods of steady-state water circulation. At times in its history the basin was sufficiently saline to deposit extensive subaqueous halite and gypsum units. The bromide content of the halite in this lower part is consistently large and indicates continuous salt deposition without intermittent dissolution. In the upper part of the San Andres, the interbedded salt, anhydrite, carbonate, and mud units are far more discontinuous than their equivalents in the lower part, indicating a change in the depositional setting. The bromide content of the halites in this upper part of the San Andres are depleted, indicating halite dissolution and reprecipitation. The upper part of the formation was deposited in smaller water bodies which were sensitive to fluctuating inflow volumes and at times shrank to the point where the former lagoon perhaps became a coastal sabkha and coastal salina complex (Fig. 5.12). The increasing importance of meteoric water in this environment can be seen in the greater contents of manganese, iron, and terrestrial organic matter. This same meteoric water was a source of dissolved carbonate to be deposited as thin carbonate units in the upper San Andres.

Although drastic differences exist in the depositional setting of the upper and lower San Andres, the dolomites of both are similar, if not identical. This indicates that the formation of most of the San Andres dolomite occurred in similar brines. Bein and Land (1982) concluded that the San Andres carbonates were stabilized and dolomitized by refluxing magnesium—calcium—chloride brines in the shallow subsurface. The periodic deposition of halite throughout San Andres time explains the predominance of such brines throughout early diagenesis. During the first stages of brine reflux through the carbonates, $CaCO_3$ was dissolved and $CaSO_4$ precipitated. The solubility product of $CaCO_3$ in hypersaline waters is greater than that in normal seawater and increases with increasing salinity (Sass and Ben-Yaakov, 1977). Thus the abundant anhydrite- cemented molds in the San Andres carbonates are attributed to this early near-surface stage of brine fluxing.

After mold formation and sulfate infill the calcium and magnesium levels in the brine remained constant while the SO_4 levels fell. At the same time the carbonate concentration must have increased until dolomite saturation occurred (Bein and Land, 1982).

The degree of dolomitization in the carbonates varies; the thinner carbonate units are the most extensively dolomitized, although the thicker carbonate units on the southern seaward side of the basin are also extensively dolomitized. Presumably, the latter were dolomitized because their marine-margin position where the presence of less saline pore fluids hindered the evaporite plugging of the molds. Plugging reduced the permeability of the thicker units in the rest of the basin and so reduced the efficiency of the brine-reflux process. This formed thick, evaporite-plugged dolomitic limestones rather than relatively pure dolomites. Bein and Land (1982) invoked the lack of early diagenetic evaporite plugging of moldic porosity in areas to the south on the Northern Shelf to explain the producing reservoirs in that region. Their interpretation differs from that of many other workers who have explained the Northern Shelf dolomites by a sabkha model, but is consistent with the model presented in Chapter 3.

Smackover Formation, Gulf of Mexico

The Smackover Formation forms a subsurface arc about the Gulf of Mexico and was deposited during Upper Jurassic (Oxfordian) time in a transgressive-regressive event (Fig. 5.13). It is underlain by the siliciclastic Norphlet Formation and overlain by the Buckner Evaporite (Fig. 5.14). The Smackover is composed of shelf dolomites, shelf-margin oolitic and peloidal grainstones, and slope and basin limestones. It can be subdivided into three informal units, the lower, middle, and

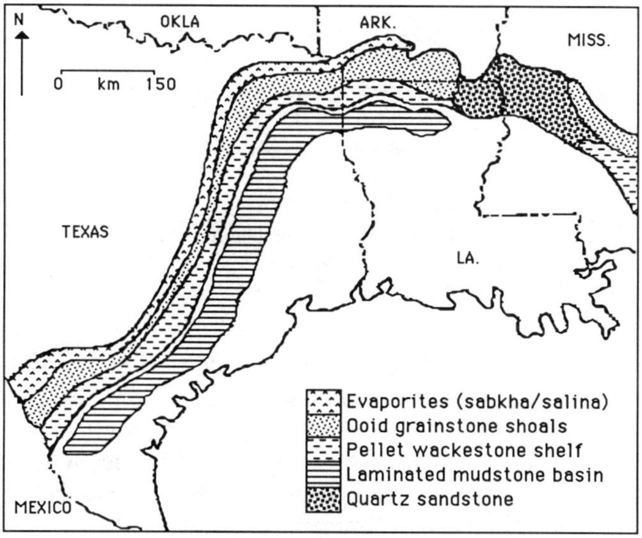

Figure 5.13 Distribution of the various facies of the Smackover Formation (Oxfordian) in the Gulf of Mexico (after Budd and Loucks, 1981).

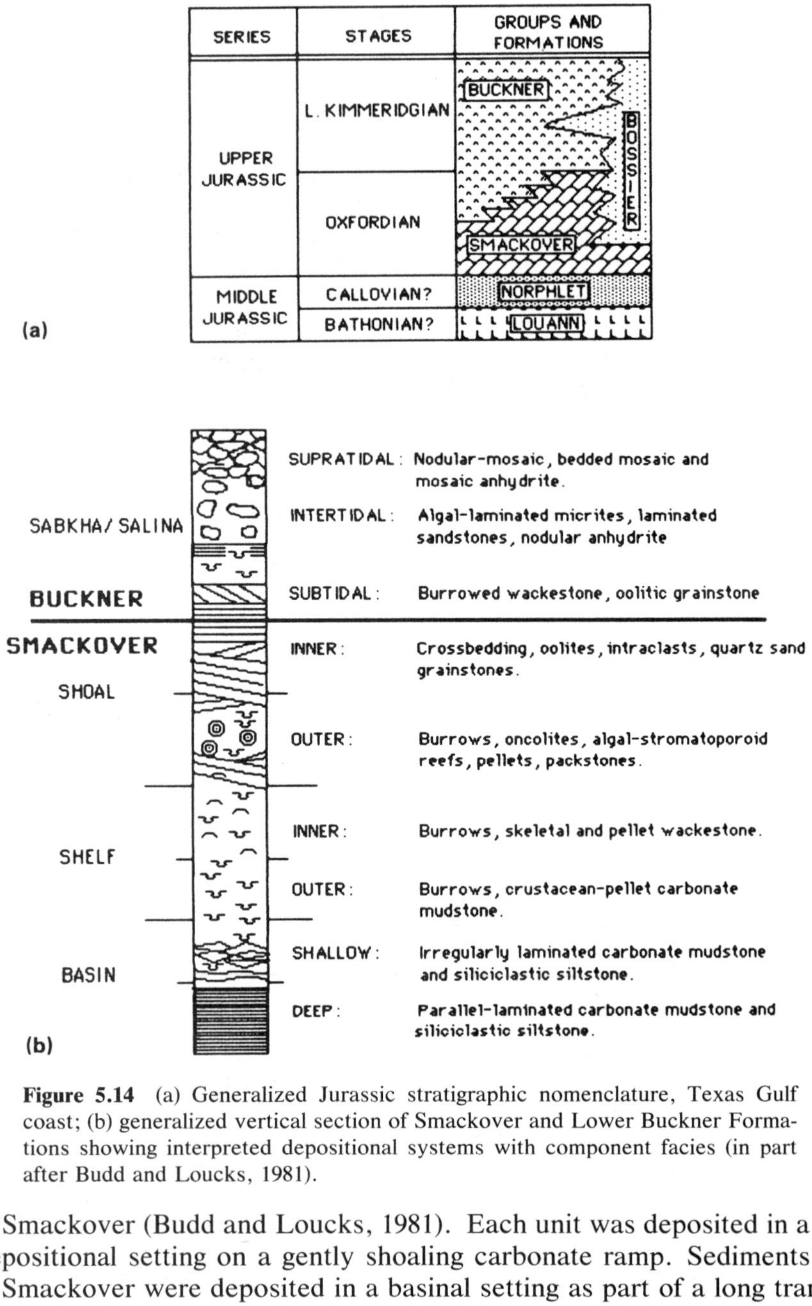

Figure 5.14 (a) Generalized Jurassic stratigraphic nomenclature, Texas Gulf coast; (b) generalized vertical section of Smackover and Lower Buckner Formations showing interpreted depositional systems with component facies (in part after Budd and Loucks, 1981).

upper Smackover (Budd and Loucks, 1981). Each unit was deposited in a different depositional setting on a gently shoaling carbonate ramp. Sediments in the lower Smackover were deposited in a basinal setting as part of a long transgressive episode. They consist of laminae of dark brown to black, organic-rich mudstones alternating with laminae of pyrite and siliciclastic silt or sand. The laminae grade upward in section from very thin parallel laminae to slightly thicker irregular laminae. The middle part of the Smackover is composed of sediments deposited on a low-energy, open-shelf setting as sedimentation caught up with relative subsidence. Two lithofacies are present: a wispy-laminated, crustacean- pellet carbonate mudstone containing a sparse pelagic fauna and a densely burrowed skeletal-pellet wackestone containing a benthic fauna. The upper part of the Smackover consists of several grain-supported lithofacies deposited in shoal settings ranging from tidal bars and channels to beaches, islands, and algal-stromatoporoid reefs (Fig. 5.15a). Burrowed pellet and oncolite packstones and algal-stromatoporoid boundstones formed moderate-energy outer shoals, and sequences of cross-bedded mixed allochem, oolite, and oolitic grainstones com-

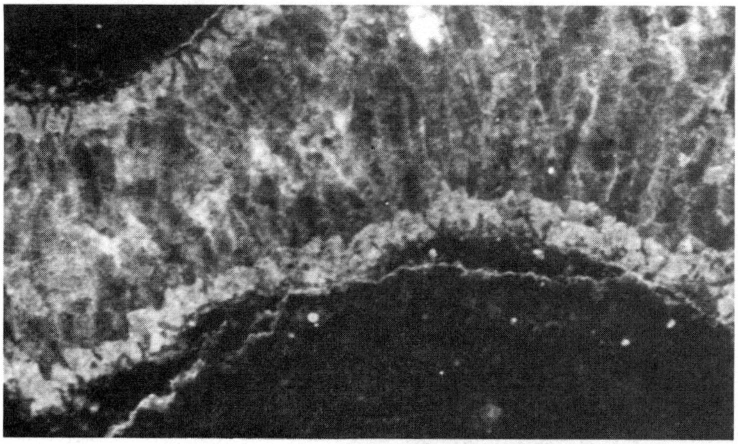

(a)

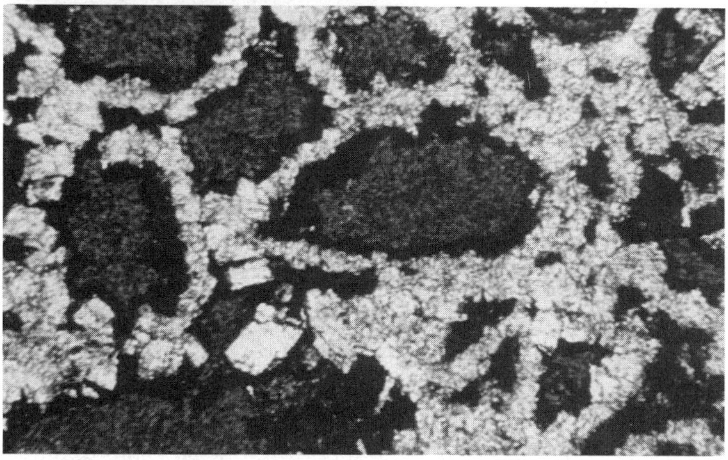

(b)

Figure 5.15 (a) Algal stromatoporoid reef facies showing an early marine rim cement phase, Hatters Pond Field, Alabama; rim is 0.1 mm thick, photographed using fluorescent light; (b) dolomitized ooid grainstone, with oil residue in pores, Hatters Pond Field, Alabama, (photographs courtesy of John Worrall).

prised the current reworked inner shoals (Fig. 5.15b). Sea level at this time was relatively constant, a stillstand, and deposition kept pace with subsidence. These upper Smackover facies create the most productive reservoirs. The Smackover is sealed by subaqueous and subaerial evaporites and redbeds of the overlying and perhaps laterally equivalent Buckner formation.

A minimum of 4% porosity and 0.1 md permeability in Smackover carbonates is considered to be reservoir grade (Becher and Moore, 1976). Economic porosity is of three main types:

1. Primary interparticle porosity
2. Leached moldic porosity
3. Intercrystalline porosity within dolomites

Most reservoirs occur in sediments deposited as grainstone shoals and sometimes in the sediments of the inner-shelf setting. The best reservoirs in Arkansas, Louisiana and eastern Mississippi are those preserving incompletely cemented interparticle primary porosity. The up-dip Smackover close to the paleoshoreline often contains leached moldic porosity due to freshwater flushing. In eastern Texas, Alabama, and Florida, the occurrence of dolomite with intercrystalline porosity enhances the reservoir potential of the Smackover grainstones (Stoudt, 1979).

Dolomitization is almost always required to form a reservoir in inner-shelf sediments and certainly enhances the reservoir characteristics of the shoal sediments. A minimum of 80% dolomite is necessary for consistently good reservoir quality, but rocks that are 100% dolomite are not always good reservoirs. Reservoirs are best developed in shoal-water grainstones which have been almost completely dolomitized with their ooids leached subsequently. Individual reservoirs can be as thick as 10 m (33 ft), with porosities as much as 26%, and permeabilities as great as 6.5 md. Although reservoir quality is often related to dolomite occurrence, a diagenetic replacement process, the distribution of dolomite, like that of the interparticle porosity and the moldic porosity, is in large part controlled by the porosity trends in the original depositional setting. In other words, although both moldic porosity and dolomitization are a result of diagenesis in the shallow and deeper subsurface, the plumbing or solution pathways allowing diagenetic solutions to flow through the Smackover were first created at the time of deposition.

Four diagenetic stages are seen in the inner shelf Smackover grainstones: marine-phreatic, shallow-meteoric, regional fluid mixing, and basinal fluid expulsion (Fig. 5.16; Budd and Loucks, 1981; Moore and Druckman, 1981; Moore, 1984; Worrall and Warren, 1986).

The *first stage*, the marine-phreatic, affected sediments in the marine environment from the sediment–water interface to a meter or so below the surface. It was primarily a stage of micrite envelope formation, micritization of grains, and the formation of the occasional submarine hardground. The effect of this first

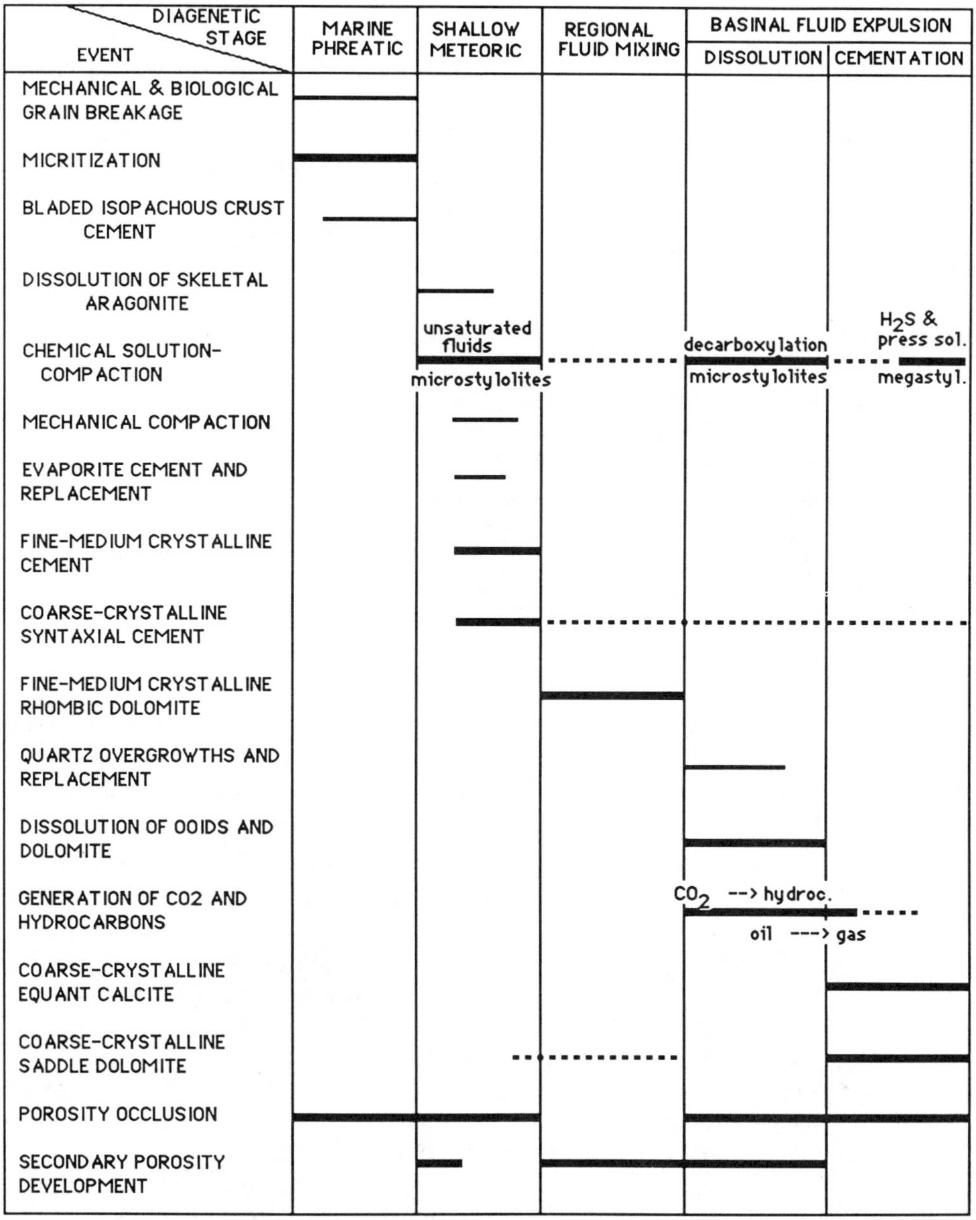

Figure 5.16 The four stages of diagenetic alteration in the Smackover Formation and their petrographic indicators (based on Moore, 1984; Moore and Druckman, 1981; Becher and Moore, 1976).

diagenetic stage on porosity and permeability was minimal, although it did eliminate some intraparticle porosity.

The *second stage,* the shallow-meteoric stage, affected sediments from the surface to several hundreds of feet in depth. Freshwater entered the Smackover either through localized groundwater sinks (islands) or as part of the regional groundwater flow system (Fig. 5.17). These waters leached aragonite allochems,

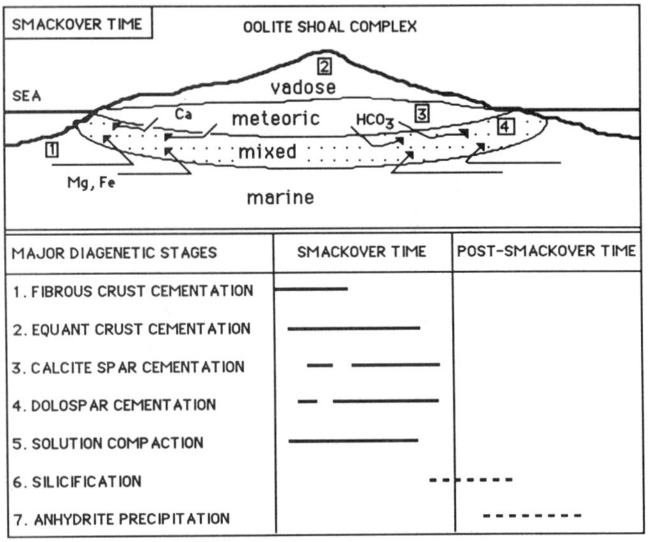

Figure 5.17 Areas of Smackover cementation, their diagenetic locality during formation and their relative timing (after Becher and Moore, 1976).

leaving behind hollow grains preserved only by their micrite envelopes. Overburben pressure caused mechanical and solution compaction (microstylolites) at this time, and solution compaction greatly reduced the porosity during this stage. Stylolitization began at this time and continued on into the deep-burial stage when it culminated in the formation of megastylolites. Fine- to medium-crystalline equant calcite also formed at this time and preceded the formation of the microstylolites. This calcite is unevenly distributed, often filling small intergranular pores, but occurring only at grain contacts in larger pores. It is probably a vadose-meteoric cement. When the same calcite completely surrounds the grains, it is probably a meteoric-phreatic cement. The early meteoric cements were minor and appeared to have had little effect on the porosity.

The *third stage,* the regional fluid-mixing stage, affected sediments from a few meters to hundreds of meters in depth in areas where meteoric fluids mixed with marine-phreatic or basinal brines. Mixing-zone dolomitization produced good intercrystalline porosity and was the dominant diagenetic process at this stage. A mixing-zone model best explains the distribution of most of the dolomite in the Smackover, as dolomite is usually concentrated in the up-dip areas (Budd and Loucks, 1981). Iron-poor, fine- to medium-crystalline, euhedral rhombic dolomite precipitated in the pore spaces and replaced allochems. Ooids in the highly dolomitized zones were not affected by solution compaction of the shallow-meteoric, indicating that dolomitization was early or that some matrix may have inhibited fluid flow and prevented dissolution. In areas of little dolomitization, the individual rhombs precipitated directly onto the allochems, and coarsely crystalline equant calcite later filled the remaining pores. At the end of this stage the intercrystalline porosity in the up-dip, highly dolomitized lithofacies ranged from high to moderate, whereas in the down-dip, poorly dolomitized lithofacies, only small amounts of interparticle porosity were still preserved. An alternative model for this stage is the brine-reflux model. But initial results show dolomitization associated with a lightening of the oxygen isotopes (Humphrey et al., 1986) implying that isotopically heavy brines were not important in the dolomitization of the shoal areas.

The *fourth stage,* the basinal fluid expulsion stage, affected Smackover sediments after a few thousand meters of burial. This stage can be divided into two substages: a dissolution-dominated substage and a later cementation-dominated substage. Both are controlled by fluids generated in the deeper parts of the basin and expelled through permeable pathways to shallower, lower-pressured strata. Such fluids probably were generated by compaction and expulsion of the diagenetic waters of deeply buried shales and brines squeezed out of deeply buried evaporites (Land and Prezbindowski, 1981). The same waters also contained hydrocarbons and H_2S generated by the thermal interaction of hydrocarbons and anhydrite (Budd and Loucks, 1981). In the dissolution stage the micritized lime ooids from the highly dolomitized areas were dissolved, but the surrounding framework of relatively stable dolomite rhombs prevented widespread solution compaction. In the down-dip dolomite-poor lithofacies, solution compaction was reactivated. CO_2-charged waters derived from the thermal decarboxylation of kerogen-rich lower Smackover may have produced this dissolution substage. In terms of reservoir development this dissolution stage is an important time as it converted relatively tight rocks into good-quality reservoirs by leaching the micritized ooids and allochems. Dissolution preceded hydrocarbon generation, as many oo-molds are filled with bitumen.

The dissolution stage was followed by the cementation stage, a time of abundant coarse-to-medium calcite precipitation accompanied by the formation of minor amounts of saddle dolomite. The cementation substage probably occurred at temperatures of 60 to 120 °C during the initial stages of hydrocarbon migration in the deep subsurface. Porosity in the dolomite-poor down-dip areas was totally occluded at the end of this cementation stage. Probably, exceptions can be found to the complete loss of porosity in the down-dip regions. Moore and Druckman (1981) observed that the Smackover in some areas contains evidence of yet another dissolution substage where dissolution created porosity at the expense of the late-stage equant calcite spar. The last diagenetic process in this late stage of diagenesis is the growth of megastylolites, probably by the enlargement of the earlier formed microstylolites. These megastylolites cut across all grain and cement types and were probably a result of continually increasing overburden pressure in conjunction with the generation of thermally derived H_2S.

Recent work by Worrall and Warren (1986) in Hatter's Pond field has shown that at least in this area, a late stage of coarse-grained dolomite formation also occurred. This coarse-grained dolomite makes up the base of the Smackover Formation. It displays carbon and oxygen isotopic signatures too negative to have formed in the mixing-zone. It is a late stage dolomite formed as hot basinal(?) brines flowed out of the Louann Salt Basin through the underlying Norphlett Sand.

Dolomite in the Miette and Other Buildups, Alberta

Dolomite occurrence in the Miette buildup has been used to argue for extensive late-stage dolomitization of a carbonate sequence (Mattes and Mountjoy, 1980). The Miette buildup or reef complex is one of a series of Upper Devonian buildups that formed in the Alberta basin after an early Frasnian rise in sea level. They are composed of a succession of distinctly layered biostromal limestones and dolomites with associated calcilutites, calcarenites, and minor amounts of calcirudite, surrounded by argillaceous carbonate foreslope deposits and basinal shales (Fig. 5.18). The stromatoporoids formed local wave-resistant shoals or biostromes, especially at or near the margins of the buildups. Most of these Upper Devonian buildups in the Alberta Basin are dolomitized extensively, with the notable exception of isolated reefs such as Redwater, Willingdon, the Golden Spike, Ancient Wall, and Miette, where the dolomite is far more localized. Some authors have explained the distribution of dolomite as due to penecontemporaneous or early diagenesis (Andrichuk, 1958; Maiklem, 1971). Others, most notably Illing (1959) and Mattes and Mountjoy (1980), found the patterns of dolomitization better explained by late-stage dolomitization, especially in the Leduc reefs. To explain the patterns of dolomitization, Illing invoked the late diagenetic, deep-burial movement of hot fluids through porous geologic reefs. The fluids were shale compaction waters, moving up-dip and vertically along the aquifers of porous reef carbonates, dolomitizing as they moved. Reefs that possessed high initial porosity and maintained permeability despite early diagenesis were dolomitized extensively. Isolated reefs and offshoots that were not part of the plumbing conduits were not dolomitized. Mattes and Mountjoy (1980) found support for Illing's model in their study of the Miette buildup.

Dolomite in the Miette buildup is between 25 and 35% of the rock volume and is confined essentially to buildup margins. The degree of dolomitization within the margin varies from negligible to complete and appears to be related directly to the permeability of the precursor carbonate. The dolomite is basically of two types, a fine- to medium-crystalline sucrosic dolomite or a dolomitic limestone containing scattered dolomite rhombs. The sucrosic dolomite shows microvuggy intercrystalline porosity and is commonly associated with vuggy and

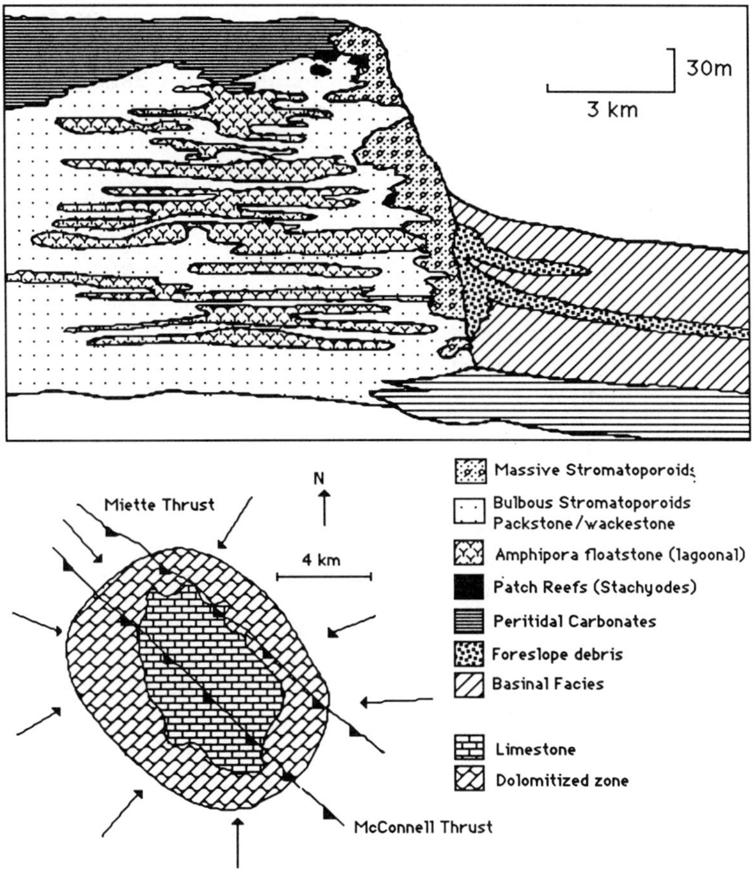

Figure 5.18 Cross section of the facies present in the dolomitized margin of the Devonian Miette buildup and the locality of dolomite distribution (schematic plan view) (after Mattes and Mountjoy, 1980).

leached porosity. The dolomitic limestones are often dolomitized mudstones and are more common in basin interior settings, the sucrosic dolomite being more common about the buildup margin. Regardless of the two dolomite types there are five crystal habits in the dolomites of the Miette buildup:

1. Diffuse micocrystalline dolomite
2. Mosaic dolomite
3. Pressure solution related dolomite
4. White sparry dolomite
5. Late stage baroque or saddle dolomite

These dolomite types formed in the following paragenetic sequence: first type 1, an early prelithification(?) dolomite; then type 2, a late-burial dolomite which forms at least 70% of all the dolomite in the buildup and has replaced large volumes of precursor limestone matrix; then type 3, a dolomite which forms along pressure-solution interfaces; then type 4, forming coarse crystalline pods at the buildup margin; and finally, type 5, forming isopachous crusts which line the leached void spaces and fractures associated with brecciation and fracturing at the buildup margin. Four of the five dolomite types and at least 90% of all the dolomite in the buildup formed during the late-burial stage of diagenesis (Mattes and Mountjoy, 1980). However, it was the primary porosity which was the dominant control over dolomite selectivity and distribution, this primary permeability restricted dolomite to the margin of the buildup. The magnesium needed to form the dolomite flowed into the buildup by three main mechanisms: (1) migration of brines from adjacent basinal muds as they are compacted, (2) pressure solution, (3) mixing of near surface fluids with deep burial brines along fracture-controlled conduits. The latter mechanism is probably responsible for the precipitation of the white sparry saddle dolomite.

The late diagenetic interpretation of the Miette dolomite distribution is a result of a surface study. Subsurface reservoir equivalents, based on detailed geochemical and petrographic studies, are not yet documented. The use of this late-stage diagenetic model involves one major flaw. It suffers from a lack of a source for magnesium capable of contributing enough dolomite to dolomitize extensive volumes of limestone. However, dolomite is a diagenetic mineral continually reequilibrating to the changes in the surrounding pore fluids as it is buried. It may well be that late stage dolomitization introduces very little new dolomite into any given volume of rock, but it may substantially alter the fabric of the final product. The same processes of cannibalization and reequilibration of dolomite mineralogy and chemistry may explain other examples of burial dolomite where the volume of magnesium is otherwise too small to explain the amount of late stage dolomite.

SUMMARY

Dolomite is an enigmatic mineral still not understood in terms of its origin and distribution. However, it is better understood today than 20 years ago. Today six major models exist to help explain subsurface dolomite distribution (Table 5.1). Probably the greatest leap forward in understanding the origin of dolomite and dolomitization controls came in the last five years with the discovery that dolomite is a mineral whose chemistry and lattice structure changes with burial and diagenesis. Many different types of dolomite are known, just as different types of calcium carbonate exist. Some forms of dolomite are more soluble than others. When the time is reached where the controls of these differences in solubility are understood, then modeling of dolomitization will be able to predict reservoir controls.

Controls on dolomitization change as the limestone under consideration passes through various depositional and diagenetic realms. Because of the variability of deposition and diagenesis within a basin, dolomitization models are more useful today on the scale of permeability trends within individual fields rather than predicting reservoir occurrence in a basin. Using whichever model is relevant to the sequence under study, a dolomitizing solution must have flowed along the most permeable pathway present at that time. At its simplest, dolomite distribution is controlled by the permeability of the sequence at the time of dolomitization. However, to predict the distribution of dolomite in the subsurface, one must know the depositional setting and the associated permeabilities for early dolomites and yet be able to superimpose the basinal plumbing to depths where connate waters are flowing to understand later dolomites.

CHAPTER 6

Salt Structures, Sediments, and Hydrocarbons

INTRODUCTION

The movement of thick salt beds into salt structures has created hydrocarbon reservoirs around the world. Better-documented examples come from Germany, the Gulf of Mexico, and the Middle East, but reservoirs associated with salt structures can be found in southern Russia, the North Sea, southern France, Utah, and the Atlantic Seaboard of Canada. Traps are formed by anticlinal closure over diapirs, salt pillows, and turtle structures, as well as fault-line and unconformity traps lying above and adjacent to salt structures. Some oil traps in Germany even occur below mushroom-shaped diapirs where the salt acts as both trap and seal.

Salt structure is an all-embracing term covering both nonpiercement and piercement features. Piercement describes a structure that appears to intrude the overlying strata, nonpiercement structures deform but do not intrude the overburden. Another word for a piercement structure is *diapir* from the Greek *diaperin,* meaning "to pierce." Most economically interesting diapirs are composed of salt, but diapirs can also be composed of shale, serpentinite, and any other material that possesses a lower relative density than that of the overlying strata. This chapter deals only with salt-related structures.

Salt structures are most often found in active rifts or along previously rifted continental margins, both areas of sediment dumping and high initial geothermal gradients. Bedded evaporites are commonly deposited in such areas during the incipient ocean phase of continental rifting, and many juvenile continental margins were floored with a thick layer of bedded salt (Evans, 1978; Kinsman, 1975a,b). The proto-Atlantic rift, now found as buried remnants along the eastern margin of the Americas and the western margin of Africa, is a well-studied example. As the rift widened and the basin floor cooled and subsided, the bedded evaporites of the early rift were buried by several kilometers of shelf sediment. The associated loading initiated salt movement and subsequent diapirism. Diapir growth then persisted for millions of years, as long as subsidence and burial continued to affect the thick, mobile salt unit. Depending on the thickness and lateral extent of the mother salt bed, salt spines may project 10 or more kilometers up to the surface. Wherever salt structures approached the surface, they created topographic highs which subsequently controlled sediment deposition and reservoir formation in their vicinity.

Not all salt structures still contain salt. Fossil salt structures outcrop along the rift axis of the Adelaide Geosyncline in South Australia (Lemon, 1985). Structures were last active in the Proterozoic and the salt has long since been flushed from the structures. Fossil diapirs are now composed of carbonate and siliceous breccia blocks originally derived from the associated sediments as the salt flowed toward the surface. The diapirs show syndepositional sedimentation features and geometries associated with modern salt structures. Some of the larger structures have been modeled by Lemon as diapirs whose accompanying lopsided sedimentation was controlled by basement faults.

Thick salt units have been laid down during times of continental collision as well as during rifting. For example, the Upper Miocene (Messinian) evaporites of the Mediterranean were laid down as Africa collided with Eurasia (Chapter 4). Continued collision remobilized this, and earlier Mesozoic salt, so that extensive diapiric provinces exist now in and about the Mediterranean Sea (Halbouty and El-Baz, 1985). In the western Mediterranean the diapirs are very similar in occurrence and geometry to the Gulf of Mexico. Such diapirs have been identified in the Northern Balearic Basin, the Balearic abyssal plain, in the region south of the Balearic Islands, in waters off the western coast of Sardinia, and in the Ligurian Sea. In the eastern Mediterranean, salt structures can be found on the Nile Cone and in the Antalya, Cilicia, Herodotus, and Levantine basins. In Israel mobile salt structures subcrop along the outer Mediterranean shelf and along the transform of the Dead Sea. Mobile salt structures occur in the Adana and Iskendurun basins north and northeast of Cyprus. In addition, thick salt and evaporites of Triassic age exist in offshore and onshore Tunisia, and onshore in Italy, Spain, Portugal, and western Greece.

This chapter describes how various salt structures grow, some of their recognizable characteristics, and how mobile salt features often control sedimentation and reservoir occurrence in their vicinity. Aspects of this chapter are covered in excellent papers and books by Trusheim (1960), Halbouty (1979), Woodbury et al. (1980), Carter and Hansen (1983), and Jackson and Seni (1983), Seni and Jackson (1983a,b), Jackson and Seni (1984), Jenyon (1986a) and the references therein. To understand how a salt structure grows requires an understanding of how and when salt moves.

SALT MOVEMENT

Salt's low density and rheidity make it particularly susceptible to movement at shallow depths. Halite attains a density of 2.2 gm/cm^3 within a few meters of burial, it then changes only slightly with further burial. During deposition, density of the overlying sediment is less than that of the halite unit but increases with depth as it compacts and dewaters. By depths of 0.5 to 0.9 km (2000 to 6000 feet) it exceeds that of the underlying salt, making the salt buoyant. Also, the low rheidity of salt, that is, the ease with which it flows as a viscous solid, means salt flows readily under burial conditions where the surrounding sediment is still undeformed. The viscosity of Gulf of Mexico salt is 10^{14} to 10^{17} poise, depending on moisture content and temperature. The viscosity of the present overburden is about 10^{21} poise. This viscosity contrast of at least four orders of magnitude means that whenever salt flow begins, the salt moves rapidly into a growing salt structure.

Temperature, density contrast, salt thickness, viscosity, and salt moisture content control the rapidity of salt flow. Figure 6.1 shows recent experimental data on the deformation of dry rock salt. A strain value of 15 on the top abscissa indicates that a rock stretches at a rate of 10^{-15} of its length per second, a slow strain rate equivalent to rock deformation rates in active orogenic belts. A value of 9 is 1 million times faster and represents the strain rates in active salt glaciers in

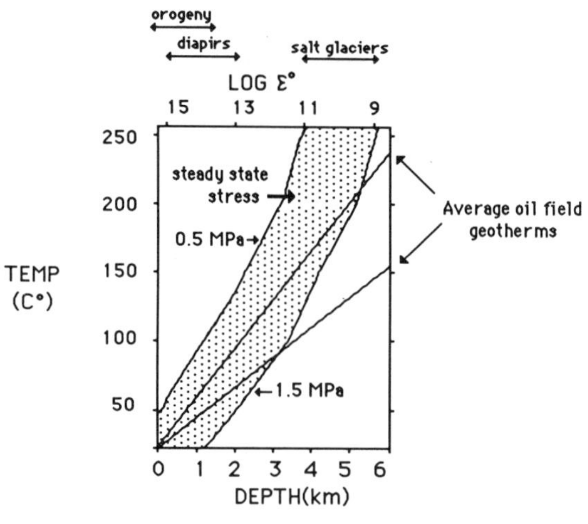

Figure 6.1 Steady-state strain rate as a function of temperature (depth) calculated from steady-state stresses. The solid straight lines are maximum and minimum geothermal gradients from the Gulf of Mexico and the Arabian Gulf (in part after Carter and Hansen, 1983; Jackson and Seni, 1984).

Iran (1.6×10^{-13} to 1.8×10^{-8} per second; Wenkert, 1979). The solid lines on the figure graph temperature versus strain rate and the shaded area between is the measured stress field for dry salt. The field of geothermal gradients from the Arabian Gulf and the Gulf of Mexico have been superimposed on the figure (shaded area). The overlap of the geothermal field with the strain curve shows that salt can deform at much lower pressures and temperatures than was previously thought possible. If diapirs in the Gulf of Mexico grow with a minimum strain rate of 10^{-15} per second, then with a minimum stress difference of 0.5 MPa, the salt could flow at temperatures of 57°C. This corresponds to a present depth of 900 m (3000 ft). Under the same strain, dry salt subjected to a stress difference of 1.0 MPa will flow at 25°C or surface conditions. The other conclusion drawn from the curve is that increases in temperature with burial encourage ductile flow in salt.

Carter and Hansen worked only with dry salt; in the real world no natural salt is completely dry. Subsurface salt always contains pore waters plus brine inclusions and is flushed by pore fluids from nearby compacting sediments. Water in salt drastically effects the rheology of salt and makes it weaker and more susceptible to flow. Salt glaciers (outcropping diapirs) in Iran always flow faster after a rain; the Dashti glacier at Kuh-e-Namak flows at 0.5 m/day during the few weeks of the annual wet season but flows little if any during the dry season when the glacier expands and contracts in response to changes in surface temperatures (Wenkert, 1979). Urai et al., (1987) showed that water need not enter a salt body from outside to be a factor in salt flow. They demonstrated that even the minute quantities of brine caught up as inclusions and intergranular films can have profound effects. Brine-inclusion-bearing samples showed marked weakening at low strain rates; a result not seen in previous experiments using dry salt. At shallow depths any entrained moisture allows salt to flow faster and earlier. Water or brine not only exerts controls on the fabric of bedded salt but maintains a control on salt even when it deforms into salt structures.

As salt flows into a salt structure it stretches and deforms, leaving behind a memory of the stress regime in the various structures and fabrics preserved in the diapir. The significance of salt flow fabrics as indicators of the various stages of diapir growth is outside the sedimentological emphasis of this book. The interested reader is referred to an excellent paper by Talbot and Jackson (1987) and references therein. Their benchmark paper showed that fine-scale grain-shape fabrics record only the late stages of the strain history and that the actual strains, resulting from superposed deformations, are recorded by larger strain markers—bedding, nodules, inclusions, veins—which have much longer strain memories.

There appears to have been no minimum depth for salt movement in many evaporite basins. The Gulf of Mexico is such an area of syndepositional salt flow. Once salt was mobilized it moved at the same time the surrounding sediment was deposited (Seni and Jackson, 1983a, b). Not all salt movement is shallow and spontaneous; bedded salt can be found at depths of more than 900 m (3000 ft) in many other areas, such as the Michigan Basin. Salt can move only when it has been triggered. The triggering mechanism can operate at shallow depths (<900 m) or at much greater depths. Shallow triggers are usually important in initiating salt flow during or soon after deposition of the salt unit, deeper triggers can initiate salt flow even after the unit has been buried for millions of years. Deep salt flow is triggered by tensional or compressional stresses, or by burial of salt to the critical depth where it becomes plastic. At the present time deep-seated mechanisms related to compression are triggering salt flow in the Zagros Mountains (Woodbury et al., 1980).

Triggers are commonly tied to some sort of irregularity in the basin located in the mother salt bed (source layer), in the overburden, or in the underlying section as basement faults (Fig. 6.2). Irregularities in the salt bed include lateral changes in thickness, density, viscosity, or temperature and are often related to facies changes in the salt unit (Fig. 6.2a). A change from an area dominated by halite (2.2 gm/cm^3) to an area dominated by carnallite—sylvite (1.7 − 2.0 gm/cm^3) gives a density and a viscosity contrast. In a rift graben the thickness change from the thick central zone of bedded evaporites to a thinner margin is another irregularity triggering salt flow. Still other irregularities can be fault imposed; these faults can be deep-seated and involve basement rocks (Jenyon, 1986a, b), or thin-skinned where the fault soles out in the salt layer (Fig. 6.2b).

Irregularities can also be fold induced, with the salt layer acting as a zone of detachment. In a subsiding basin the salt slips down slope into the deeper portion of the basin to create the fold-induced thicks and thins in the buckled frontal zone of a gravity slide (Fig. 6.2c). The Mexican Ridges in the Gulf of Mexico are examples of gravity slides. Gravity spreading is another mechanism for creating fold-induced thickness irregularities (Fig. 6.2c). The Sigsbee Scarp is a good example where salt domes are forming in the toe of a huge thrust nappe of Jurassic Louann salt as it spreads southward into the deep Gulf of Mexico. Mesozoic salt now covers Quaternary and Tertiary sediments to a horizontal distance measured in tens of kilometers (Humphris, 1979).

Regional compression can also create buckling and folding in the salt layer and so instigate salt structures (Fig. 6.2c). Active fold-induced diapirs outcrop in the Zagros Mountains of Iran, where Precambrian Hormuz salt is currently extruded as salt glaciers. The glaciers flow out of outcropping salt domes which rise as gleaming hills up to 1300 meters (3900 ft) above the surrounding desert. They are solid rivers of salt which flow for kilometers away from the feeder stock (Kent, 1979). In the Zagros Mountain region the African and Eurasian plates are overriding one another and have been colliding since the mid-Miocene. This has created a huge bellows squeezing salt to the surface. The current phase of salt extrusion began in the late Tertiary, but this latest episode of extrusion began long after diapirs of Hormuz Salt first reached the surface. Hormuz salt surfaced in the Arabian Gulf region during the Jurassic to mid-Cretaceous (Ala, 1974). At that time it influenced sedimentation patterns of many of the concurrent sediments and associated reservoirs (personal observation). Today, the folding in the Zagros Mountains drives salt to the surface at the staggering rate of 170 km/m.y., a rate so

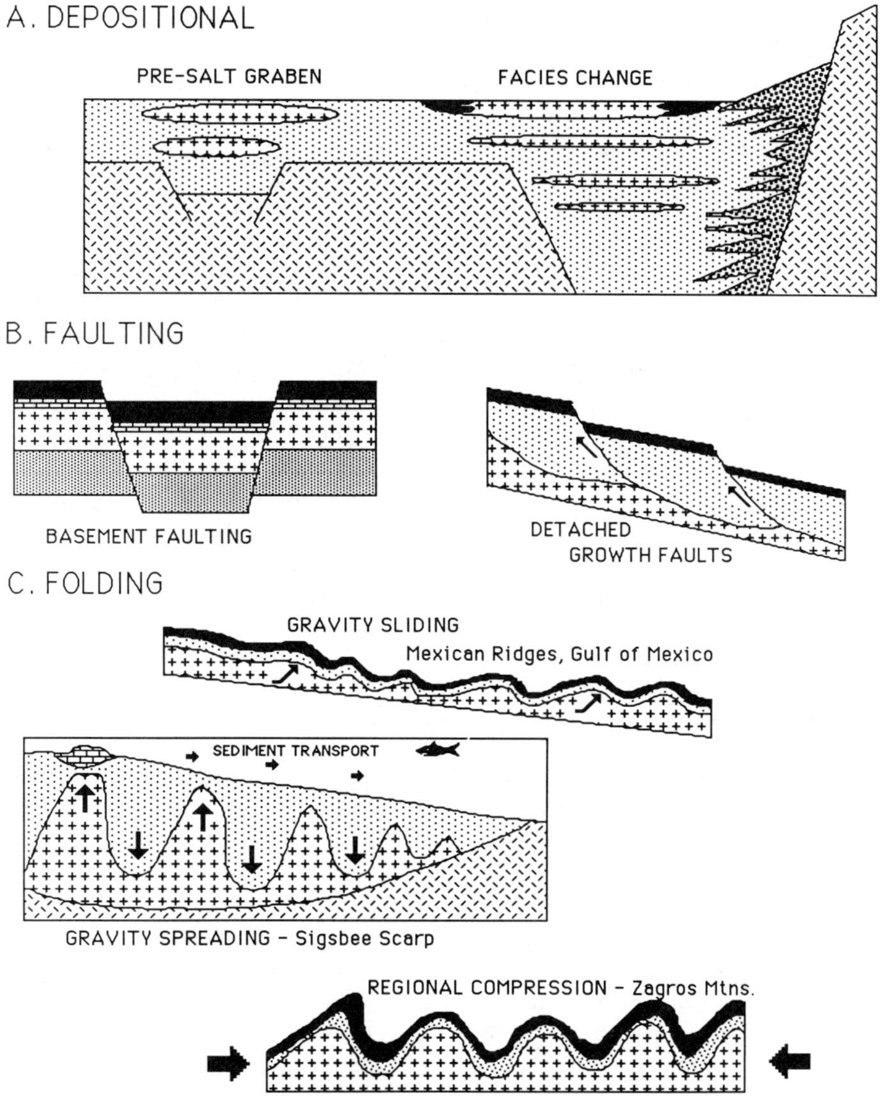

Figure 6.2 Mechanisms that trigger salt movement: (a) depositional triggers related to facies changes in lithology, thickness, and density; (b) faulting irregularities which may involve basement rocks or may sole out in the salt layer; (c) fold-induced triggers initiated by gravity slides, gravity spreading, and regional compression.

fast that no known depositional rate can possibly keep pace (Table 1.4). At Kuh-e-Namak (Mountain of Salt) in Dashti province in Iran, no sediment is laid down atop the actively flowing salt of the glacier, until a distance of some kilometers out from the zone of extrusion.

GROWTH OF SALT STRUCTURES

Once triggered, a salt structure can grow by a number of different mechanisms. The depth of burial usually controls which mechanism or group of mechanisms exerts most influence on salt flow. Mechanisms are divided into two groups, a shallow group controlling salt flow at shallow depths of less than 600 m (2000 ft) and a deeper group which is most important below such depths.

At shallow depths (<600m) three major mechanisms are known:

1. *Differential loading*. Differential loading on top of the salt layer often occurs in areas of reefs and shoals or where deltas and/or submarine fans are prograding as depositional lobes over an underlying salt layer (Fig. 6.3a).
2. *Gravity glide folding*. This causes the passive rise of salt into growing anticlines. In the Gulf of Mexico this mechanism commonly operates beneath the deep waters of the continental slope and margin (Fig. 6.2c).
3. *Buoyancy*. This mechanism usually dominates at depths of >600 m (Fig. 6.3b), but can operate at shallower depths under denser overburden. Such conditions can be found beneath well-cemented carbonate layers and reefs where densities often exceed 2.2 gm/cm^3 only tens of meters below the surface.

At depths greater than 600 to 900 meters (2,000 to 3,000 ft) the effects of differential loading are not so important. There the major mechanisms are:

1. *Buoyancy*. This occurs when sediments atop the salt are buried and compacted to where the overburden density exceeds that of the salt creating a density inversion (Fig. 6.3b). The depth of inversion varies according to the type of overburden. In the Gulf of Mexico the density inversion occurs around 600 m subbottom, although in overpressured shales it may not be reached until 900 m. Bishop (1978), a strong proponent of differential loading as the major mechanism of salt growth, argues that density inversion may not occur until 5.5 km (18000 ft).
2. *Differential loading*. Differential loading continues to drive salt movements at depths greater than 600 m in many depotroughs (Fig. 6.3a).
3. *Gravity spreading*. Salt spreading and flow beneath a prograding continental margin causes diapirism on the continental slope at subbottom depths greater than 600 m as salt rises into growing anticlines. The Sigsbee Scarp is the type example of this mechanism (Humphris, 1979).
4. *Thermal convection*. This process increases the rheidity of salt below depths of 3 km and probably increases the rate of movement of deep salt (Fig. 6.3c).

Halite is an unusual mineral, as it can become slightly less dense with deep burial. At a depth of 5 km (17000 ft) under a geotherm of 30°C/km halite expands

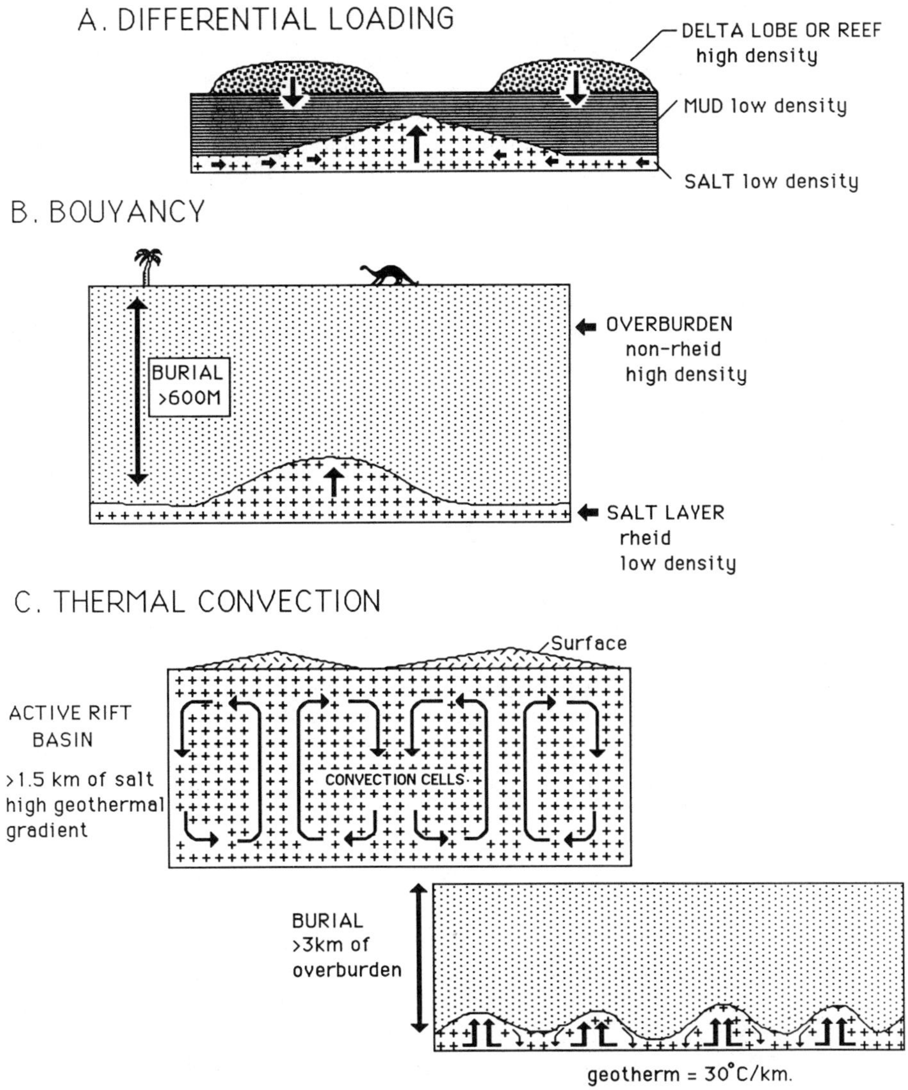

Figure 6.3 Mechanisms of salt structure growth: (a) differential loading; (b) buoyancy or density inversion caused by the burial to a depth where the overburden is denser than the underlying salt; (c) thermal convection in areas of high geothermal gradient salt can flow at much shallower depths (in part after Seni and Jackson, 1984).

2% due to heat and contracts 0.5% due to pressure. In other words, it becomes less dense with burial in contrast to all other sedimentary minerals (Jackson and Seni, 1984). Under such a geothermal regime salt flow should be induced thermally at depths of more than 1.5 km (5000 ft). In areas with elevated near-surface temperatures and high geothermal gradients such as the African Rift, thermally induced salt movement can be important in the near surface. In Ethiopia's Danakil depression the volcano-shaped salt mounds associated with hot springs may well be the tops of convecting plumes of salt. Gussow (1968) argued that the Iranian salt glaciers are catastrophic extrusives emplaced in a similar way to a lava flow. He thought that the salt was emplaced hot from depth and now the salt glaciers have cooled and salt is no longer flowing. More recent work by Kent (1979, 1987) and Talbot (1979) indicated that the salt is still flowing and that hot intrusion is not the major mechanism of salt emplacement. Regional compression is much more important in driving salt to the surface.

Whatever mechanisms control salt movement, a salt structure grows by moving upward relative to the surrounding sediment. Crests of many salt structures are shallow through much of their growth history and often influence sedimentation in the adjacent and surrounding sediment. The tops of shallow salt structures remain at about the same position in earth space while the roots of the structures are built downward. This process is called "downbuilding," a concept first hypothesized by Barton (1933). Syndepositional sediments sink down in subsiding inter-salt-structure troughs, down past the adjacent growing salt mass. Sediment fills in the trough as it replaces the underlying salt which moves up into the adjacent growing salt structure. The subsiding areas are sometimes called salt withdrawal basins or rim synclines and are marked by syndepositional sediment thicks. In the last 5 to 10 years a growing body of evidence has been published on the syndepositional effects of salt flow on sedimentation above the crest of growing salt structures as well as in the troughs (Humphris, 1979; Bishop, 1978). It supports downbuilding as the mode of growth of many shallow salt structures, although some geologists still believe that salt masses ram their way to the surface by intruding the overlying sediments in a fashion similar to a volcanic plug. Reality is probably somewhere between the two extremes: when salt movement was initiated soon after the deposition of the mother salt bed, episodes of downbuilding were separated by times of intrusion or quiescence; when salt flow began millions of years after the mother salt bed was deposited, the initial flow was intrusive, once the crest of the salt structure was near the surface downbuilding became the dominant mechanism.

The best way to distinguish which mechanism dominated in a particular salt structure is to isopach or seismically map the thickness of units around the root of a salt structure. Downbuilding creates sediment thicks associated with salt withdrawal and thins near the crest of the growing salt structure (Fig. 6.4). Intrusive processes do not affect greatly the thickness of the units overlying the mother salt bed. Often, one can time the salt movement using a seismic section by looking for the first sediment-thick subjacent to a salt structure—this is the time that the salt first started to flow into a structure. Using this method in a number of salt provinces, it appears that some salt units have been deeply buried before salt flowed into a near surface position by intrusive movement (personal observation; portions of the Zechstein in the Danish North Sea and the Hormuz in the Trucial

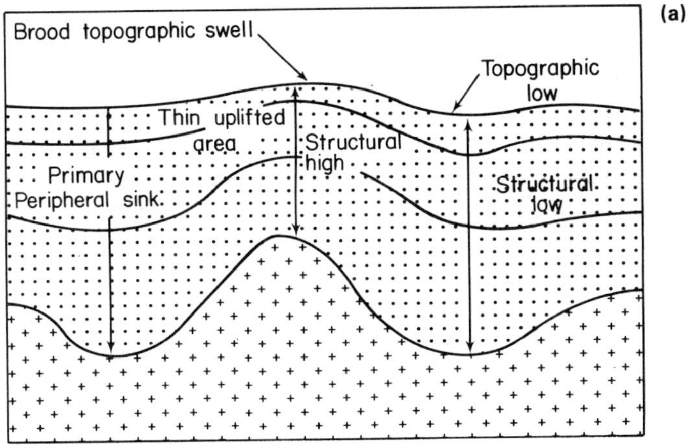

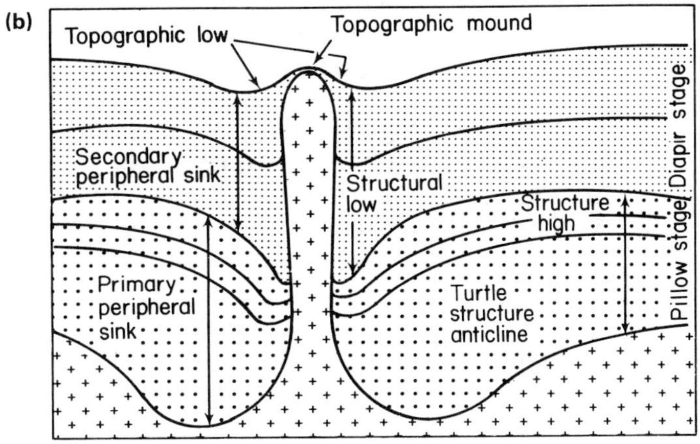

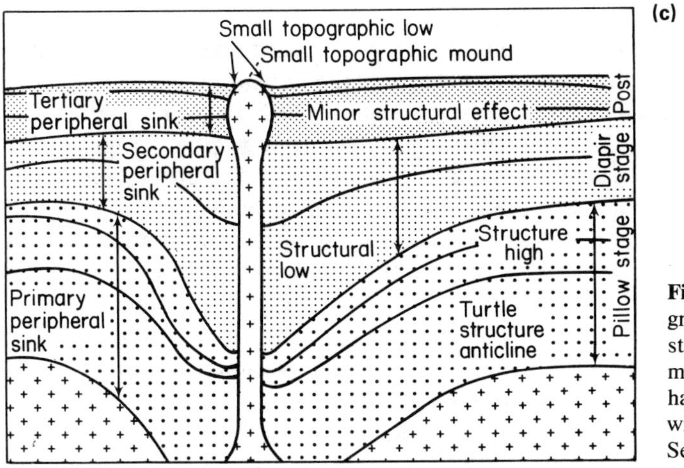

Figure 6.4 Stages of salt structure growth: (a) pillow stage; (b) diapir stage—associated with the development of turtle structures and overhangs; (c) post diapir stage-associated with the development of cap rock (after Seni and Jackson, 1983a).

States of the Arabian Gulf). Other salt units moved during the deposition of the immediately overlying unit (e.g., the Smackover Formation in the Gulf of Mexico). Once flowing salt gets near or at the sediment depositional surface, it usually maintains its position in earth space until it is cut off from the mother salt bed. Downbuilding is often the dominant process during this time of near-surface salt emplacement, although episodes of short-lived intrusion at the salt crest can interrupt the longer-term downbuilding mechanism.

As a salt structure grows, it passes through three stages:

1. Salt pillow stage
2. Salt diapir stage
3. Post diapir stage

The successive growth stages are seen in Fig. 6.4. Not all salt structures will grow to stage 3, and many reservoirs have been drilled above salt pillows.

The *pillow stage* is a nonpiercement stage; sediments above the pillow are thin over a broad equidimensional area with maximum thinning over the crest of the structure (Fig. 6.4a). The area above the pillow is often a topographic high, while the area between the pillows is often a topographic low or primary rim syncline created by salt withdrawal. Rim synclines are often areas of preferential sedimentation, that is, sediment thicks. In the East Texas Basin in the Gulf of Mexico, the percentage of stratigraphic or seismic-stratigraphic thinning ranges from 10 to 100% in units above pillows composed of the Jurassic Louann Salt (Seni and Jackson, 1983a,b).

By the *diapir stage,* a piercement stage, the rate of salt movement is so rapid that sediment is usually not preserved over the crest of the structure. Salt in the crest of the diapir is near or at the surface throughout the diapir stage; it is continually dissolved in the shallow subsurface, or extruded as a salt glacier (Fig. 6.4b). In the East Texas basin, about 70% of the salt in the central diapir province was calculated to have been lost, of which 12% was lost by subsurface dissolution, and 88% by extrusion and erosion (Seni, 1984). The maintenance of the diapir crest at or near the surface means that the diapir stage is characterized by many unconformities and thinning of strata in the immediate vicinity of the diapir. On the continental rise and slope off North Carolina the topographic irregularities created by diapiric growth have generated associated zones of slumping and shallow faulting (Cashman and Popenoe, 1985). Depressions form in areas adjacent to diapirs above areas of salt withdrawal near the salt neck. These secondary rim synclines are located closer to the salt structure than the primary rim synclines of the pillow stage. They are equidimensional to elongate in plan; equidimensional rim synclines indicate that salt flowed into the diapir from all directions. In the East Texas Basin, the secondary rim synclines contain sedimentary units up to 215% thicker than their off-structure equivalents (Seni and Jackson, 1983b).

Overhangs also form during the diapir stage; they occur where part or the whole of the shallower portion of a diapir extends laterally beyond the cross sectional area of the diapir root (Figs. 6.17 and 6.18). Mechanisms that form salt diapir overhangs are poorly understood, in the past they were often interpreted as buried salt glaciers. The modelling of Talbot and Jackson (1987) suggests mature salt structures can form overhanging crestal bulbs (balloon or mushroom shapes) well below the land surface or level of neutral buoyancy. Such bulbs are due to toroidal circulation of salt below the crest of the salt structure. Jenyon (1987) used seismic data of the North Sea Zechstein basin to show that some overhangs in the area formed in the later stages of diapir emplacement by the lateral injection of salt beneath relatively thin near-surface sediments. He also showed that the salt in the overhang could be later dissolved, but the effects of the injection were preserved as depositional patterns in the overburden sediments, specifically infill of a two-stage collapse graben.

Turtle structures also form in the diapir stage. They grow as the locus of maximum salt withdrawal moves closer and closer to the growing salt mass. This means that the rim syncline also migrates toward the salt structure. As the area of salt withdrawal migrates closer and closer to the stem of the salt structure, the associated zone of subsidence moves closer and closer to the salt neck. Thus the area of primary rim syncline of the pillow stage becomes a zone of relatively thicker sediment (Fig. 6.4b). Continued salt withdrawal and subsidence around the edges of the primary rim syncline creates a turtle structure anticline. The crest above a turtle structure can be a syndepositional high controlling sedimentation (Fairway Field, East Texas basin).

The last stage of salt structure growth is the *post diapir stage*. By this time the salt supply is dwindling as the volume of salt flowing into the foot of the salt mass decreases (Fig. 6.4c). The crest of the structure is once again buried, but still near the surface. It now feels the effects of meteoric waters as a caprock carapace forms by dissolution and alteration to encase the upper few hundred meters of the structure. The buried crest means that sediment can once again be deposited over the crest of the structure, and small, localized tertiary rim synclines can be seen adjacent to the diapir. The outcropping salt domes of Yas Island and Jebel Dhanna in the Arabian Gulf are post-diapir stage structures (Kent, 1987).

The relationship between salt withdrawal, salt structure growth, and sediment deposition is a good example of a feedback system, and, as shown in the next section, salt movement controls sedimentation while the sedimentation regime is controlling salt structure development.

MODERN SALT STRUCTURES

Gulf of Mexico

Feedback between salt structure growth and sedimentation is seen along the continental margin of Texas and the other Gulf coast states (Fig.6.5). In the Texas coastal zone about 56% of the post-diapir-stage structures possess a positive relief

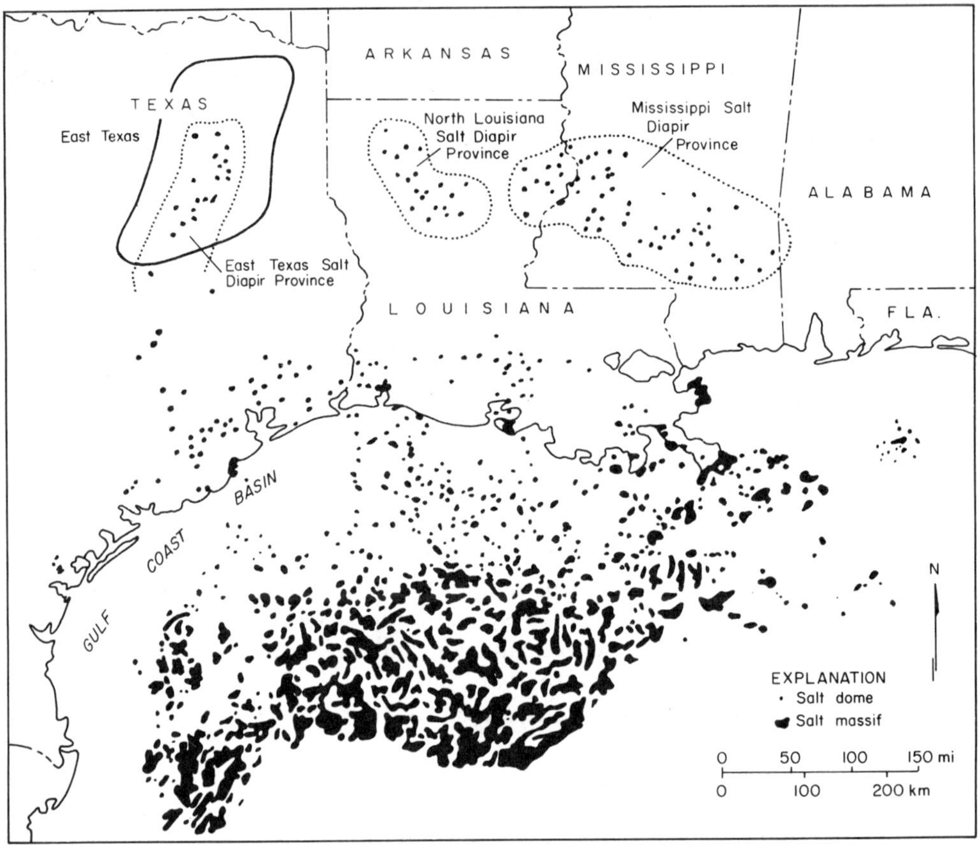

Figure 6.5 Distribution of salt structures in the Gulf of Mexico (after Martin, 1976).

of more than 5 ft (1.5 m) at the ground surface above the structure (Seni and Jackson, 1983a). This relief strongly influences Holocene sedimentation, as salt structures always occur in sand-poor areas or along the edges of sand belts. Topographic highs above coastal salt masses have channeled sand into the areas between the salt stocks, which are themselves capped by muddy sediments. In the Houston-Galveston area Plio-Pleistocene sands form stacked channels between the salt structures. Areas above the crests of the structures have been mud-rich throughout this time. The lack of relief over the remaining 44% of the structures in the coastal zone is due to the greater depth of burial of the structures, to the cessation of any upward growth, and to the dissolution of the upper portions of the salt structures.

In the offshore Gulf of Mexico numerous active salt structures are growing along the boundary of the continental shelf and the upper continental slope (Fig 6.5). Most of the hills and knolls on the continental slope are underlain by salt structures (Trippet, 1981). Deeper in the Gulf waters abyssal plain seamounts are cored by salt structures (e.g., Sigsbee Knolls). The Pleistocene sediments above these structures have been raised 150 to 300 m above the surrounding seafloor by salt movement (Ewing and Ewing, 1962). Diapirs, especially shallow piercement diapirs in the Gulf of Mexico, tend to grow in pulses (Woodbury et al., 1980). Short, shallow upthrusting episodes are superimposed on salt masses whose predominant growth mode is downbuilding. Prior to upthrusting, the diapir crest may be covered by a few tens of meters of sediment. Once a new pulse of salt moves into the diapir, it drives salt to the surface, where it is extruded as a salt glacier. The salt quickly dissolves on or just below the seafloor to leave a zone of dissolution atop the diapir. Seismically this is often seen as a zone of collapse or a graben on the structural crest (Lohmann, 1979). Growth pulses uplift the overlying strata, which are then removed by erosion, creating the numerous unconformities above and on the flanks of a diapir. As a general rule, sediment is not preserved above the crest of an active salt structure.

Along the well-studied northwest Gulf of Mexico, diapirs are more common on the upper continental slope than on the continental shelf (Humphris, 1979). Depositional loading by lobes of siliciclastic sediment is moving the Jurassic Louann Salt into salt ridges and then into diapirs from the top of these regional ridges (Fig. 6.6). Below regional tension faults, many diapirs coalesce at depth into salt ridges and walls (Trippet, 1981). The initial movement was basinward gravity sliding, forming low-amplitude, deep-seated salt ridges and pillows. Subsequent differential loading beneath fluviodeltaic and submarine fan lobes moved salt from the pillows into diapirs. Regionally, the ridges are aligned nearly perpendicular to the shelf edge, their distribution a result of salt squeezing out from under local depositional basins or individual depositional prisms. Local ridge orientation is related to a general pattern of feedback between deposition and diapirism in the Gulf sediments. As the shelf progrades, the older diapirs on the shelf influence the direction of the stream flow. When the streams prograde out to the shelf edge, the delta builds out on to the upper slope. The weight of this thick sediment prism activates salt movement around the edges of sediment wedges (Fig. 6.2c). This, in turn, creates a depositional basin as sediment ponds against the edges of the growing salt ridges and diapirs. The basin subsequently fills with additional sediment, which then spills over the edge of the shelf to form a new sediment wedge and so repeats the cycle. This style of shelf progradation is called marginal accretion (Trippet, 1981).

Marginal accretion has been responsible for as much as 80 km (50 miles) of post-Pliocene shelf edge progradation along the northwest Gulf of Mexico. Salt movement on the present upper slope seems to have reached a maximum level in the Pleistocene, whereas at present the lower slope is in the early stages of activity (Martin, 1976; Humphris, 1979). Movement of the Jurassic Louann Salt has controlled shelf sedimentation in the Gulf of Mexico from the Jurassic until today. Jurassic salt pillows and diapirs were locally important in controlling the depositional style of the Norphlett and Smackover Formations, two units which directly

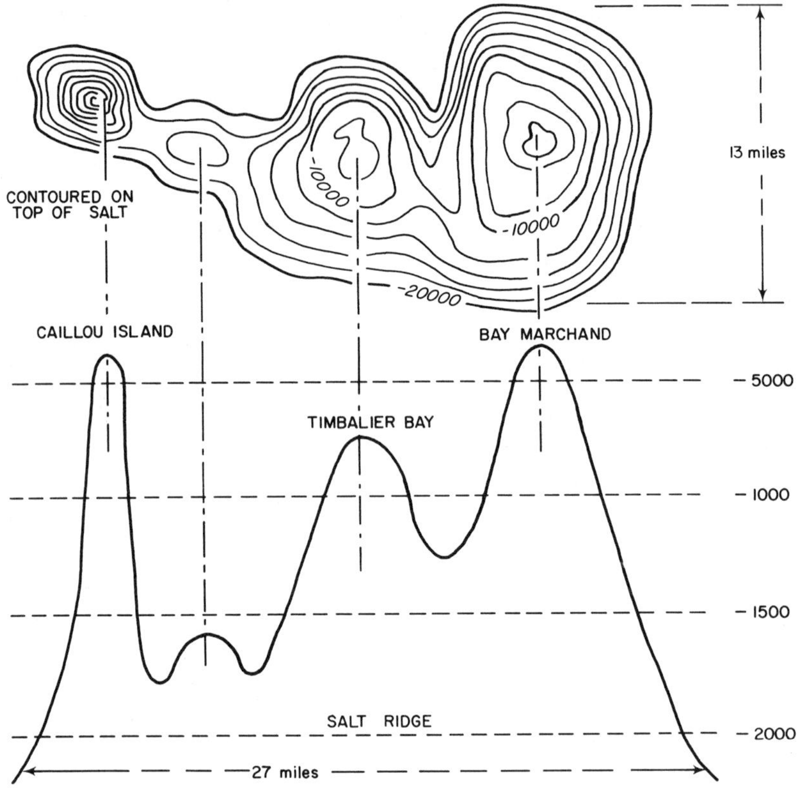

Figure 6.6 Plan and cross-sectional view of a salt ridge which has developed into a series of diapirs: the Caillou Island, Timbalier Bay, and Marchand structures, Lafourche Parish, Louisiana (after Halbouty, 1979).

overlie the Louann Salt (Fig. 5.14a). As the continental margin has prograded basinward for the last 150 million years, so has the locus of most active salt movement moved farther and farther into the Gulf.

Salt movement in the Gulf not only affects siliciclastics but has created conditions on the upper slope which favor carbonate reef deposition in an otherwise deep water setting (Fig. 6.7). The effects of salt movement on carbonate deposition are best seen in the Flower Garden Banks in the northwest Gulf of Mexico, where the rising hummocks above salt structures have developed into coralgal reefs (Rezak, 1977, 1985; Fig. 6.8). On the south (seaward) side, the West Flower Garden Bank rises from a depth of 136 m (450 ft) and is capped by a living reef which has grown to 20 m (65 ft) below the sea surface. The reef marks the northernmost extent of coralgal reef growth in the Gulf of Mexico. The living reef, a *Diploria-Monastrea-Porites* boundstone assemblage, rises from a depth of

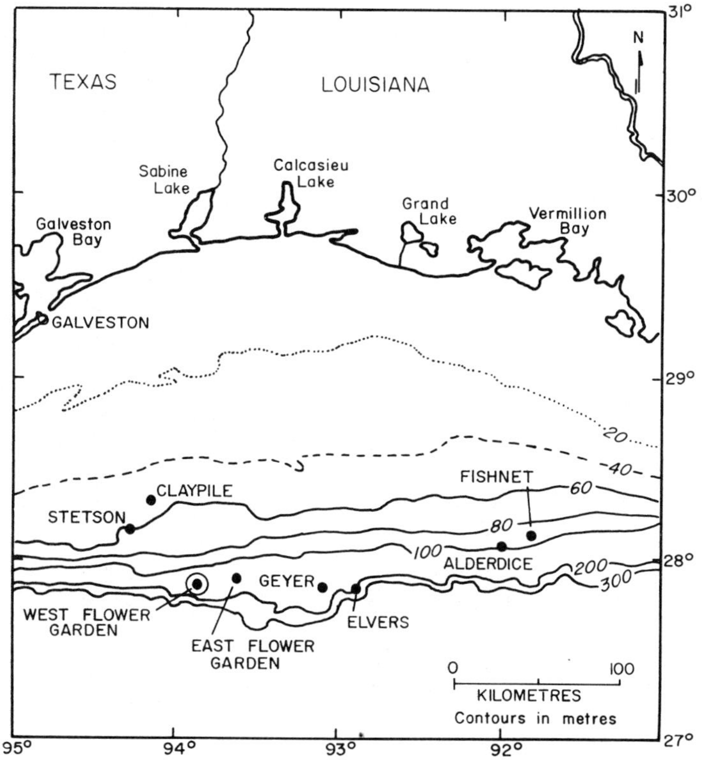

Figure 6.7 Location of major banks related to salt structures on the continental slope of the northern Gulf of Mexico (from Rezak and Bright, 1981). Note the major shoals atop salt structures, including West Flower Garden Bank, are located on the upper continental slope.

45 m (150 ft) with forereef slopes ranging from 45° to vertical. Below 45 m the reef sediments are deposited in successively deeper, concentric bands. Immediately below the living reef at depths of 45 to 51 m (150 to 167 ft) is a band of coral debris. Next comes a zone of algal nodules at depths of 51 to 73 m (167 to 240 ft). Below this at 73 to 91 m (240 to 300 ft) is an *Amphistegina* (benthonic foram) zone. Finally, well away from the reefs, at depths greater than 91 m (300 ft), is the quartz-planktonic foram zone characteristic of the upper slope of the Gulf of Mexico.

These carbonate buildups on the Flower Garden Banks are there only because of the shoaling effects of salt movement. In the East Flower Garden Bank a brine lake was created by salt dissolution in the central part of the structure (Rezak and Bright, 1981). Collapse over the crest of the structure has generated the graben which houses the lake. Healthy growing corals surround the lake except where a canyon cuts across the bank to the deep open seafloor. Halokine-

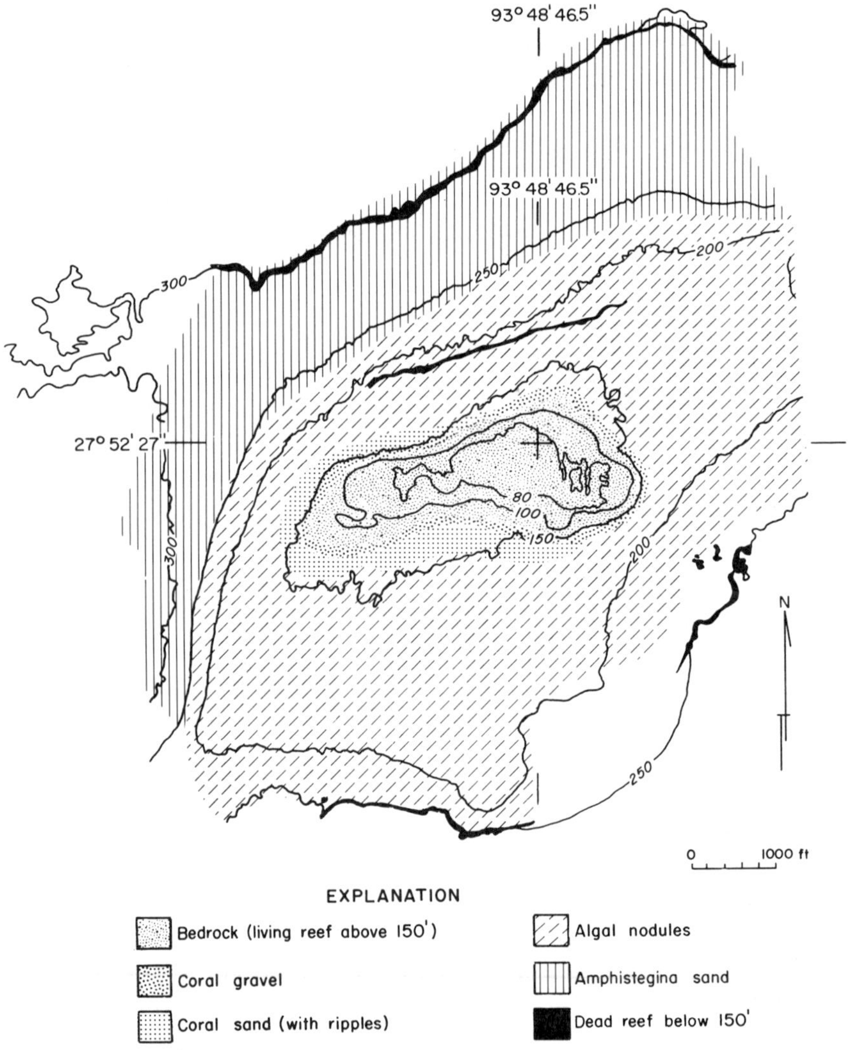

Figure 6.8 Distribution of major carbonate facies on the West Flower Garden Bank, Gulf of Mexico (after Rezak, 1977) showing concentric depth-related control of facies and an earlier Pleistocene reef at 150 ft depth.

sis beneath the Flower Garden Banks has converted an otherwise muddy siliciclastic platform into a region suitable for carbonate deposition. Shoaling has lifted the seafloor into a clear, well-lit environment and created a potential carbonate reservoir in what is otherwise a muddy, siliciclastic province.

Arabian (Persian) Gulf

The Arabian Gulf in the Middle East is another area where growing salt structures are modifying Holocene deposition in what is otherwise a carbonate ramp setting. Today the Arabian Gulf is a shallow epicontinental sea, a large portion of which is underlain and intruded by Precambrian Hormuz Salt (Fig.6.9: Purser,

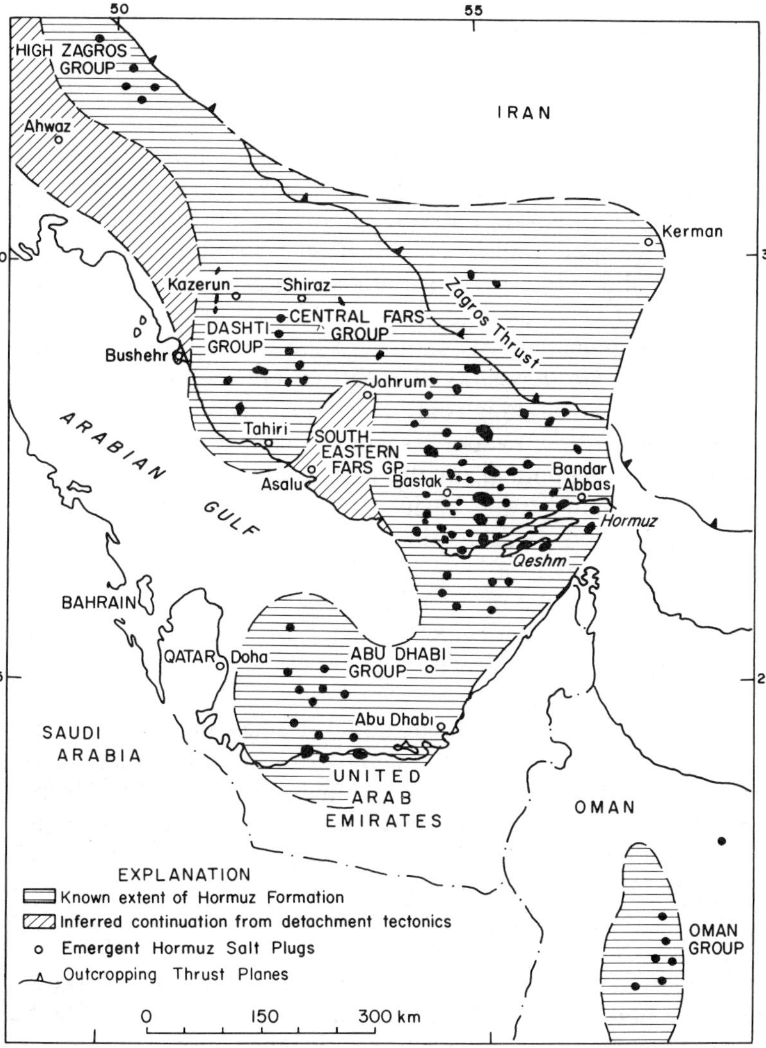

Figure 6.9 Distribution of Hormuz-Formation-related salt structures in the Arabian Gulf (after Kent, 1979).

1973a, b; Kent, 1979, 1987). To date, little work has been done on the interaction between Holocene carbonate deposition and salt movement, but when studied in detail this area will become the "type area" for modeling feedback between carbonate deposition and halokinesis.

The salt-cored bathymetric highs of the Arabian Gulf can be divided into three classes based on relative position offshore and associated depth in the Gulf (Purser, 1973b):

1. Outer – basin center
2. Intermediate – basin
3. Coastal

The basin center highs can be subdivided further into two types. First are the highs with crests below the wave base; these are capped by grainstones of perforate foraminifera and surrounded by the lamellibranch muddy sands and muds of the open seafloor (Fig. 6.10a). Second are the basin center highs whose crests are above the wave base; these are capped by porous and permeable coralgal reefs and associated grainstones (Fig. 6.10b). Both types of high display concentric sediment patterns similar to the buildups in the Gulf of Mexico. At the other end of the depositional spectrum are the coastal highs (Fig. 6.10c). Sediments on the coastal highs are pelletal sands and carbonate muds with restricted faunas, algal mats, evaporites, and sabkha sequences. Sediment distribution is asymmetric, dependent on the interactions of wind and waves, and so is less predictable than on the basin center highs. In the coastal zone the bioclastic grainstones tend to accumulate on the leeward side of the high. Some of the coastal highs, such as Yas Island, are rimmed by shallow depressions, which favor the accumulation of finer-grained sediments compared to the surrounds (Fig. 6.11). Salt-rafted blocks carried up from depth litter the immediate area around the salt structure, and because of their coarse size tend to remain close and downwind of the high.

The multiplicity of depositional settings above salt structures in the Arabian Gulf gives an insight into the geometries of carbonate reservoir sands about ancient salt structures. Basin center highs surrounded by deep water do not show the effects of widespread current reworking. These structures shed shallower water sediments directly into deep sub-wave base depths. Facies patterns tend to form concentric rings, with the best sorted sands or reefs at or near the crest of the structure. For its ancient equivalent one would drill on or near the crest of the structure. Intermediate and coastal highs are surrounded by waters of moderate to shallow depths, and now the sediment is often reworked by tides and other currents both across the high and in the areas surrounding the high. Often, this type of structure generates a down current sediment tail and a strong asymmetry to any reef or sand buildup on the structure itself. The sediment tails can be quite thick; off Dalma island the downwind tail attains thicknesses of 20 m (Fig. 6.10c). Drilling for porous units on the ancient equivalent of this type of structure would

(a)

(b)

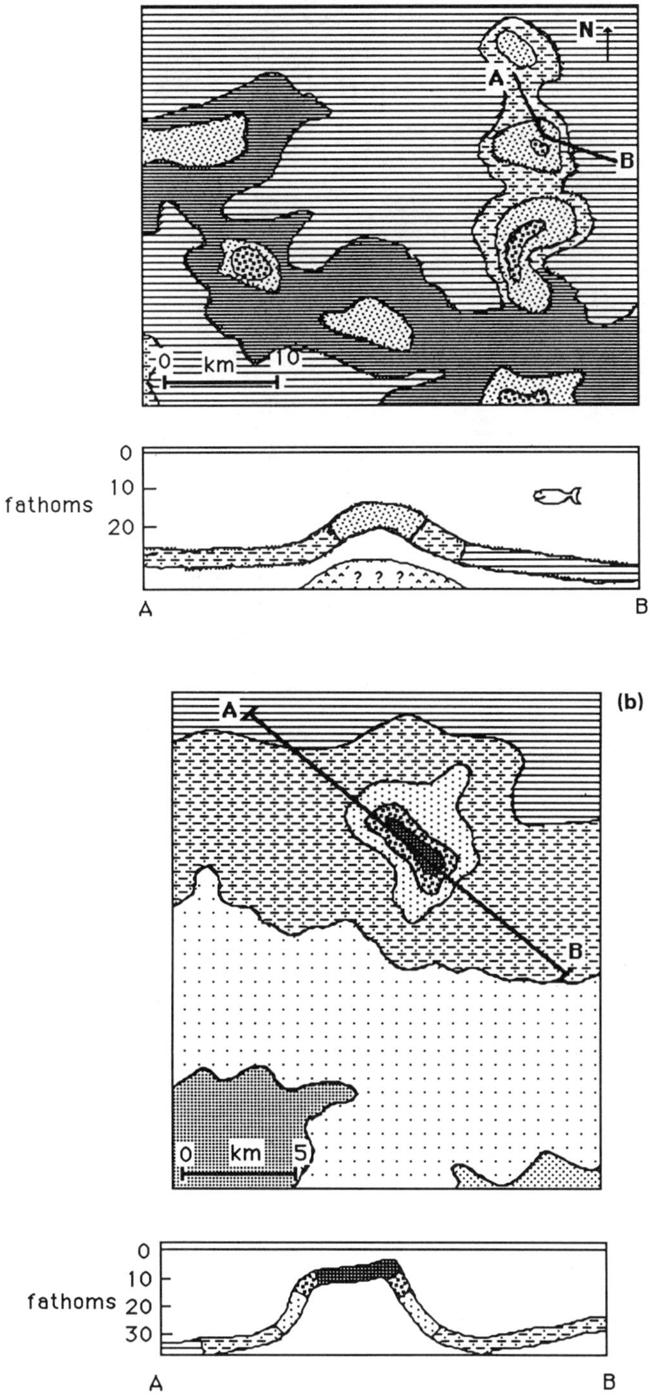

Figure 6.10 Sediment distribution about salt structures in the Arabian Gulf: (a) sediment patterns on highs east of Halul Island, northeast of Doha (note the concentric facies patterns); (b) sediment patterns on Abu Thama shoal north of the Qatar Peninsula (note the capping coralgal sand/reef); (c) sediment pattern on Halat Dalma shoal (Fig. 6.11 for locality) with an asymmetric sediment pattern and the reef on the seaward side (after Purser, 1973b).

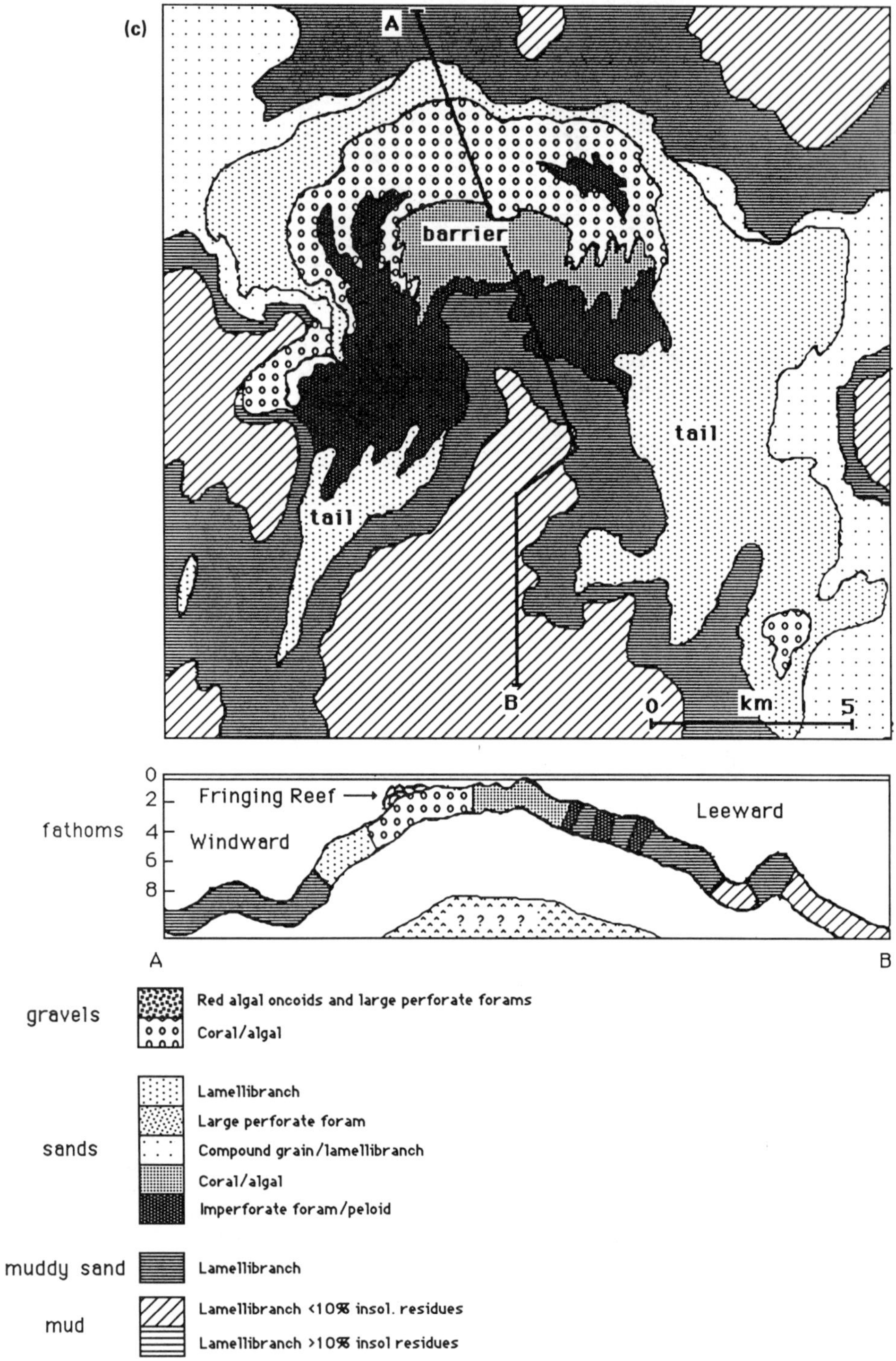

Figure 6.10 (Continued)

involve drilling both on or near the crest of the structure as well as expecting and drilling sand trends which cut across the structural crest. Twenty-meter-thick sands encased in muds and sabkha evaporites are not beyond the resolution of modern high-frequency seismic. Sediment tails about Arabian Gulf salt structures often trend in the same direction as the prevailing winds of the region (Fig. 6.11).

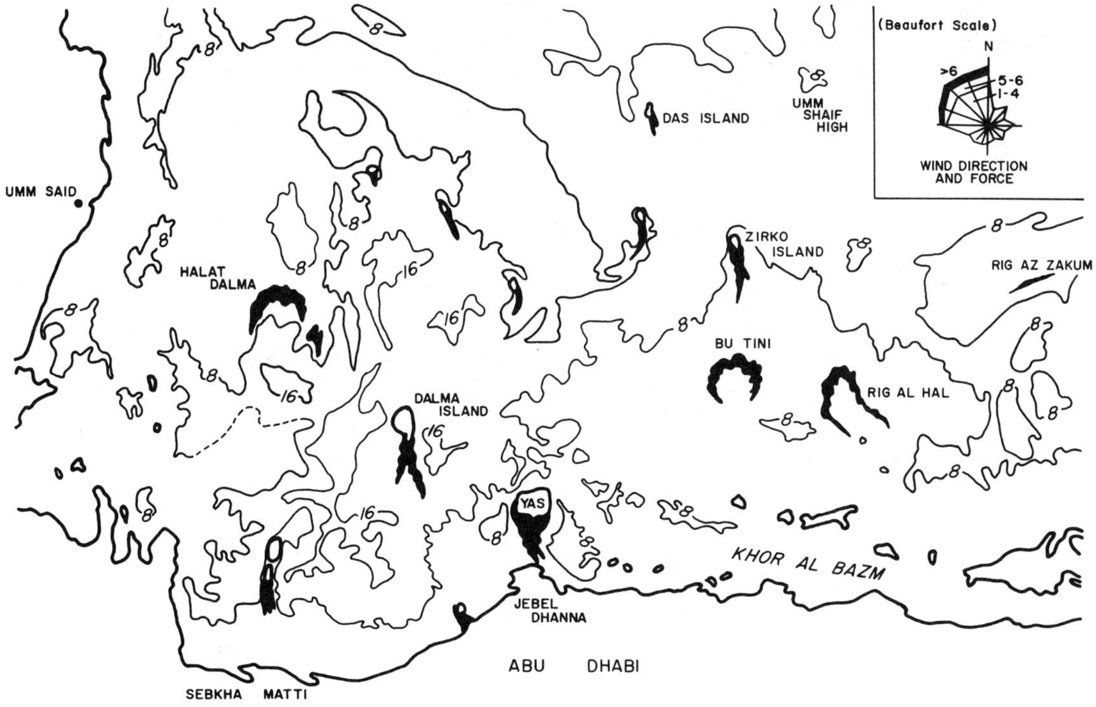

Figure 6.11 Relationships between wind direction and the orientation of sediment tails in the southeast Arabian Gulf (after Purser, 1973b).

This is a useful predictive tool when exploring for sediment tails atop ancient salt structures in shelf settings. Once the position and trend of an ancient grainstone unit is established about an individual salt structure, grainstone orientation and position of potential reservoirs about other diapirs in the region can be predicted. In coastal settings in the Arabian Gulf the sediment trend is often modified by coastal transport processes. Leeward sediment tails often extend from the high to the shoreline (Yas Island), generating a progressive shallowing between the high and the mainland. Shallowing increases the velocity of tidal currents across the shoal and deposits oolite shoals, trending at an oblique angle to the trend of the sediment tail (Fig. 6.12).

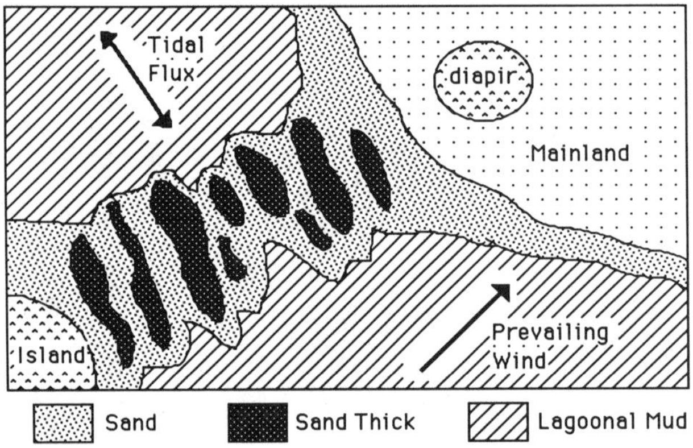

Figure 6.12 Schematic of the interactive effects of wind and tide on sediment distribution. Longshore and tidal currents are redistributing sand on wind aligned shoals.

MODELS OF SEDIMENTATION

Salt structures can grow beneath depositional settings ranging from the continental to the deep marine, although they are more common on the continental margin. To understand sedimentation associated with salt structure growth the setting must be classified in relation to the stage of salt growth and the regional depositional setting (Fig. 6.13). Remember that for many salt structures the area above the crest of the structure is a topographic high for all growth stages: the pillow, diapir, and post-diapir stages. This controls sedimentation in various depositional settings.

In shelf settings the sediments deposited atop a growing salt pillow are often deposited above the wave base and in the photic zone (Fig. 6.13). In a siliciclastic province this may lead to the development of current-deposited, well-winnowed sandstones above the pillow structure. In a carbonate province, current-dominated skeletal and oolitic grainstones as well as reefs can form above the pillow. In the intervening areas between the pillows, the quieter-water, low-permeability, shelf muds are deposited. In a continental setting, the high areas above a pillow crest are sites of muddy sedimentation. In a siliciclastic province, the areas between the pillows are often composed of stacked fluviodeltaic sands; high-porosity, high-permeability units pinching out about the edge of the rim syncline. In a carbonate province the areas above the pillow are mud-rich supratidal flats or sabkhas, often with thick porous grainstones wedging out into the sabkha. Subaerial exposure of the muddy sediments of the pillow crest can sometimes lead to meteoric diagenesis, dolomitization, and associated porosity enhancement on the crest. In the deep marine setting of the continental slope, the submarine carbonate and siliciclastic sands tend to stack up between the pillows, while the pillow crests are covered with muds and hemipelagic oozes.

By the diapir stage, a piercement stage with active salt extrusion, little or no sediment is preserved over the crest of the structure, whatever the setting (Fig. 6.13). Pulse-like diapir growth creates numerous unconformities and sediment

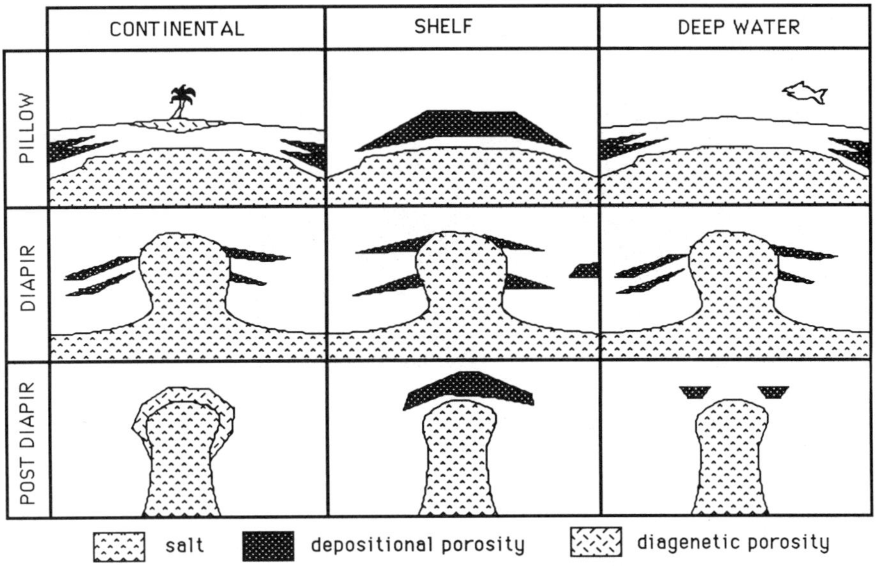

Figure 6.13 Schematic cross sections showing the control of salt structures on the distribution of depositional and diagenetic porosity. Porosity distribution changes with the stage of salt structure development and the depositional setting, continental versus shelf versus deep water.

thins adjacent to the topographic high of the diapir. Fluviodeltaic and submarine sands stack up in the secondary rim synclines. In a shelf setting, the sediments deposited in the secondary rim syncline are muddy, although coarser-grained sediment may be deposited adjacent to the diapir. Coarse-grained shelf sediment and/or reefs can form on areas above turtle structures. Such structures form between the diapirs in areas above the primary rim syncline of the preceding pillow stage.

By the post-diapir stage, the salt once again is just below the surface and shoal water facies can once again be deposited over the crest of the structure. In fact, the various sediment types developed in the pillow stage are once again developed over the post-diapir-stage structure, although on a much smaller scale. In the Arabian Gulf the reefs and sabkhas of Yas island and Jebel Dhanna are forming on topographic highs above post-diapir-stage structures. The sand pinchouts of the Texas coastal plain mentioned earlier in this chapter are also due to shoaling above post-diapir-stage structures.

This time of dwindling salt supply is also the time of caprock formation. Caprock is a porous limestone/anhydrite carapace found atop many shallow inactive diapirs (Figs. 6.14a and 6.17). It is most common and thickest on domes which are currently close to the surface but can be found on domes buried as deep as 9000 ft (3000 m). The cap is usually 300 to 450 feet (100 to 150 m) thick but can be thicker than 1000 ft (300 m). Caprock is thickest over the crest of the structure

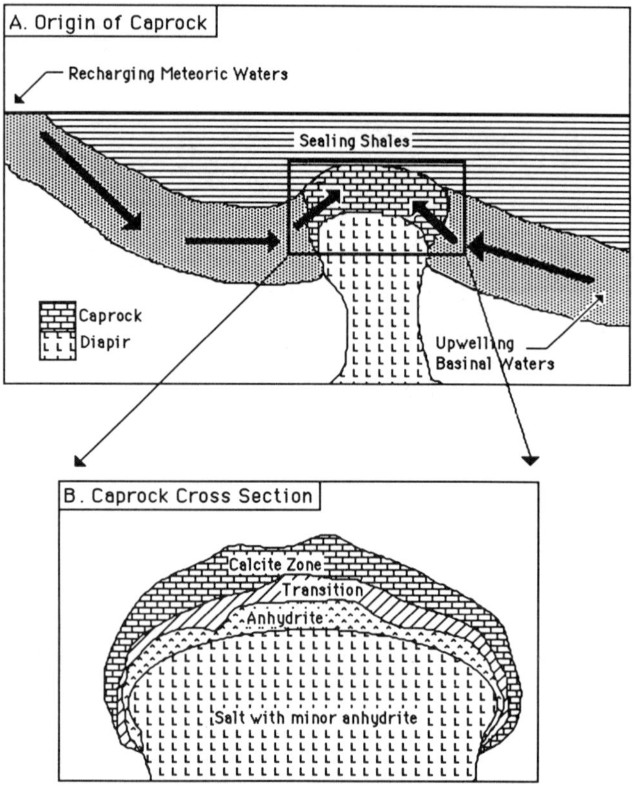

Figure 6.14 Caprock: (a) regional setting showing the importance of regional groundwater flow in forming caprock; (b) distribution of the various caprock facies.

and thins toward the flanks. Occasionally, the steep vertical walls of the diapir stem are also mantled with caprock. At depth the caprock is composed of granular anhydrite and minor halite; at shallower depths the anhydrite grades up into gypsum and limestone with some accessory minerals (Fig. 6.14b). Most workers agree that the anhydrite caprock forms as groundwaters dissolve the anhydrite-bearing halite of the diapir stem, leaving behind the less soluble anhydrite (Galloway et al., 1983). Gypsum then forms by the hydration of the anhydrite. The calcite portion of the caprock forms as anaerobic sulfate-reducing bacteria break down the $CaSO_4$ phase into $CaCO_3$ and H_2S. Sulfur often forms in caprocks and many Gulf of Mexico diapirs contain economic sulfur deposits, some of which have been mined. Some caprocks also contain metal sulfides.

Salt structures can affect the thermal history and pore fluid chemistry of a sedimentary basin. Salt and anhydrite are two of the most thermally conductive sedimentary rocks (Table 6.1). In the Gulf of Mexico coastal plain the thermal conductivity of the diapiric rock salt is approximately four times that of the surrounding sedimentary rocks (Smith and Dees, 1982). This contrast leads to modified temperature isoclines and can lead to drastic effects on the geothermal

TABLE 6.1 THERMAL CONDUCTIVITY RANGES OF VARIOUS SEDIMENTS (AFTER SELLEY, 1985)

Lithology	Thermal Conductivity ($Wm^{-1}{}^{\circ}C^{-1}$)
Anhydrite	5.5
Halite	5.5
Dolomite	5.0
Limestone	2.8–3.5
Sandstone	2.6–4.0
Shale	1.5–2.9
Coal	0.3

gradient in the vicinity of a salt structure. A salt spine or diapir poking up through the sediment pile acts as a heat-conducting rod. At the top of the spine it is a heat source and emits heat to the surrounding sediments. At its base it is a heat sink and draws heat from the surrounds. Such conditions are ideal for generating convective flow cells which move diagenetic waters and hydrocarbons up and down the sediment column, causing widespread alteration in the surrounding sediments. Such convection may help explain the "Mississippi Valley" type of mineralization seen in many salt structures about the Gulf of Mexico (Price et al., 1983). Salt structure provinces often show geothermal gradients which are much higher than would be expected from the associated sediment thicknesses in the areas between the structures. Organic-rich sediments will mature much faster in such areas, and this factor should be accounted for when studying the thermal maturity of a diapiric basin.

The brine chemistry of much of the northern Gulf of Mexico is influenced by salt structures; much of the high salinity pore fluid in the northern Gulf region is derived from diapir dissolution (Hanor, 1987; Bennett and Hanor, 1987). Dissolution of the upper part of the Iberia and Welsh Dome salt domes in the Louisiana Gulf coast is supplying ions to the surrounding strata, while driving large-scale fluid convection in the surrounding sediments. In Iberia the circulation cell is driven by density inversion created during the dissolution of halite at the crest of the diapir. This creates dense saline brines at shallow depths which then flow out and down, away from the dome. Flow rates in the convecting cell may be as high as meters per year.

Salt movement has affected sedimentation in the Gulf of Mexico and the Arabian Gulf for tens of millions of years at a time. Many of the giant fields in these areas are related directly or indirectly to salt movement. An understanding of sedimentation models based on the Quaternary sedimentation can now be used to better interpret the controls on reservoir distribution in the Mesozoic and Tertiary reservoirs in these areas.

SALT STRUCTURE RELATED RESERVOIRS

Oil and gas reservoirs associated with salt structures are as varied as the shapes of the salt structures that cause them. Practically all known types of hydrocarbon traps have been found in halokinetic provinces (Halbouty, 1979; Woodbury et al., 1980; Seni and Jackson, 1983c). Figure 6.15 is a schematic summary of all known

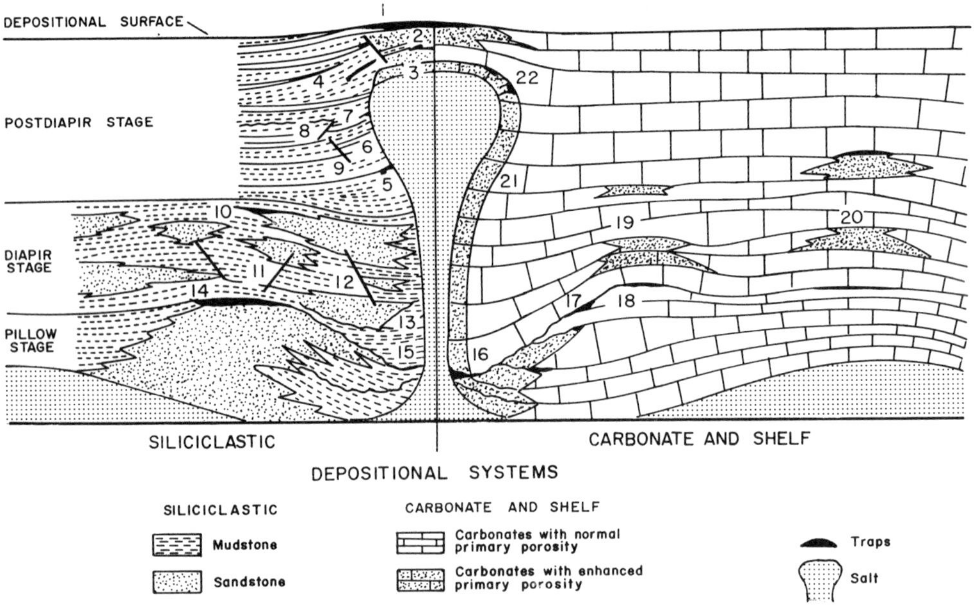

Figure 6.15 Schematic of the distribution of potential reservoirs in carbonates (right-hand side of figure), and siliciclastics (left-hand side of figure) during the three stages of salt structure growth (after Seni and Jackson, 1983c). *Trapping mechanisms:* 1, combination trap in sand in anticline over crest of dome; 2, graben fault trap over dome; 3, porous cap rock; 4, stratigraphic trap in flank sand pinch-out; 5, structural trap beneath overhang; 6, structural trap uplifted and buttressed against salt stock; 7, unconformity trap; 8, fault trap downthrown away from salt stock; 9, fault trap downthrown toward salt stock; 10, combination trap in sand from up-dip pinchout of porous facies in sink; 11, fault trap in sand over turtle structure; 12, fault trap in sand in peripheral sink; 13, stratigraphic trap in sand from domeward pinchout of porous facies in sink; 14, combination trap in sand at crest of turtle structure; 15, unconformity trap in sand over crest and flanks of precursor pillow; 16, unconformity trap in carbonates from enhanced porosity over crest and flanks of pillow; 17, combination trap in carbonates from pinchout of enhanced porosity zone on distal flanks of precursor pillow; 18, structural trap in carbonates over crest of turtle structure; 19, combination trap in carbonates from enhanced porosity due to paleotopography over pillow; 20, combination trap in carbonates from enhanced porosity due to paleotopography over raised saddle between peripheral sinks; 21, combination trap in carbonates from enhanced porosity near dome due to paleotopography and buttressing against salt stock; 22, combination trap in carbonates over crest of salt stock from enhanced porosity due to paleotopography.

potential hydrocarbon traps associated with salt structures. All of these traps can be explained by a combination of structural deformation, faulting, and depositional variations all acting in conjunction with salt movement and basin subsidence. Salt structures not only form traps by postburial salt movement, but they substantially modify sedimentation patterns in the overlying sequences. In this section we deal with reservoir formation within the framework of the three types of salt structure:

1. Pillows and non piercement structures
2. Diapirs and piercement structures (also turtle structures)
3. Post-diapir structures

As discussed in previous sections, growing salt structures often create topography at the surface. The effect of this paleotopography on the position of highly porous units depends on the depositional setting and which salt structure type is active.

Salt Pillows (Figs. 6.15 and 6.16)

The areas above salt pillows are topographic highs, while the areas between pillows are the topographic lows of the primary rim synclines. Fluviodeltaic, coastal, and submarine slope deposits are usually sand poor over the crests of salt pillows (Fig. 6.13). Most of the coarser-grained sediments in these systems were deposited between the pillows as various types of channel fill and beach deposits in the primary rim syncline (salt-withdrawal area). In continental settings the pinchout of the high-porosity sand occurs on the flanks of a salt pillow usually well down-dip from the pillow crest. Occasionally, reservoirs can form in the muddy sediments of the pillow crest in a continental setting. Crest reservoirs are often due to diagenesis associated with subaerial exposure and/or unconformities. In arid coastal areas the pillow crest could have been covered by sabkhas, environments often tied to the development of early dolomite in pelleted mudstones with the associated formation of good intercrystalline porosity (Chapter 2).

In shelf settings, high-porosity sediments form over the shoaling pillow crest (Fig. 6.13). Boundstones, skeletal and oolitic grainstones, and well-winnowed siliciclastics are found on such highs. For example, studies by Mancini and Benson (1980) and Ahr and Palko (1981) of the Jurassic Smackover Formation reservoirs in southwest Alabama and Louisiana show that reservoirs often occur in oolitic-pisolitic grainstones located over the pillow crests (Figs. 6.16a and b). On the down-dip side the grainy reservoir facies is replaced by muddy subtidal facies of the primary peripheral sink and passes up-dip into restricted marine carbonates and supratidal anhydrites of the overlying Buckner Formation (Fig. 5.14a). The pillow crest reservoirs are now extensively dolomitized (mixed waters) and contain evidence of earlier syndepositional erosion and meteoric diagenesis. Thus sediment atop the pillow crest was never far below sea level and was at times

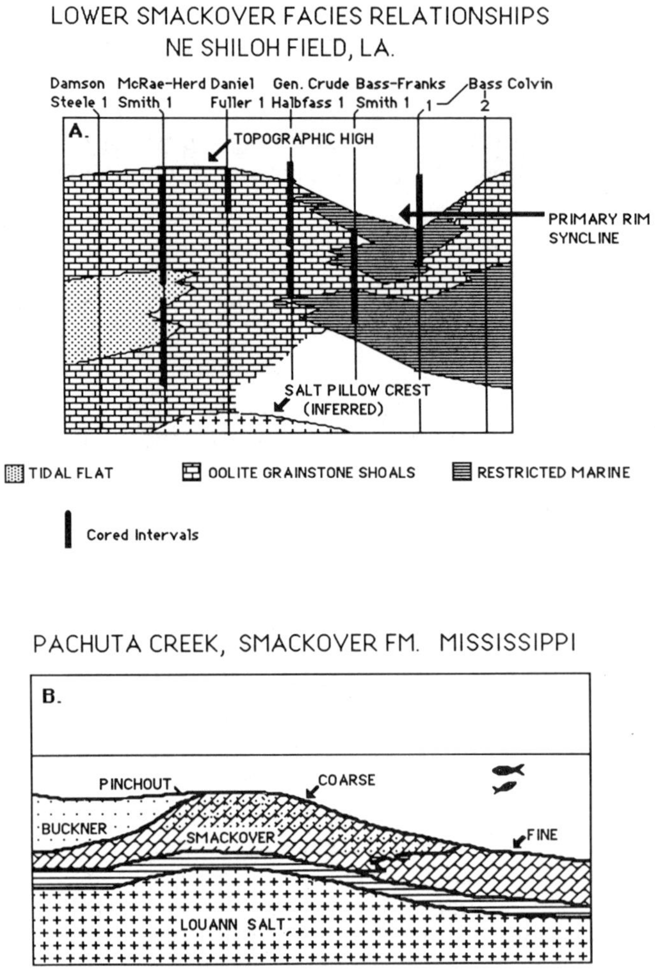

Figure 6.16 Smackover reservoirs: (a) carbonate depositional facies in northeast Shiloh Field, Louisiana showing thick oolite shoals over the pillow crest and muds in the rim syncline (after Ahr and Palko, 1981); (b) schematic cross section of facies development in Pachuta Creek Field, Mississippi.

subaerially exposed. Jurassic stromatolite and coral stromatoporoid reefs are found in the Smackover atop other salt pillows, such as the Walker Creek Field, Arkansas, in the Hico Knowles Field, Louisiana (Baria et al., 1982), and Hatter's Pond in southwest Alabama (research in progress). The buildups are 3 to 40 m (10 to 130 ft) thick, commonly elongate, and several square kilometers in plan. The reef rubble zones have reservoir-quality porosity throughout the Smackover trend but reef boundstones are economic only when extensively leached, dolomitized, and fractured. Such diagenesis has created commercial production from reefs with mean porosities of 15% and permeabilities of 20 md. The Hatters Pond Field is rather interesting in that the reef is tight and cross-cuts the reservoir trend. The distribution of the reef was controlled by the position of the Jurassic high above a growing salt pillow, while the reservoir structure was created during a Cretaceous faulting associated with the diapiric stage (Worrall and Warren, 1986). The reservoir dolomites lie well below the reef crest and above a late stage dolomite unit.

The porosity trends in sediments above a salt pillow are often depositional but are usually complicated by closures related to faulting subsequent to burial. Such faulting is due to tension created by the subsidence of sediments around the growing salt structure and dissolution of salt at the pillow crest (Woodbury et al., 1980). The fault closures form as the pillow crest maintains its position in earth space and the surrounding rim syncline subsides. Examples of fault closure forming reservoirs atop pillows can be seen in the West Bay Field in Louisiana and the Heidelberg Field in Mississippi (Halbouty, 1979). The significance of crestal faulting is discussed further in the next two sections.

Diapirs

Most of the traps associated with the diapir stage are flank traps. During the diapir stage the salt is very near to, or at the surface, and any overlying sediment is periodically eroded. Flank traps are of two interrelated types (Halbouty, 1979):

1. Created by sedimentation and erosion contemporaneous with salt movement
2. Created by the relative upward movement of the salt

In siliciclastic settings and carbonate continental slopes, the major syndepositional traps occur in sand-rich secondary rim synclines and associated pinchouts onto the diapir. An example of this type of reservoir is the West Saratoga Field, Hardin County, Texas. There, multiple Yegua sand reservoirs shale out as they pass into the Saratoga Salt Dome (Galloway et al., 1983). The secondary peripheral sink continued to subside as the diapir grew, resulting in steep dips in the sand units as they shale out into the diapir (Fig. 6.17). Such reservoirs are characterized by thick oil columns which may approach 300 m (1,000 ft). Efficient gravity segregation aided by water influx results in high recovery efficiencies. In carbonate shelf settings grainstone shoals and reefs may rim the diapir; with burial these porous units will form a rim around the diapir stem. A good example of this type of grainstone trap associated with a diapir is the West Purt Field in the East Texas Basin, where a mollusk peloid grainstones and rudstone reservoir (Rodessa Limestone, Lower Cretaceous) abuts the northwest side of the Brushy Creek Diapir (Lomando et al., 1984).

The pulselike nature of salt diapirism means that whatever the depositional setting most sediments deposited above a diapir are uplifted and eroded. The resulting unconformities truncate flanking coarser grained deposits, which may then be sealed by subsequent deposition and so create unconformity traps. The dip of such unconformity traps is as steep as the dip of the shaled-out sands just described in the Saratoga Dome. Unconformity traps also have thick oil columns but with limited lateral extent away from the salt stem. A good example occurs

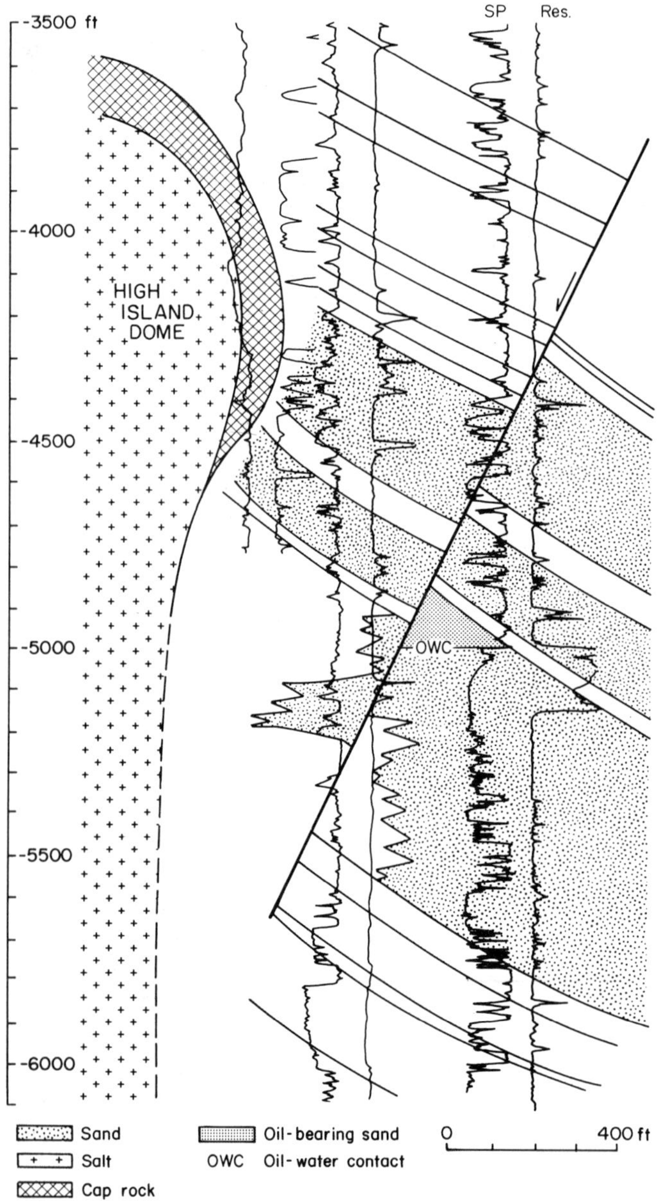

Figure 6.17 Cross section of the south flank of the High Island Dome, Chambers County, Texas (after Galloway et al., 1983). The trap in this case is due to peripheral faulting.

along the north side of Nash Dome in Texas, where steeply dipping Frio sands (Oligocene) are truncated by multiple dome-flank unconformities. In the Gulf of Mexico the two types of steeply dipping sand reservoirs (unconformity and flank sands) stack up along the sides of many diapirs. They yield greater amounts of hydrocarbons per acre than any other type of trap in the Gulf of Mexico, regardless of geologic structure (Halbouty, 1979).

The other major type of flanking traps associated with the diapir stage form when the salt actually pierces (intrudes) the overlying sediment column and the structure generates multiple faults. There are three types of traps formed in this fashion:

1. Sands terminating under the salt or caprock overhang
2. Sands buttressed against the salt stock
3. Simple and complex fault traps

Traps often occur below the salt or caprock overhang of mushroom-shaped domes. Prolific production comes from beneath the overhangs of Barber Hill Dome in Chambers County, Texas; High Island Dome in Galveston County, Texas; Batson Dome in Hardin County, Texas; and the Hanigsen Salt Dome in the Zechstein Basin of northwest Germany (Figs. 6.17 and 6.18). Sands below the Barber Hill overhang have produced more than 100 million barrels of oil. Reservoirs also form where a pierced sand unit abuts the salt stock. The position of flank traps at the edge of a subsiding peripheral sink and adjacent to the salt stock means that they are often modified by faulting (Fig. 6.17; Halbouty, 1979).

Faulting often controls the volume of entrapped hydrocarbons in the various types of flank traps. Faults possess radial, peripheral, or tangential patterns relative to the salt stock. In the Gulf of Mexico where such faults have downfaulted, uplifted, or overthrust sands against impervious shales, reservoirs have formed (Bryan Salt Mound, Brazoria County, Texas). Of course, fault traps can form at any stage of salt structure growth, Peach Point Field, Brazoria County is associated not with diapirs but with salt ridges and pillows (Halbouty, 1979).

Chalk reservoirs above salt structures, such as Danian chalks in Ekofisk, North Sea, often had reservoir properties enhanced by the movement of subadjacent salt. Tension was created in the potential reservoir above the growing salt structure leading to microfracturing (D'Heur, 1980). It is now widely believed that oil migrated up these fractures and that when the oil column in the fracture system was 50 to 100 m the buoyancy force was sufficient to overcome the high entry pressures inherent in narrow pore throats. Hardman and Kennedy (1980) postulated that this process can explain why high-relief features in the North Sea chalk/salt structures have better oil saturations and why low-relief traps can have water-wet matrix porosity. Early entry of oil into matrix pores and the associated overpressures then retarded the normal sequence of chalk lithification with burial by inhibiting chemical diagenesis and reducing overburden pressure on the matrix grains (D'Heur, 1980). These processes were most effective in reservoirs above the center of the salt structure and with time migrated out into the flank reservoir; hence the reduction of porosity and increase in water saturation radially and the concave-upward oil water contact at West Ekofisk.

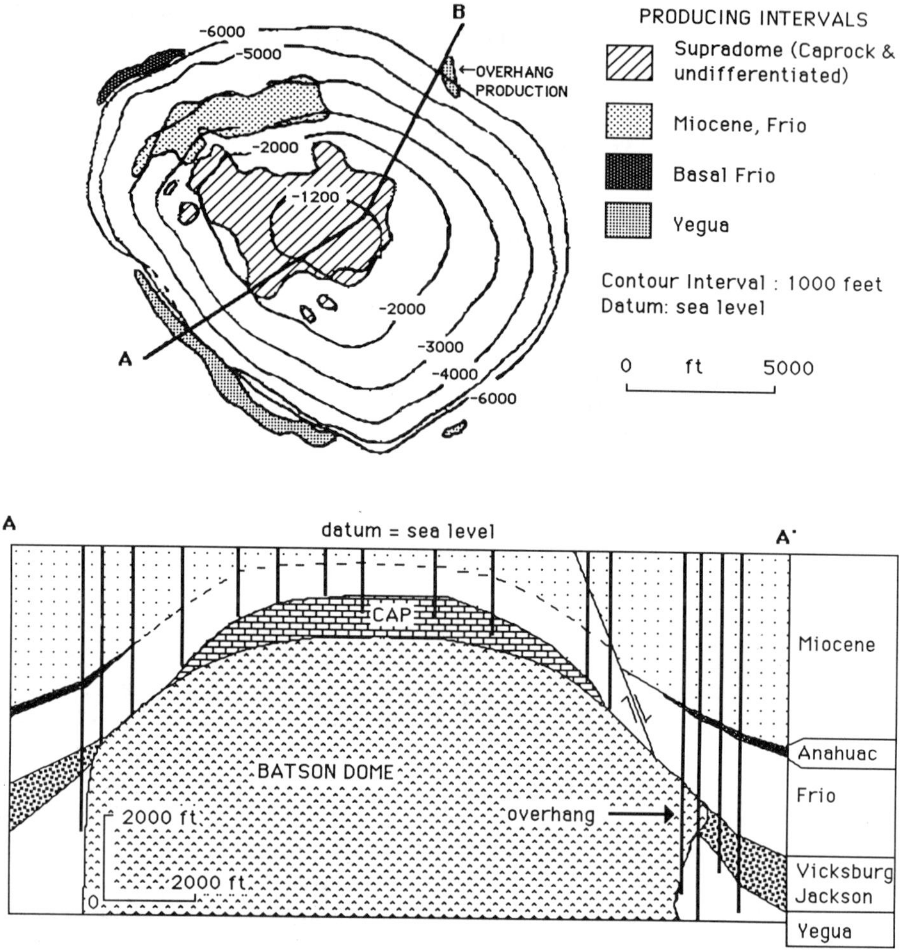

Figure 6.18 Structure map and cross section of Batson Field, Hardin County, Texas: the caprock and supradome production are comingled; note the overhang trap atop the Yegua sands (after Galloway et al., 1983).

The diapir stage is also the time when turtle structure anticlines form in the salt-withdrawal basins between diapirs. In a shelf setting the region above a turtle is often a topographic high where high-energy sands and reefs are deposited atop an anticlinal drape, an excellent setting for the subsequent accumulation of hydrocarbons. A good example is the giant Fairway Field in Lower Cretaceous James Reef Limestones in Henderson and Anderson counties, East Texas (Terriere, 1976, Achauer, 1983). The reservoir is a rudist-reef complex above a shoaling turtle structure, situated on a saddle between the salt-withdrawal areas of Boggy Creek, Brushy Creek, and La Rue Domes (Fig. 6.19a). Postdepositional faulting truncated the northwestern and northeastern margins of the trap, which is now a southeast-plunging nose (Fig. 6.19b). Reservoir porosity is largely secondary and is higher in mechanically deposited rudist grainstone belts rather than in the reef proper (Fig. 6.19c). This is the case in many rudist associated reservoirs in the Gulf of Mexico and the Middle East. The growth of a turtle structure converted a muddy area of shelf limestones into a shoal area suitable for rudist reef growth.

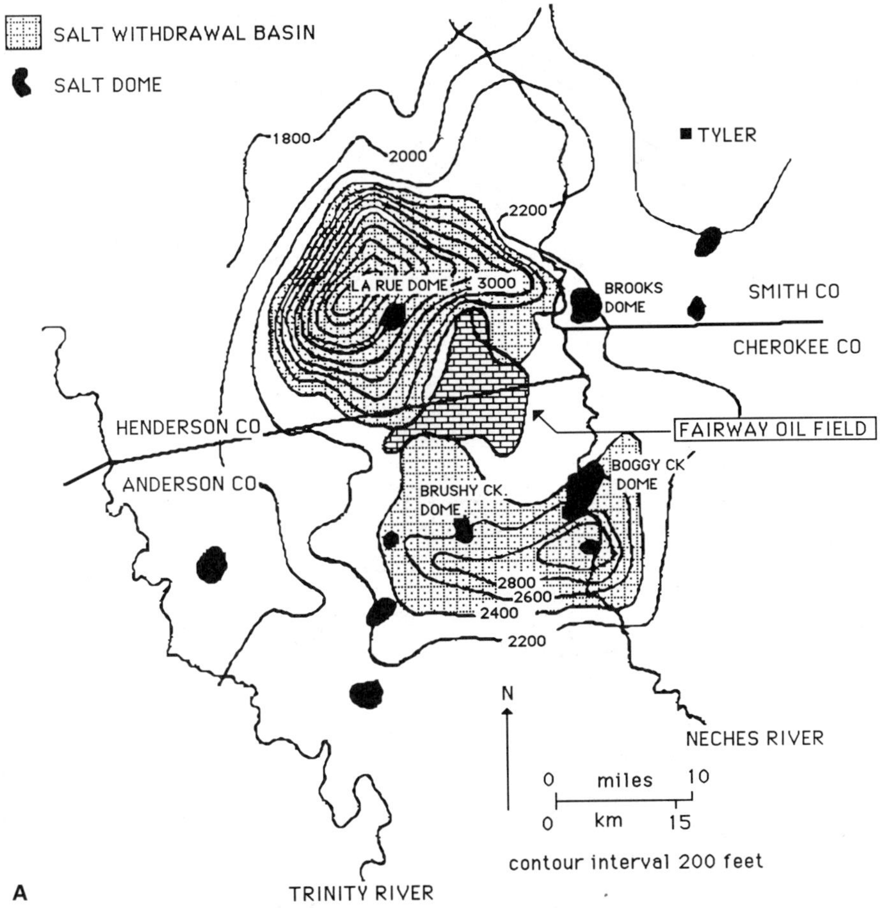

Figure 6.19 Fairway Field, East Texas basin (after Galloway et al., 1983). (a) Isopach map of Glen Rose Group (contours in feet) showing the muddy thicks in the salt withdrawal basin, the Fairway Field (limestone pattern) and salt domes. Fairway Field formed atop a turtle structure anticline between salt domes. (b) Structure map of Fairway Field, Henderson and Anderson Counties, Texas. The field trap is a faulted southeastplunging nose. (c) Facies map and interpreted depositional setting in the James Limestone. The main reef core lies in the northwestern part of the field and satellite reefs to the southeast and southwest are composed of corals, stromatoporoids, algae, and rudists. Carbonates sands and gravels were deposited in the southern part of the field; while elsewhere, muddy sands are the dominant lithology.

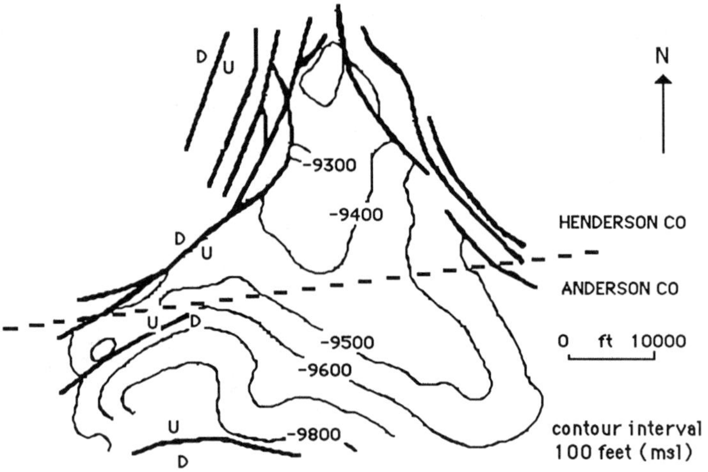

B

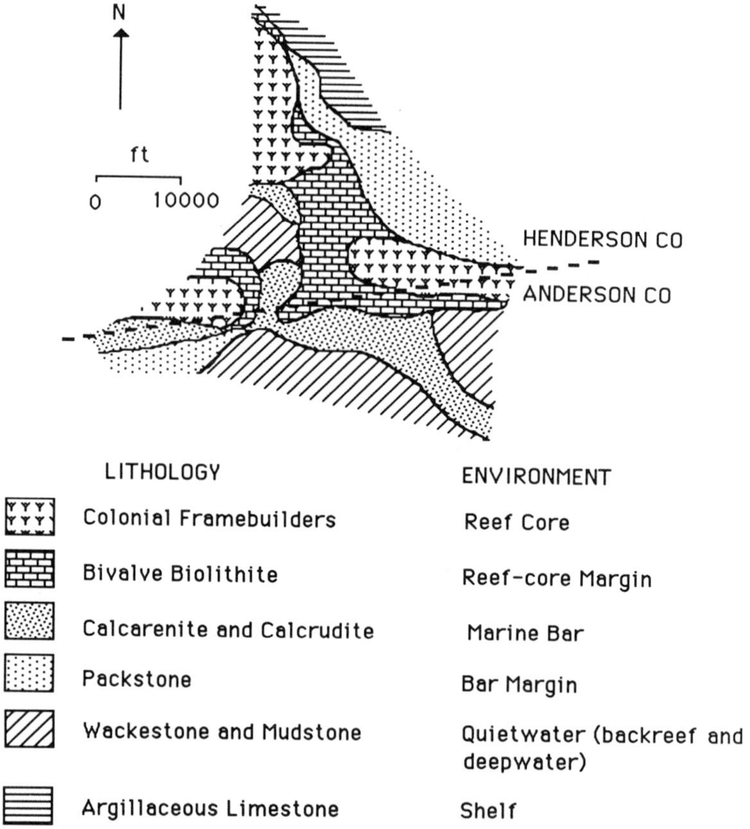

C

Figure 6.19 (Continued)

Post-diapir Stage

By this time salt flow into the structure is dwindling and will eventually cease. The crest of the structure is shallow and underlies a small topographic high (Fig. 6.4c). Reservoirs formed in this stage are similar to those formed during the pillow stage but closures are much smaller. They include combination traps in the anticline above the structure's crest as well as graben fault traps over the dome. Above the post-diapir-stage structures of the shelf and slope settings in the Gulf of Mexico and the Arabian Gulf are the Holocene reefs and grainstones of the Flower Garden Banks and Yas Island. Ancient equivalents are the reef-associated reservoirs capping Damon, Nash, and West Columbia domes in Brazoria County, Texas (Cantrell et al., 1959). There the porous coral-rich facies of the Oligocene Anahuac Formation caps the salt domes; off the crest the Anahuac passes into the more normal, deeper-water *Heterostegia* limy shales and shales. The limestone unit thins and shoals over the post-diapir crest (e.g., Anahuac in Fig. 6.18).

Caprocks also host hydrocarbons in many Gulf of Mexico domes. The porosity in caprocks can be as high as 40%, with porosity types ranging from pinpoint to vuggy to cavernous. Caprock reservoirs are shallow and so were attractive targets in the early heyday of drilling in Texas. Many famous early Texas reservoirs were in caprocks; examples include Spindletop (the most famous), Sour Lake, and Humble (Fig. 6.18). Caprock reservoirs tend to be relatively small; the most difficult thing about working them was their lack of predictability.

The early history of Gulf Coast oil shows an interesting aside. Oldtime drillers found that drilling graveyards led to a significant number of successes. Earlier this century the high success rates associated with "graveyard-tology" meant that church elders and many still-living relatives developed a sudden if somewhat belated attachment to long-since-departed great aunt Maude. Bitter legal battles were fought over the percentage of royalties that one would receive for part of, or even a distant ownership of, a small piece of "Boot Hill." A reason exists for these early success rates. Many early Texas fields were salt-structure associated, with the shallow reservoirs (caprock-associated) lying directly over the crest of the structure. The surface expression of an underlying salt structure is a small topographic high or hill, and since John Wayne was a whippersnapper, graveyards have been "up on Boot Hill." The first oil discovered in southern Louisiana in 1901 came from a similar salt-formed topographic feature known as Jennings Hill about 100 km west of Baton Rouge.

SALT STRUCTURES IN TECTONICALLY ACTIVE PROVINCES

There are more than a dozen fold and thrust belts that have at some time, at least in places, moved across a layer of evaporites, usually including salt (Davis and Engelder, 1987). Such movement in a zone of horizontal compression often results in "thin-skinned" mountain belts. Deformation in the overlying rocks can produce inhomogeneity in the overburden as it is deposited, as well as stress instability in the salt bed; both processes required to initiate salt flow and diapirism. Even when there is no development of diapirs in these areas, such as in the Jura and the Appalachian Plateau, the folds contain considerable anticlinal thickening of the salt, and associated synclinal thinning. Salt diapirism usually follows fault planes in the overlying rock. The deformation of the overlying rocks tends to have a geometry controlled by the regional state of stress, and that geometry is inherited by the resulting salt structures. The close correspondence between the elongate salt structures of the North Sea and northern Germany clearly demonstrate this relationship. Probably the best studied area of salt influenced thrusting is the Zagros Mountain region of the Arabian Gulf.

In Iran this tectonic regime formed some of the largest oil fields in the world. Folding concurrent with deposition means that the lower members of the Gachsaran Formation (seal) overlap and wedge out onto folds in the Asmari Limestone (reservoir; Ala, 1982 and references therein). This led to the deposition of sandier units above the crests of growing thrust folds, and the sealing of these porous units by evaporites deposited within hypersaline basins generated by the collision.

The late Tertiary collision of Africa and Eurasia did three things to create conditions that were suitable for hydrocarbon trapping in Iran and Iraq (Fig. 6.20). First, it created a reservoir, the Asmari Formation of Oligocene-Miocene age. In the main producing area the Asmari is predominantly a shallow-water non-reefal micritic limestone containing minor dolomite. Matrix porosity and permeability are low and the unit owes its excellent reservoir qualities to extensive brittle fracturing induced by Plio-Pleistocene folding. Second, it sufficiently restricted the basin in the Miocene to deposit the Lower Fars or Gachsaran Formation, a basin-wide evaporite seal (Kashfi, 1983,1985). The presence of this unit with its impermeable anhydrite at the base is critical in sealing the Asmari reservoirs. The erosion of the Lower Fars seal has destroyed potential traps along strike (to the northwest and southeast) of the main producing Asmari folds. And third, in the southeast portion of the Zagros Belt, collision acted as a huge bellows to the underlying Eocambrian Hormuz salt which by flowing up toward the surface further exaggerated the structural closure of the folds. Zagros folds are world renowned for their huge amplitude and areas of closure; along-strike fold dimensions often exceed 160 km. These dimensions together with brittle fracture in the Asmari limestone reservoir have created fields such as Gach Saran and Bibi Hakimeh which have estimated ultimate recoveries of oil of 12 and 8 billion barrels, in a province which has recoverable reserves estimated at 87 billion barrels of oil and 514 trillion cubic feet of gas, and produces 6 million barrels of oil per day (Fig. 6.20; Burwood, 1984).

Although the presence of an evaporitic seal overlying a reservoir is always good news for sealing hydrocarbons, a thick underlying salt unit deformed into salt structures is not always so useful in terms of accumulating large quantities of

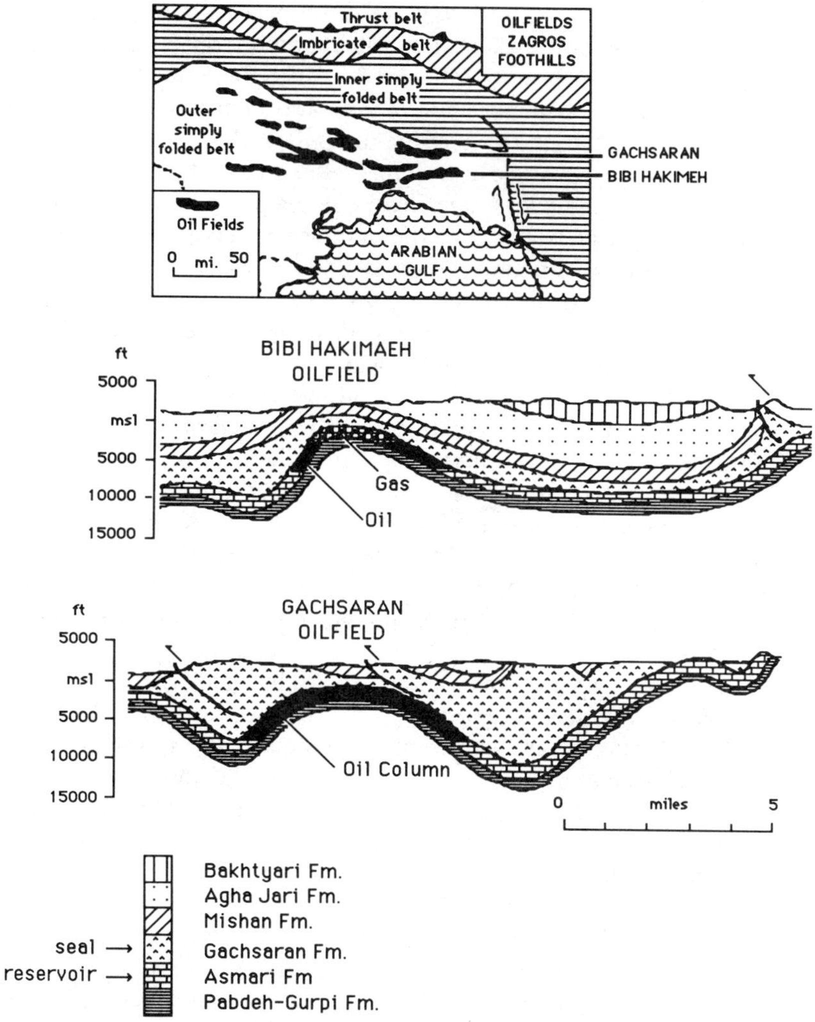

Figure 6.20 Northwest Arabian Gulf showing location and cross sections from two of the major producing fields, Bibi Hakimaeh and Gachsaran (after McQuillan, 1985).

hydrocarbons. Alongstrike changes in the tectonic style of the Zagros Mountains demonstrate this point. The northwest part of the Zagros belt is the most prolific hydrocarbon area and is characterized by a thick and relatively deep section of Asmari, folded and sealed beneath Gachsaran/Fars evaporites. Deformation is intensive but regional, due to lateral compression, some fracturing of the reservoir is probably related to deep seated basement faults (Ala, 1982; McQuillan, 1985). There is no underlying Hormuz Salt in this area. Oil fields in the outer Zagros belt were relatively easy to define and develop by stepping out along anticlinal axes. To the southeast the Zagros belt is characterized by many salt domes sourced in the underlying Hormuz Salt. Most of the fields in the south are gas fields (Ala, 1982). There the Asmari is thinner and often outcrops in breached anticlines. The style of deformation in this area is uplift with a greater vertical component compared to areas to the west, perhaps related to withdrawal of thick salt into adjacent salt diapirs and glaciers. During the latest period of vertical uplift many hydrocarbons previously trapped in the Asmari either escaped through fractures in the exposed Asmari or underwent secondary migration, possibly becoming trapped on the western side of the Oman Mountains (Kashfi, 1983). In the southeast Zagros the gravity faults developed in association with salt intrusions, probably allowed the further migration and escape of hydrocarbons. It is unlikely hydrocarbon accumulations will be found in this area with the same anticlinal style of the northwest Zagros. The presence of a thick Hormuz section in this part of the Arabian Gulf may well have prevented the transferal of the fracture system from the basement to the Asmari, the same fracture system which McQuillan (1985) uses in explaining the prolific production of the giant fields in the thrust belts further to the Northwest. Many large structural accumulations may still remain to be discovered in the area offshore from the Southern Zagros but exploration here should perhaps be for stratigraphic or combination traps about halokinetic structures.

To the southeast of the Zagros Mountains there is another collision belt, the Oman Mountains (Lippard et al., 1982; Robertson and Woodford, 1983). The Oman Mountains are outcropping evidence of an earlier Mesozoic collision, an ocean-continent subduction belt. In particular the NNE-SSW trending Dibba or Oman line has long been recognized as an important tectonic trend in the Gulf. Murris (1980) interprets the Dibba Line as an ocean/continental transform fault. In terms of the present discussion of evaporite-associated reservoirs the Dibba Line is not of primary exploration interest as Hormuz salt was not caught up in this collision belt. However this Mesozoic collision may have created a different tectonic grain to major fold structures (anticlinal traps) in onshore regions in the southeast of the Gulf. It may have reactivated many salt features sourced in the Hormuz salt so that they once again affected sedimentation styles on the Mesozoic seafloor. To the west closer to the collision belt it may have created stress and fracture fields in the basement with stress vectors related to the collision trend. This is one possible explanation for the well developed NNE-SSW orientation of most of the onshore fields in the United Arab Emirates.

In the Romanian section of the Carpathians Mountains of eastern Europe is another collisional thrust-fold belt where hydrocarbon accumulation is intimately related to the presence of basin-wide evaporites, in this case of Paleogene and Miocene age (Fig. 6.21; Paraschiv and Olteanu, 1970). Most of the reservoirs are Pliocene lacustrine sands, interbedded with fine-grained marls. The latter may well have been both source and seal. Suitable Pliocene sands with sufficient

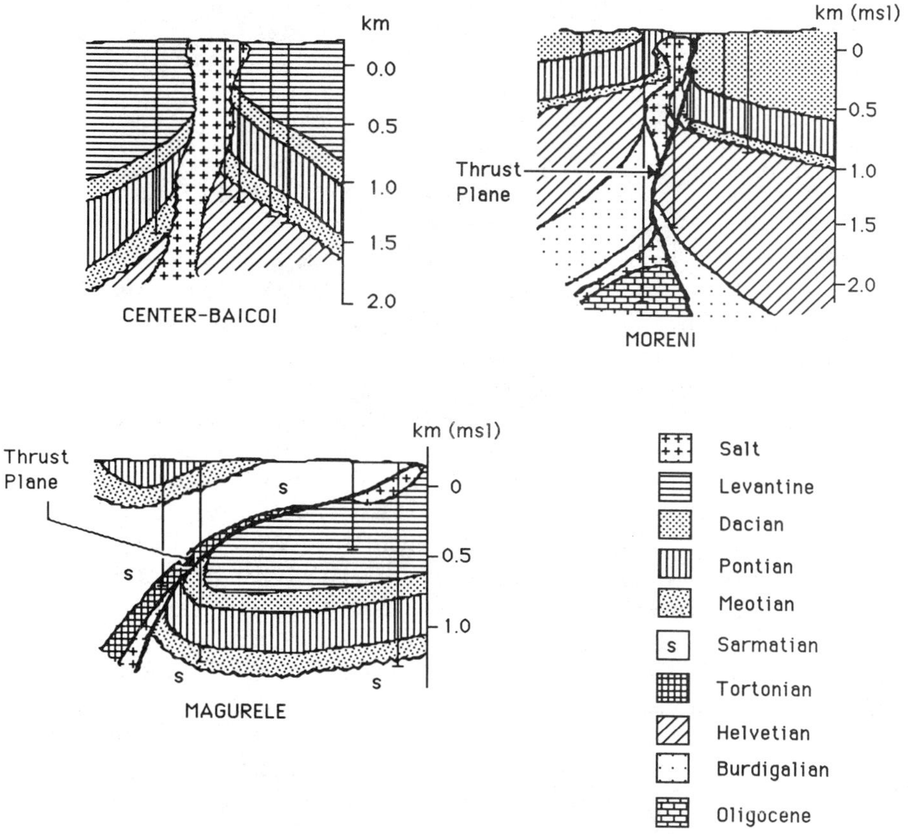

Figure 6.21 Cross sections of three salt-structure-generated oil fields in the Romanian Carpathians (after Paraschiv and Olteanu, 1970).

thicknesses and favorable reservoir facies only occur in the Ploiesti area at the convexity of the Carpathian Bend. Like the Zagros Mountains this hydrocarbon province is another salt structure or halokinetic province where the original evaporites have been squeezed into a compressive deformational setting by large scale, mainly thrust movements. Tectonism has mobilized Tertiary salt into stocks, sills, dyke-like masses and other irregular shapes whose original injection pathways were controlled by faults, folds and fault/fold intersections. Such irregular masses of salt often squeeze up along thrust faults and other planes of weakness to seal adjacent deformed porous units (Fig. 6.21).

Evaporites play a varied role in generating traps in collision belts. They act as seals, often deposited in the early stages of collision when the creation of foreland depressions restricts surface water inflow. They act as decollement surfaces which aid in the creation of anticlinal traps. And finally they can flow and be squeezed into diapiric structures which generate structural traps. Such salt structures tend to follow the tectonic trends in a basin and so their position in the subsurface is often predicted by a thorough knowledge of the regional structural setting.

SALT SOLUTION FEATURES

Salt removal or dissolution does not just occur on the crests of diapirs but can occur where bedded salt is later reexhumed and flushed by meteoric water or seawater. The foundering of the overlying rocks into voids created by dissolution of the evaporites creates *solution breccias* composed of clasts of the overlying sequences. Dissolution of the evaporite also leaves behind a thin layer of nonsoluble residue from within the evaporite bed itself. Carbonate residues created by leaching of evaporites layer are described by various authors as *dissolution boxworks or "rauhwacke"*. As the name suggests a dissolution boxwork is characterized by angular void shapes. Often the boxwork shape reflects the shapes of former evaporite crystals. Such boxworks have been documented as early syndepositional dissolution textures in Holocene lakes in southern Australia where bedded gypsum evaporites about the edges of coastal lakes are cannibalized by the inflow of less saline groundwaters (Chapter 3). Dissolution boxworks formed by subsurface dissolution long after the evaporite bed was deposited do not often reflect the original evaporite mineral outline, rather they outline the shape of the evaporite block as it was dissolved. Angularity in the dissolving blocks tends to outline fractures and joints rather than the original crystal outlines of the evaporites composing the bed, these depositional textures were destroyed by burial diagenesis (Chapter 1).

The dissolution boxwork layer is not volumetrically important in ancient strata, rather it is the blocks of the overlying sequence which subside and collapse into the zone left behind by evaporite dissolution. Often evaporite-solution breccias, forming at or near the land surface, pass down-dip into equivalent evaporite units (P_1 and P_2 in Chapter 3). When an evaporite-solution breccia is buried it is a laterally extensive, tabular unit at the stratigraphic position of the preexisting evaporite. The basal contact is usually smooth and coincides with the basal surface of the former evaporite. The upper contact is commonly gradational upward, first through a sequence of breccia, then through breccia in which remnants of the original stratification is present, then into highly fractured and then into less fractured rock and finally through intact strata where the only reflection of the dissolution of an underlying evaporite unit is some undulation of the bedding. The upper contact of the evaporite-solution breccia is often irregular and breccia-filled pipes (solution dolines) may extend up above the main breccia horizon for tens to hundreds of meters. Thickness of the main breccia layer and the fractured zone is dependant on a number of factors including; original evaporite thickness, lithology of strata above the evaporite layer, amount and type of strata interbedded with the evaporite bed, and the depth of burial of the evaporite undergoing dissolution.

Evaporite-solution brecciation and fracturing of overlying strata has formed evaporite-solution reservoirs in many areas around the world (Stanton, 1966; Beales and Oldershaw, 1969). Reservoirs in these areas of salt solution tend to occur in

1. The brecciated zone created by salt dissolution
2. Fractured rocks overlying areas of dissolution
3. Pseudo-reef crests

Deformation of the overlying strata due to differential solution and associated subsidence can also create potential reservoirs as part of the dissolution process; the end result is often a carbonate "pseudo-reef" in an area with no remaining salt (Anderson and Hunt, 1964). This type of reservoir has been described from the Devonian Nisku (Birdbear) Formation of northeastern Montana where the oil occurs in steep-sided, flattopped structures less than 1 mile in diameter (Swenson, 1967). They are not reefs but are the end result of multiple stages of dissolution of the bedded salts of the Devonian Prairie salt, which lies beneath the Nisku (Fig. 6.22). The first episode of salt dissolution in the Prairie Evaporite took place during deposition of the Souris River Formation. Leaching of this salt created local lows that were filled by Souris River carbonates. When dissolution ceased the Duperow, Nisku, and Mississippian carbonates were laid down as layer cakes. Then at some post-Mississippian time, widespread dissolution of the Prairie salts began anew. As the underlying salt dissolved the overlying sequences settled down into the space left by the dissolved salt. Local highs in the Nisku, which are now reservoirs, are actually compactional features created by differential drape about sediment thicks. Thicks were created initially by the infill of an earlier stage of localized salt dissolution.

North American fields formed by this process include Flat Lake, Fryburg, Hummingbird, the Medoria-Scoria Complex, and Outlook. They stretch from Saskatchewan to southwestern North Dakota. Reservoirs created by salt leaching are described in the North Sea by Jenyon and Taylor (1983) and Lohmann (1979), and in North America by Anderson (1966), Parker (1967), and Smith and Pullen, (1967). In every case the most successful exploration strategies were based on precise stratigraphic correlation and detailed isopachs of the potential reservoir unit.

The mother salt bed may not be completely dissolved; in some cases dissolution has differentially altered the edge of a salt bed. In such cases there are often residual outliers of salt left along the edge of a dissolving salt bed. Potential reservoir structures are difficult to resolve unless migrated seismic data are available (Sheriff, 1980, p. 36). When dissolution is early, the sediments often thicken above the areas of dissolution and thin in the areas over the remaining salt outlier. These overlying sediments can develop the same types of facies patterns that

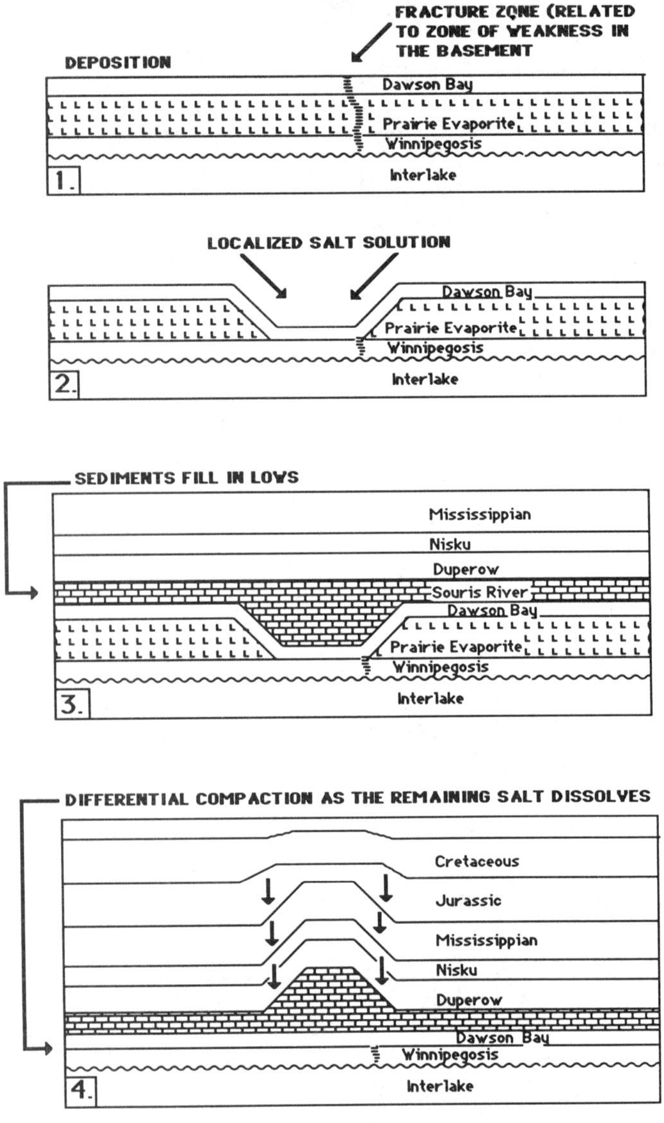

Figure 6.22 Model of Nisku Formation reservoirs showing they formed by salt solution in northeastern Montana (after Swensen, 1967).

form above salt pillows. Sometimes the salt dissolves in stages and so creates sequential thickened beds up section as the locus of salt dissolution shifts. In Fig. 6.23, two areas of sediment thickening are created by salt dissolution; the first forms in stage 2, the second in stage 4. This and later differential compaction can create reservoirs above salt remnants (high in stage 2) or by later differential compaction (stages 3 and 4).

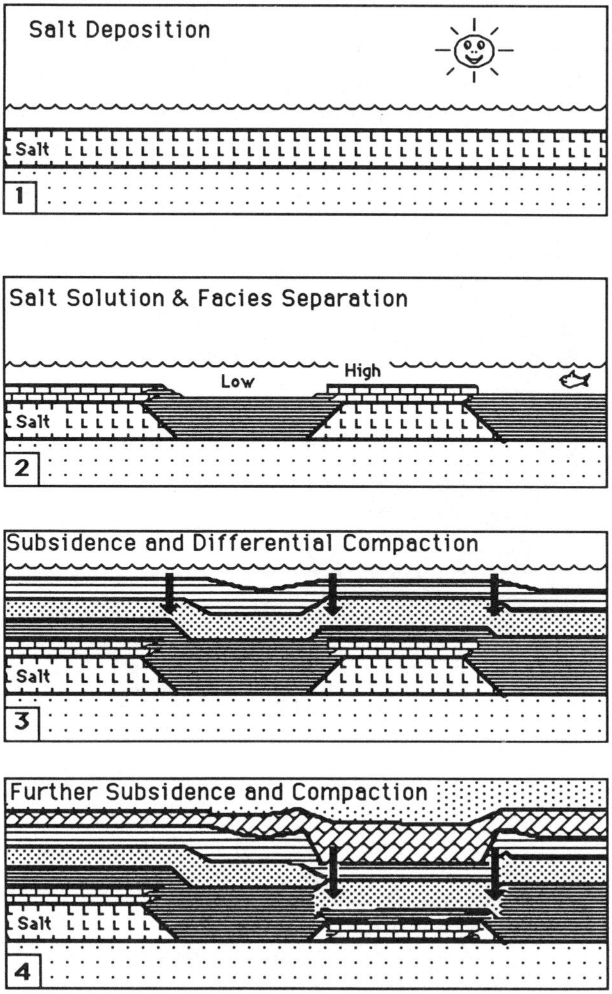

Figure 6.23 Formation of potential reservoirs by sequential salt dissolution.

SUMMARY

In this chapter I have presented a series of evaporite-associated plays which evolve as bedded salt evolves from a salt pillow to the post-diapir-stage structure or through its exposure and dissolution creates reservoirs by its absence. The working geologist must be aware of the complex interplays among hydrocarbon accumulations, bedded evaporites, and salt structures. For example, in an area characterized by salt pillows, one would not expect overhang traps, but in an area characterized by diapirs, one may expect deeper potential reservoirs which formed during the pillow stage. One should not use accepted models for salt structure development without first testing them against the known geology of the area where one is working. Like the rest of our knowledge of evaporites, our understanding of salt structures is in its infancy. The most detailed knowledge comes from siliciclastic reservoirs in the Gulf of Mexico. As knowledge of reservoir control grows in other salt-affected basins, such as in the Middle East, the North Sea, and South America, models of reservoir controls are sure to change.

References

ABDINE, A. S., 1981, Egypt's petroleum geology: Good grounds for optimism: World Oil (December), p. 99–106.

ACHAUER, C. W., 1982, Sabkha anhydrite: The supratidal facies of cyclic deposition in the upper Minnelusa Formation (Permian) Rozet field area, Powder River Basin, Wyoming, in Handford, C. R., Loucks, R. G., Davies, C. R. (eds.), Depositional and diagenetic spectra of evaporites—a core workshop: Core Workshop 3, Society of Economic Paleontologists and Mineralogists, p. 193–209.

ACHAUER, C. W., 1985, Facies morphology and major reservoir controls, James Atoll Reef, Fairway Field, Texas, in Roehl, P., and Choquette, P. (eds.), Carbonate petroleum reservoirs—a casebook: Springer–Verlag, New York, p. 485–494.

ADAMS, J. E., and RHODES, M. L., 1960, Dolomitization by seepage refluxion: American Association of Petroleum Geologists Bulletin, v. 44, p. 1912–1920.

AHARON, P., KOLODNY, Y., and SASS, E., 1977, Recent hot brine dolomitization in the "Solar Lake", Gulf of Elat: Isotopic, chemical, and mineralogical study: Journal of Geology, v. 85, p. 27–48.

AHMED, S. S., 1972, Geology and petroleum prospects in eastern Red Sea: American Association of Petroleum Geologists Bulletin, v. 56, p. 707–719.

AHR, W. M., 1973, The carbonate ramp: an alternative to the shelf model: Transactions of Gulf Coast Association of Geological Societies, v. 23, p. 221–225.

AHR, W. M., and PALKO, G. J., 1981, Depositional and diagenetic cycles in Smackover limestone-sandstone sequences, Lincoln Parish, Louisiana: Transactions of Gulf Coast Association of Geological Societies, v. 31, p. 7–17.

ALA, M. A., 1974, Salt Diapirism in southern Iran: American Association Petroleum Geologists Bulletin, v. 58, p. 1758–1770.

ALA, M. A., 1982, Chronology of trap formation and migration of hydrocarbons in Zagros Sector of Southwest Iran: American Association of Petroleum Geologists Bulletin, v. 66, p. 1535–1541.

ALDERMAN, A. R. and SKINNER, H. C. W., 1957, Dolomite sedimentation in the southeast of South Australia: American Journal of Science, v. 255, p. 561–567.

ALSHARHAN, A. S. and KENDALL, C. G. St. C., 1986, Precambrian to Jurassic rocks of Arabian Gulf and adjacent areas: Their facies, depositional setting, and hydrocarbon habitat: American Association of Petroleum Geologists Bulletin, v. 70, p. 977–1002.

AMIEL, A. J. and FRIEDMAN, G. M., 1971, Continental sabkha in Arava Valley between Dead Sea and Red Sea: Significance for origin of evaporites: American Association of Petroleum Geologists Bulletin, v. 55, p. 581–592.

ANDERSON, R. Y., 1982, Orbital and climatic control of Permian evaporite deposition, Delaware Basin, Texas and New Mexico (abs.): 95th Annual Meeting Geological Society of America, New Orleans, La., Oct. 1982.

ANDERSON, R. Y. and KIRKLAND, D. W., 1966, Intrabasin varve correlation: Geological Society of America Bulletin, v. 77, p. 241–256.

ANDERSON, R. Y., DEAN, W. E., KIRKLAND, D. W., and SNIDER, H. I., 1972, Permian Castile varved anhydrite sequence, West Texas and New Mexico. Geological Society of America Bulletin, v. 83, p. 59–86.

ANDERSON, S. B., 1966, A look at the petroleum potential of southwestern Dakota: North Dakota Geological Survey Report of Investigation 42.

ANDERSON, S. B., and HUNT, J. B., 1964, Devonian salt solution in north central North Dakota: Third International Williston Basin Symposium Proceedings, p. 93–104.

ANDRICHUK, J. M., 1958, Stratigraphy and facies analysis of Upper Devonian reefs in Leduc, Stettler and Redwater areas, Alberta: American Association of Petroleum Geologists Bulletin, v. 42, p. 1–93.

ANGELIER, J., 1985, Extension and Rifting: the Ziet region, Gulf of Suez: Journal of Structural Geology, v. 7, p. 605–612.

ARAKEL, A. V., 1980. Genesis and diagenesis of Holocene evaporitic sediments in Hutt and Leeman lagoons, Western Australia: Journal of Sedimentary Petrology, v. 50, p. 1305–1326.

ARTHURTON, R. S., 1973, Experimentally produced halite compared with Triassic layered halite rock from Cheshire, England: Sedimentology, v. 20, p. 145–160.

AYRES, M. G., BILAL, M., JONES, R. W., SLENZ, L. W., TARTIR, M., and WILSON, A. O., 1982, Hydrocarbon Habitat in Main Producing Areas, Saudi Arabia: American Association of Petroleum Geologists Bulletin, v. 66, p. 1–9.

BABEL, M., 1987, Giant gypsum intergrowths from the Middle Miocene evaporites of southern Poland: Acta Geologica Polonica, v. 37, p. 1–20.

BADIOZAMANI, K., 1973, The Dorag dolomitization model—application to the middle Ordovician of Wisconsin: Journal of Sedimentary Petrology, v. 43, p. 965–984.

BAKER, P. A. and BURNS, S. A. J., 1985, Occurrence and formation of dolomite in organic-rich Continental Margin sediments: American Association of Petroleum Geologists Bulletin, v. 69, p. 1917–1930.

BAKER, P. A. and KASTNER, M., 1981, Constraints on the formation of sedimentary dolomite: Science, v. 213, p. 214–216.

BARBER, P. M., 1981, Messinian subaerial erosion of the Proto-Nile delta: Marine Geology, v. 44, p. 253–272.

BARIA, L. R., STOUDT, D. I., HARRIS, P. M., and CREVELLO, P. D., 1982, Upper Jurassic reefs of Smackover Formation, United States Gulf Coast: American Association Petroleum Geologists Bulletin, v. 66, p. 1449–1482.

BARLEY, M. E., DUNLOP, J. S. R., GLOVER, J. E., and GROVES, D. I, 1979, Sedimentary evidence for an Archean shallow-water volcano-sedimentary facies, eastern Pilbara Block, Western Australia: Earth and Planetary Science Letters, v. 43, p. 74–84.

BARTON, D. C., 1933, Mechanics of formation of salt domes with special reference to Gulf Coast salt domes of Texas and Louisiana: American Association Petroleum Geologists Bulletin, v. 17, p. 1025–1083.

BEALES, F. W., and OLDERSHAW, A. E., 1969, Evaporite-solution brecciation and Devonian carbonate reservoir porosity in western Canada: American Association of Petroleum Geologists Bulletin, v. 53, p. 503–512.

BECHER, J. W., and MOORE, C. H., 1976, The Walker Creek Field—A Smackover diagenetic trap: Transactions of Gulf Coast Association of Geological Societies, v. 26, p. 34–56.

BEGIN, Z. B., EHRLICH, A., and NATHAN, Y., 1974, Lake Lisan the Pleistocene precursor of the Dead Sea: Geological Survey of Israel Bulletin 63, 30 pp.

BEHRENS, E. W., and LAND, L. S., 1972, Subtidal Holocene dolomite, Baffin Bay, Texas: Journal of Sedimentary Petrology, v. 42, p. 155–161.

BEIN, A., and LAND, L. S., 1982, San Andres carbonates in the Texas Panhandle: sedimentation and diagenesis associated with magnesium-calcium-chloride brines: Bureau of Economic Geology, University of Texas, Report of Investigation 121, 48 p.

BENNETT, S. S. and HANOR, J. S., 1987, Dynamics of subsurface salt dissolution at the Welsh Dome, Louisiana Gulf Coast, in Lerch, I. and O'Brien, J. J. (eds.), Dynamical geology of salt and related structures, Academic Press Inc., New York, p. 653–677.

BIJU-DUVAL, B., LETOUZEY, J., and MONTADERT, L., 1978, Variety of margins and deep basins in the Mediterranean, in Watkins, J. S., Montadert, L., and Dickerson, P. W. (eds.), Geological and geophysical investigations of continental margins, American Association of Petroleum Geologists Memoir 29, p. 293–317.

BISCHOFF, J. L., ROSENBAUER, R. J., and SMITH, G. I., 1985, Uranium series dating of sediment from Searles Lake: Differences between continental and marine climate records: Science, v. 227, p. 1222–1224.

BISHOP, R. S., 1978, Mechanism for the emplacement of piercement diapirs: American Association of Petroleum Geologists Bulletin, v. 62, p. 1561–1583.

BLATT, H., 1982, Sedimentary Petrology: W. H. Freeman and Company, San Francisco, 564 pp.

BRICE, S. E., COCHRAN, M. D., PARDO, G., and EDWARDS, A. D., 1982, Tectonics and Sedimentation of the South Atlantic rift sequence: Cabinda, Angola, in Watkins, J. S. and Drake, C. L. (eds.), Studies in continental margin geology, American Association of Petroleum Geologists Memoir 34, p. 5–18.

BUDD, D. A. and LOUCKS, R. G., 1981, Smackover and Lower Buckner Formations, South Texas: Depositional systems on a Jurassic carbonate ramp: Bureau of Economic Geology, University of Texas, Report of Investigation 112, 38 p.

BURWOOD, R., 1984, Carbonate source rocks for six million barrels of oil per day—Zagros fold belt, southwestern Iran (abs.), in Palacas, J. G. (ed.), Petroleum geochemistry and source rock potential of carbonate rocks, American Association of Petroleum Geologists Studies in Geology 18, p. 206.

BUSH, P., 1973, Some aspects of the diagenetic history of the sabkha in Abu Dhabi, Persian Gulf, p. 395–407, in Purser, B. H. (ed.), The Persian Gulf: Holocene carbonate sedimentation and diagenesis in a shallow epicontinental sea; Springer–Verlag, New York, 471 p.

BUTLER, G. P., 1966, Early diagenesis in the recent sediments of the Trucial Coast of the Persian Gulf: Ph.D. thesis, University of London, 251 p.

BUTLER, G. P., 1969, Modern evaporite deposition and geochemistry of coexisting brines, the sabkha, Trucial Coast, Arabian Gulf: Journal of Sedimentary Petrology, v. 39, p. 70–89.

BUTLER, G. P., 1970, Holocene gypsum of the Abu Dhabi sabkha, Trucial Coast: An alternative explanation of origin: 3rd Symposium on Salt, Northern Ohio Geological Society, v. 1, p. 120–152.

BUTLER, G. P., HARRIS, P. M., and KENDALL, C. G. St. C., 1982, Recent evaporites from the Abu Dhabi Coastal Flats, in Handford, C. R., Loucks, R. G., Davies, C. R. (eds.), Depositional and diagenetic spectra of evaporites—a core workshop: Core Workshop 3, Society of Economic Paleontologists and Mineralogists, p. 33–64.

CANTRELL, R. B., MONTGOMERY, J. C., and WOODARD, A. E., 1959, *Heterostegina* reef on Nash and other piercement salt domes in northwestern Brazoria County, Texas: Transactions of Gulf Coast Association of Geological Societies, v. 9, p. 59–62.

CARBALLO, J. D. and LAND, L. S., 1984, Holocene dolomitization of supratidal sediments by active tidal pumping, Sugarloaf Key, Florida (abs.): American Association of Petroleum Geologists Bulletin, v. 68, p. 459.

CARTER, N. L., and HANSEN, F. D., 1983, Creep of rock salt: Tectonophysics v. 92, p. 275–333.

CASHMAN, K. V. and POPENOE, P., 1985, Slumping and shallow faulting related to the presence of salt on the continental slope and rise off North Carolina: Marine and Petroleum Geology, v. 2, p. 260–271.

CASTENS-SEIDELL, B., 1984, The anatomy of a modern marine siliciclastic sabkha in a rift valley setting; northwest Gulf of California tidal flats, Baja California, Mexico, Ph.D. dissertation, Johns Hopkins University, Baltimore, Maryland.

CATALANO, R, and D'ARGENIO, B., 1982, Guida alla Geologia Della Sicilia Occidentale: Societá Geologica Italiana - Guide Geologiche Regionali, 1st Centenario della Societá Geologica Italiana, Palermo, Italy, 1982, 155 p.

CATALANO, R., RENDA, P. and SLACZKA, A., 1976, Redeposited gypsum in the evaporite sequence of the Ciminna Basin, Sicily: Memorie della Societá Geologica Italiana, v. 16, p. 666–676.

CHUBER, S., and PUSEY, W. C., 1972, Cyclic San Andres facies and their relationships to diagenesis, porosity and permeability in the Reeves Field, Yoakum County, Texas, in

Elam, J. G., and Chuber, S. (eds.), Cyclic sedimentation in the Permian Basin, West Texas Geol. Soc., p. 135–150.

Cita, M. B., 1983, The Messinian salinity crisis in the Mediterranean: A review, in Berckhemer, H., and Hsu, K. J., (eds.), Alpine Mediterranean geodynamics review: American Geophysical Union; Geodynamics Series, v. 8, p. 113–140.

Clark, D. N., 1980, The sedimentology of the Zechstein-2 Carbonate Formation of eastern Drenthe, the Netherlands, in; Fuchtbauer, H., and Peryt, T. (eds.), The Zechstein Basin with emphasis on carbonate sequences. E. Schweizerbart'sche Verlagsbuchhandlung, Stuttgart, West Germany, p. 131–65.

Clark, D. N., 1986, The distribution of porosity in Zechstein carbonates, in Brooks, J., Goff, J. C., and Van Hoorn, B., (eds.), Habitat of Palaeozoic gas in N.W. Europe, Geological Society Special Publication 23, p. 121–149.

Clayton, R. N., Jones, B. F. and Berner, R. A., 1968, Isotope studies of dolomite formation under sedimentary conditions: Geochimica et Cosmochimica Acta, v. 32, p. 415–432.

Clemmey, H., and Badman, N., 1982, Oxygen in the Precambrian atmosphere: An evaluation of the geological evidence: Geology, v. 10, p. 141–146.

Curtis, R., Evans, G., Kinsman, D. J. J., and Shearman, D. J., 1963, Association of dolomite and anhydrite in the recent sediments of the Persian Gulf: Nature, v. 197, p. 6779–6800.

Davis, D. M. and Engelder, T., 1987, Thin-skinned deformation over salt, in Lerch, I., and O'Brien, J. J. (eds.), Dynamical geology of salt and related structures, Academic Press Inc., New York, p. 301–337.

Dean, W., 1978a, Trace and minor elements in evaporites: In Dean, W., and Schreiber, B. C. (eds.), Marine evaporites, Society of Economic Paleontologists and Mineralogists short course notes 4, p. 86–104.

Dean, W. E., 1978b, Salinity cycles: evidence for subaqueous deposition of Castile Formation and lower part of Salado Formation, Delaware Basin, Texas and New Mexico, in Austin, G. S. (compiler), Geology and mineral deposits of Ochoan rocks in Delaware Basin and adjacent areas, New Mexico Bureau of Mines and Mineral Resources Circular 159, p. 15–20.

Dean, W. E., and Anderson, R. Y., 1974, Trace and minor element variations in the Permian Castile Formation, Delaware Basin, Texas and New Mexico, revealed by varve calibration: in 4th International Symposium on Salt, North Ohio Geological Society, v. 1, p. 275–285.

Debenedetti, A., 1982, The problem of the origin of salt deposits in the Mediterranean and of their relations to other salt occurrences in the Neogene formations of the contiguous regions: Marine Geology, v. 49, p. 91–114.

Deffeyes, K. S., Lucia, F. J. and Weyl, P. K., 1965, Dolomitization of Recent and Plio-Pleistocene sediments by marine evaporite waters on Bonaire, Netherlands Antilles, in: Pray, L. C., and Murray, R. C. (eds.), Dolomitization and limestone diagenesis—a symposium; Society of Economic Paleontologists and Mineralogists Special Publication 13, p. 71–88

Dellwig, L. F., 1955, Origin of the Salina Salt of Michigan: Journal of Sedimentary Petrology, v. 25, p. 83–110.

DELLWIG, L. F. and EVANS, R., 1969, Depositional processes in the Salina Salt of Michigan, Ohio and New York: American Association of Petroleum Geologists Bulletin, v. 53, p. 949–956.

DEWEY, J. F., PITMANN, W. C. III., RYAN, W. B. F., and BONNIN, J., 1973, Plate tectonics and the evolution of the Alpine System: Geological Society of America Bulletin, v. 84, p. 3137–3180.

D'HEUR, 1980, Chalk reservoir of the West Ekofisk Field, in The Sedimentation of the North Sea Reservoir Rocks, Geilo, Norsk Petroleumforening, Section X.

DIMROTH, E., and KIMBERLY, M. M., 1976, Precambrian atmospheric oxygen: Evidence in the sedimentary distributions of carbon, sulfur, uranium and iron: Canadian Journal of Earth Sciences v. 13, p. 1161–1185.

DOW, W. G., 1971, Theory of Red River reservoir development in Williston Basin (abs.): American Association of Petroleum Geologists Bulletin, v. 55, p. 536.

DRAPER, J. J. and JENSEN, A. R., 1976, The geochemistry of Lake Frome, a playa lake in South Australia: Bureau of Mineral Resources Journal of Australian Geology Geophysics, v. 1, p. 83–104.

DUNLAP, W. H. 1967, San Andres oil exploration in the Cato-Slaughter trend of southeastern New Mexico, in the oil and gas fields of southeastern New Mexico, 1966 supplement—a symposium, Roswell Geological Society, Roswell, New Mexico, p. 21–24.

DUNLOP, J. S. R. J., 1978, Shallow-water sedimentation at North Pole, Pilbara, Western Australia: Publications of the Geology Department Extension Service, University of Western Australia, v. 2, p. 30–38.

EARDLEY, A. J., 1938, Sediment of Great Salt Lake, Utah: American Association of Petroleum Geologists Bulletin, v. 22, p. 1305–1411.

ELLIOTT, L. A. AND WARREN, J. K., in review, Stratigraphy and depositional environment of the lower San Andres Formation in the Levelland-Slaughter trend and equivalent outcrops: American Association Petroleum Geologists Bulletin.

ELLIOTT, T., 1986, Siliciclastic shorelines, in Reading, H. G. (ed.), Sedimentary environments and facies, 2nd Edition, Elsevier, New York, p. 155–188.

ERICKSSON, K. A., and WARREN, J. K., 1983, A paleohydrological model for early Proterozoic dolomitization and silicification: Precambrian Research, v. 21, p. 299–321.

EUGSTER, H. P., 1969, Inorganic bedded cherts from the Magadi area, Kenya: Contributions Mineralogy and Petrology, v. 22, p. 1–31.

EUGSTER, H. P., 1980, Lake Magadi, Kenya, and its precursors, in Nissenbaum A. (ed.), Hypersaline brines and evaporitic environments, Developments in Sedimentology 28, Elsevier, New York, p. 195–232.

EUGSTER, H. P. and CHOU, I-Ming, 1973, The depositional environment of Precambrian bedded iron-formations: Economic Geology, v. 68, p. 1144–1168.

EUGSTER, H. P. and HARDIE, L. A., 1978, Saline Lakes, in Lerman A. (ed.), Lakes—chemistry, geology, and physics: Springer-Verlag, New York, p. 237–293.

EUGSTER, H. P. and JONES, B. F., 1978, Behavior of major solutes during closed-basin brine evolution: American Journal of Science, v. 279, p. 609–631.

EVANS, G., SCHMIDT, V., BUSH, P., and NELSON, H., 1969, Stratigraphy and geologic history of the sabkha, Abu Dhabi, Persian Gulf: Sedimentology, v. 12, p. 145–159.

EVANS, R., 1978, Origin and significance of evaporites in basins around Atlantic margin: American Association of Petroleum Geologists Bulletin, v. 62, p. 234.

EWING, M. and EWING, J., 1962, Rate of salt dome growth: American Association Petroleum Geologists Bulletin, v. 46, p. 708–709.

FELMY, A. and WEARE, J. H., 1986, The prediction of borate mineral equilibria in natural waters: Application to Searles Lake, California: Geochimica et Cosmochimica Acta, v. 50, p. 2771–2783.

FERGUSON, J., BURNE, R. V., and CHAMBERS, L. A., 1982, Lithification of peritidal carbonates by continental brines at Fisherman Bay, South Australia, to form a megapolygon/spelean limestone association: Journal of Sedimentary Petrology, v. 52, p. 1127–147.

FOLK, R. L. and LAND, L. S., 1975, Mg/Ca ratio and salinity: Two controls over crystallization of dolomite: American Association of Petroleum Geologists Bulletin, v. 59, p. 60–68.

FOX, B. W., 1963, How many James reef pools lie hidden in the Fairway region?: Oil and Gas Journal, June 10, 1963, p. 246–252.

FRIEDMAN, G. M., 1980, Dolomite is an evaporite mineral: evidence from the rock record and from sea-marginal ponds of the Red Sea, in Zenger, D. H., Dunham, J. B., and Ethington, R. H. (eds.), Concepts and models of dolomitization, Society of Economic Paleontologists and Mineralogists, Special Publication 28, p. 71–88.

FRIEDMAN, G. M., AMIEL, A. J., BRAUN, M., and MILLER, D. S., 1973, Generation of carbonate particles and laminates in algal mats—Examples from sea-marginal hypersaline pool, Gulf of Aqaba, Red Sea: American Association of Petroleum Geologists Bulletin, v. 57, p 541–557.

FUCHTBAUER, H. and PERYT, T., 1980, The Zechstein Basin with emphasis on carbonate sequences, Developments in Sedimentology, E. Schweizerbart'sche Verlagsbuchhandlung, Stuttgart, West Germany, 328 p.

FULLER, J. G. C. M. and PORTER, J. W., 1969, Evaporite formations with petroleum reservoirs in Devonian and Mississippian of Alberta, Saskatchewan and North Dakota: American Association of Petroleum Geologists Bulletin, v. 53, p. 909–926.

GALLOWAY, W. E., EWING, T. E., GARRET, C. M., TYLER, N., and BEBOUT, D. G., 1983, Atlas of major Texas oil reservoirs: The University of Texas at Austin, Bureau of Economic Geology special publication, 139 p.

GARBER, R. A., 1980, The Sedimentology of the Dead Sea: Ph.D thesis, Rensselaer Polytechnic Institute, 170 p.

GARBER, R. A. and FRIEDMAN, G. M., 1983, Coated grains along the Dead Sea shore, in Peryt, T. (ed.), Coated Grains, Springer Verlag, New York, p. 163–168.

GARBER, R. A., LEVY, Y. and FRIEDMAN, G. M., in preparation, The Dead Sea: Modern evaporite basin in a deep basin

GARFUNKEL, Z., ZAK, I., and FREUND, R., 1981, Active faulting in the Dead Sea Rift: Tectonophysics, v. 62, p. 37–52.

GAVISH, E., 1974, Geochemistry and mineralogy (sic) of a recent sabkha along the coast of Sinai, Gulf of Suez: Sedimentology, v. 21, p. 397–414.

GAVISH, E., 1980, Recent sabkhas marginal to the southern coasts of Sinai, Red Sea, in Nissenbaum, A.(ed.), Hypersaline brines and evaporitic environments, Elsevier, Amsterdam, p. 233–251.

Glennie, K. W., 1986, Development of NW Europe's Southern Permian gas Basin, in Brooks, J., Goff, J. C., and Van Hoorn, B., (eds.), Habitat of Palaeozoic gas in NW Europe, Geological Society Special Publication 23, p. 3–22.

Glennie, K. W. and Buller, A. T., 1983, The Permian Weissliegend of NW Europe: the partial deformation of eolian sands caused by the Zechstein transgression: Sedimentary Geology, v. 35, p. 43–81.

Gornitz, V. M. and Schreiber, B. C., 1981, Displacive halite hoppers from the Dead Sea: Some implications for ancient evaporite deposits: Journal of Sedimentary Petrology, v. 51, p. 787–794.

Gratton, P. J. F., and Lemay, W. J., 1969, San Andres oil east of the Pecos in San Andres limestone, a reservoir for oil and water in New Mexico, in Summers, W. K., and Kottlowski, F. E. (eds.), New Mexico Geological Society, Special Publication 3, p. 37–43.

Gregg, J. M. and Sibley, D. F., 1984, Epigenetic dolomitization and the origin of xenotopic dolomite texture: Journal of Sedimentary Petrology, v. 54, p. 908–931.

Gussow, W. C., 1968, Salt diapirism: Importance of temperature, and energy source of emplacement, in Braunstein, J., and O'Brien, G. D. (eds.), Diapirism and diapirs: American Association of Petroleum Geologists Memoir 8, p. 16–52

Halbouty, M. T., 1979, Salt domes, Gulf region, United States and Mexico, 2nd ed., Gulf Publishing, Houston, Texas, 561 p.

Halbouty, M. T., and El-Baz, F., 1985, Circum-Mediterranean has potential: American Association of Petroleum Geologists Explorer v. 6, no.6, p. 28–30.

Hallam, A., 1981, Facies interpretation and the stratigraphic record: W. H. Freeman and Co., San Francisco, 291 p.

Halley, R. B., 1977, Ooid fabric and structure in the Great Salt Lake and the Geologic record: Journal of Sedimentary Petrology, v. 47, p. 1099–1120.

Handford, C. R., 1981a, A process-sedimentary framework for characterizing recent and ancient sabkhas: Sedimentary Geology, v. 30, p. 255–265.

Handford, C. R., 1981b, Coastal sabkha and salt pan deposition of the lower Clear Fork Formation (Permian), Texas: Journal of Sedimentary Petrology, v. 51, p 761–778.

Handford, C. R., 1982, Sedimentology and evaporite genesis in a Holocene continental sabkha playa basin—Bristol Dry Lake, California: Sedimentology, v. 29, p. 239–253.

Handford, C. R., Kendall, A. C., Prezbindowski, D. R., Dunham, J. B., and Logan, B. W., 1984, Salina-margin tepees, pisoliths and aragonite cements, Lake MacLeod, Western Australia; Their significance in interpreting ancient analogs: Geology, v. 12, p. 523–527.

Hanor, J. S., 1987, Kilometer-scale thermohaline overturn of pore waters in the Louisiana Gulf Coast: Nature, v. 327, p. 501–503.

Hanshaw, B. B., Back, W. E., and Deike, R. G., 1971, A geochemical hypothesis for dolomitization by groundwater: Economic Geology, v. 66, p. 710–724.

Hardie, L. A., 1967, The gypsum-anhydrite equilibrium at one atmosphere pressure: American Mineralogist, v. 52, p. 172–200.

Hardie, L. A., 1982, Reconstruction of brine compositions of ancient evaporites from their mineralogy: Application to chemistry of ancient seawater, presented at the American Association of Petroleum Geologists Annual Convention, Calgary, Alberta, Canada, 1982.

HARDIE, L. A., 1984, Evaporites: Marine or non-marine: American Journal of Science, v. 284, p. 193–240.

HARDIE, L. A., Dolomitization: A critical review of some current views: Journal of Sedimentary Petrology, v. 57, p. 166–183.

HARDIE, L. A., SMOOT, J. P., and EUGSTER, H. P., 1978, Saline lakes and their deposits: A sedimentological approach, in Matter, A. and Tucker, M. E. (ed.), Modern and ancient lake sediments: International Association of Sedimentologists Special Publication 2, p. 7–41.

HARDMAN R. F. P. and KENNEDY, W. J., 1980, Chalk reservoirs of the Hod Fields, Norway, in The sedimentation of the North Sea reservoir rocks, Geilo: Norsk Petroleumforening, Section XI.

HARVIE, C. E., EUGSTER, H. P., and WEARE, J. M., 1982, Mineral equilibria in the six-component seawater system, Na—K—Mg—Ca—SO4—Cl—H2O at 250°C, II: Compositions of the saturated solutions: Geochimica et Cosmochimica Acta, v. 46, p. 1603–1618.

HARVIE, C. E., MØLLER, N., and WEARE, J. H., 1984, The prediction of mineral solubilities in natural waters: The Na—K—Mg—Ca—H—Cl—SO4—OH—HCO3—H2O system to high ionic strengths at 25°C. Geochimica et Cosmochimica Acta, v. 44, p. 723–751.

HARVIE, C. E., WEARE, J. H., HARDIE, L. A. and EUGSTER, H. P., 1980, Evaporation of sea water: Calculated mineral sequences: Science, v. 208, p. 498–500.

HAY, R. L., 1968, Chert and its sodium-silicate precursors in the sodium-carbonate lakes of East Africa: Contributions to Mineralogy and Petrology, v. 17, p. 225–274.

HAY, R. L., 1970, Silicate reactions in three lithofacies of a semi arid basin, Olduvai Gorge, Tanzania: Mineralogical Society of America Special Paper 3, p. 237–255.

HERRMANN, A. and RICHTER-BERNBERG, G., 1955, Frühdiagenetische Störungen der Schichtung und Lagerung im Werra-Anhydrit (Zechstein-1) am Südwestharz: Zeitschrift der Deutschen Geologischen Gesellschaft, v. 105, p. 659–702.

HOLSER, W. T., 1979, Mineralogy of Evaporites, in Burns, R. G. (ed.) Marine Minerals, Mineralogical Society of America Short Course Notes 6, p. 211–235.

HOLWERDA, J. G. and HUTCHINSON, R. W., 1968, Potash bearing evaporites of the Danakil area, Ethiopia: Economic Geology, v. 63, p. 124–150.

HOVORKA, S., 1987, Depositional environments of marine-dominated bedded halite, Permian San Andres Formation, Texas: Sedimentology, v. 34, p. 1029–1054.

HOVORKA, S. D., LUNEAU, B. A., and THOMAS, S., 1985, Stratigraphy of bedded halite in the Permian San Andres Formation, Units 4 and 5, Palo Duro Basin, Texas: Bureau of Economic Geology, University of Texas, Internal Report No. OF-WTWI-1985–9.

HSU, K. J., CITA, M. B., and RYAN, W. B. F., 1973, The origin of the Mediterranean evaporites, in Ryan, W. B. F. and Hsu, K. J. (eds.), Initial reports of the Deep Sea Drilling Project, Washington, D. C., United States Government Printing Office, v. 13, p. 1203–1231.

HSU, K. J., MONTADERT, L., BERNOULLI, D., CITA, M. B., ERIKSSON, A., GARRISON,

R. E., KIDD, R. B., MELIERES, X., MULLER, C., and WRIGHT, R., 1977, History of the Mediterranean salinity crisis: Nature, v. 267, 399–403.

HSU, K. J. and SCHNEIDER, J., 1973, Progress report on dolomitization of Abu Dhabi Sabkhas, Arabian Gulf, p. 409–422, in Purser, B. H. (ed.), The Persian Gulf: Holocene carbonate sedimentation and diagenesis in a shallow epicontinental sea: Springer-Verlag, New York, 471 p.

HUMPHREY, J. D., RANSOM, K. L. and MATTHEWS, R. K., 1986, Early meteoric diagenetic control of Upper Smackover production, Oaks Field, Louisiana, American Association of Petroleum Geologists, v. 70, p. 70–85.

HUMPHRIS, C. C. Jr., 1979, Salt movement on continental slope, northern Gulf of Mexico: American Association Petroleum Geologists Bulletin, v. 63, p. 782–798.

HUNT, C. B., ROBINSON, T. W., BOWLES, W. A., and WASHBURN, A. L., 1966, Hydrologic Basin, Death Valley, California: U.S. Geological Survey Paper 494-B, 138 p.

HUTCHINSON, R. W. and ENGELS, G. G., 1970, Tectonic significance of regional geology and evaporite lithofacies in northeastern Ethiopia: Philosophical Transactions of the Royal Society, A267, 313–329.

ILLING, L. V., 1959, Deposition and diagenesis of some Upper Paleozoic carbonate sediments in western Canada: 5th World Petroleum Congress Proceedings, p. 23–52.

ILLING, L. F., WELLS, A. J. and TAYLOR, J. C. M., 1965, Penecontemporaneous dolomite in the Persian Gulf, in Pray, L. C., and Murray, R. C. (eds.), Dolomitization and limestone diagenesis: Society of Economic Paleontologists and Mineralogists Special Publication 13, p. 89–113.

IRWIN, M. L., 1965, General theory of epeiric clear water sedimentation: American Association of Petroleum Geologists Bulletin, v. 49, p. 445–459.

JACKSON, M. P. A. and SENI, S. J., 1983, Geometry and evolution of salt structures in a marginal rift basin of the Gulf of Mexico, east Texas: Geology, v. 11, p. 131–135.

JACKSON, M. P. A. and SENI, S. J., 1984, Guide to oil and gas exploration in the East Texas Basin with emphasis on subtle traps: Professional Seminar Group, Course Notes.

JENYON, M. K., 1986a, Salt Tectonics, Elsevier Applied Science Publishers, Barking Essex, U. K.

JENYON, M. K., 1986b, Some consequences of faulting in the presence of a salt rock interval: Journal of Petroleum Geology, v. 9, p. 29–52.

JENYON, M. K., 1987, The development by salt diapirs of superficial overhang features, and effects on associated sediments, in Lerch, I., and O'Brien, J. J. (eds.), Dynamical geology of salt and related structures, Academic Press, New York, p. 679–709.

JENYON, M. K. and TAYLOR, J. C. M., 1983, Hydrocarbon indications associated with North Sea Zechstein shelf features: Oil and Gas Journal, Dec. 5, 1983, p 155–160.

JONES, B. F., EUGSTER, H. P., and RETTIG, S. L., 1977, Hydrogeochemistry of the Lake Magadi Basin, Kenya: Geochimica Cosmochimica Acta, v. 41, p. 53–72.

KAHLE, C. F., 1978, Patch reef development and effects of repeated subaerial exposure in Silurian shelf carbonates, Maumee, Ohio: Geological Society America North-Central Section, Field excursions, Apr. 1978, p. 63–115.

KASHFI, M. S., 1983, Variations in tectonic styles in the Zagros geosyncline and their relation to the diapirism of salt in Southern Iran: Journal of Petroleum Geology, v. 6, p. 195–206.

KASHFI, M. S., 1985, The Pre-Zagros integrity of the Iranian Platform: Journal of Petroleum Geology, v. 8, p. 353–360.

KASTNER, M., 1970, An inclusion hourglass pattern in synthetic gypsum: American Mineralogist, v. 55, p. 2128–2130.

KATZ, A., KOLODNY, Y. and NISSENBAUM, A., 1977, The geochemical evolution of Lake Lisan–Dead Sea system: Geochimica et Cosmochimica Acta, v. 41, p. 1609–1626.

KELTS, K. and MCKENZIE, J. A., 1982, Diagenetic dolomite formation in Quaternary anoxic diatomaceous muds of Deep Sea Drilling Project Leg 64, Gulf of California, in Scientific Party, Initial Reports DSDP, U.S. Govt. Printing Office, Washington D. C., v. 64, part 2, p. 553–570.

KENDALL, A. C., 1979, Continental and supratidal (sabkha) evaporites. No. 13, in Walker, R. G. (ed.), Facies Models: Geoscience Canada Reprint Series 1, p. 145–158.

KENDALL, A. C., 1984, Evaporites: In Walker, R. G. (ed.) Facies Models: 2nd ed., Geoscience Canada Reprint Series, p. 259–296.

KENDALL, C. G. St. C., 1969, An environmental reinterpretation of the Permian evaporite/carbonate shelf sediments of the Guadalupe Mountains: Geological Society of America Bulletin, v. 80, p. 2503–2526.

KENDALL, C. G. St. C., and SKIPWITH, Sir P. A. D'E., 1969, Holocene shallow water carbonate and evaporite sediments of Kor al Bazam, Abu Dhabi, Persian Gulf: American Association of Petroleum Geologists Bulletin, v. 53, p. 841–869.

KENDALL, C. G. St. C., and WARREN, J. K., 1987, Disrupted carbonate hardgrounds found in shallow carbonate shelf sediments: The origin and setting of tepees and their associated fabrics: Sedimentology, v. 34, p. 1007–1027.

KENT, P. E., 1979, The emergent Hormuz salt plugs of southern Iran: Journal of Petroleum Geology, v. 2, p. 117–144.

KENT, P. E., 1987, Island salt plugs in the Middle East and their tectonic implications, in Lerch, I. and O'Brien, J. J. (eds.), Dynamical geology of salt and related structures, Academic Press, New York, p. 3–37.

KEZAO, C. and BOWLER, J. M., 1985, Preliminary study on sedimentary characteristics and evolution of paleoclimate of Qarhan Salt Lake, Qaidam Basin: Scientia Sinica (series B), v. 28, p. 1218–1232.

KINSMAN, D. J. J., 1963, Recent carbonate sedimentation near Abu Dhabi, Trucial Coast, Persian Gulf: Ph.D. thesis, London University, 315 p.

KINSMAN, D. J. J., 1965, Gypsum and anhydrite of recent age, Trucial Coast, Persian Gulf, in Rau, J. L. (ed.), 2nd Symposium on Salt, Northern Ohio Geological Society, v. 1, p. 302–326.

KINSMAN, D. J. J., 1969, Modes of formation, sedimentary associations and diagnostic features of shallow water and supratidal evaporites: American Association of Petroleum Geologists Bulletin, v. 53, p. 830–840.

KINSMAN, D. J. J., 1974, Calcium sulfate minerals of evaporite deposits: Their primary mineralogy: v. 1, p. 343–348: in Coogan, A. J. (ed.), 4th Symposium on Salt: Northern Ohio Geological Society, Cleveland, Ohio.

KINSMAN, D. J. J., 1975a, Salt floors to geosynclines: Nature, v. 255, p. 375–378.

KINSMAN, D. J. J., 1975b, Rift valley basins and sedimentary history of trailing continental margins, in Fischer, A. G., and Judson, S. (eds.), Petroleum and plate tectonics: Princeton University Press, Princeton, N.J., p. 83–126.

KINSMAN, D. J. J., 1976, Evaporites: Relative humidity control of primary mineral facies: Journal of Sedimentary Petrology, v. 46, p. 273–279.

KIRKLAND, D. W., and EVANS, R., 1981, Source-rock potential of evaporitic environment: American Association of Petroleum Geologists Bulletin, v. 65, p. 181–190.

KOCUREK, G., and DOTT, R. H., Jr., 1983, Jurassic paleogeography and paleoclimate of the central and southern Rocky Mountain region, in Reynolds, M. W., and Dolly, E. D.(eds.), Mesozoic paleogeography of west-central United States: Rocky Mountain Section, Society of Economic Paleontologists and Mineralogists, Denver, Colorado, p. 101–116.

KOCURKO, M. J., 1979, Dolomitization by spray-zone brine seepage, San Andres, Colombia: Journal of Sedimentary Petrology, v. 49, p. 209–214.

KUMAR, N., and FOSTER, J. D., 1982, Effect of an underlying bioherm of the San Andres (Permian) reservoir and trap, Hanford field, Gaines County, Texas: American Association of Petroleum Geologists Bulletin, v. 66, p. 2571–2583.

KUSHNIR, J., 1981, Formation and early diagenesis of varved evaporitic sediments in a coastal hypersaline pool: Journal of Sedimentary Petrology, v. 51, p. 1193–1203.

LALOU, C., NGUYEN, H. V., FAURE, H. and MOREIRA, L., 1970, Datation par la methode uranium-thorium des hauts niveaux de coraux de la depression de l'Afar (Ethiopie): Revue Geographie Physique et Geologie Dynamique, v. 12, p. 3–8.

LAND, L. S., 1973, Holocene meteoric dolomitization of Middle Pleistocene limestones, North Jamaica: Sedimentology, v. 20, p. 411–424.

LAND, L. S., 1983, Dolomitization: American Association of Petroleum Geologists Education Course Note Series, 24, 20 p.

LAND, L. S., 1985, The origin of massive dolomite: Summary and suggestions: Journal of Geological Education, v. 33, p. 112–125.

LAND, L. S., and PREZBINDOWSKI, D. R., 1981, The origin and evolution of saline formation water, Lower Cretaceous carbonates, south-central Texas: Journal of Hydrology, v. 54, p. 51–74.

LAND, L. S., SALEM, M. R. I., and MORROW, D. W., 1975, Paleohydrology of ancient dolomites—geochemical evidence: American Association of Petroleum Geologists Bulletin, v. 59, p. 1602–1625.

LAUBSCHER, H., and BERNOULLI, D., 1977, Mediterranean and Tethys: in Nairn, A. E. M., Kanes, W. H., and Stehli, F. G. (eds.), The ocean basins and margins, Plenum, New York, v. 4a, p. 1–28.

LEMON, N. M., 1985, Physical modeling of sedimentation adjacent to diapirs and comparison with late Precambrian Oratunga breccia body in central Flinders Ranges, South Australia: American Association of Petroleum Geologists Bulletin, v. 69, p. 1327–1338.

LEVY, Y., 1977, Description and mode of formation of the supratidal evaporite facies in northern Sinai coastal plain: Journal of Sedimentary Petrology, v. 47, p. 463–474.

LEVY, Y., 1980, Evaporitic environments in the Northern Sinai, in Nissenbaum, A.(ed.), Hypersaline brines and evaporitic environments, Elsevier, Amsterdam, p. 131–143.

LIPPARD, S. J., SMEWING, J. D., ROTHERY, D. A., and BROWNING, P., 1982, The geology of the Dibba zone, northern Oman Mountains; a preliminary study: Journal of the Geological Society of London, v. 139, p. 59–66.

LOGAN, B. W., 1981, Modern carbonate and evaporite sediments of Shark Bay and Lake Macleod, Western Australia: Geological Society of Australia Field Excursion Guidebook, 5th Australian Convention, 64 p.

LOGAN, B. W., and SEMENIUK, V., 1976, Dynamic metamorphism; processes and products in Devonian carbonate rocks, Canning Basin, Western Australia: Geological Society of Australia, Special Publication, 6, 138 p.

LOHMANN, H. H., 1979, Seismic recognition of salt diapirs: American Association of Petroleum Geologists Bulletin, v. 63, p. 2097–2102.

LOMANDO, A. J., BIRDSALL, T. P., and GOLL, C. L., 1984, Deposition and porosity evolution of Rodessa (Lower Cretaceous) grainstone reservoirs, examples from West Purt and Bois D'Arc fields, Texas, in Harris, P. M. (ed.), Carbonate sands—a core workshop: Society of Economic Paleontologists Mineralogists Core Workshop 5, p. 365–390.

LOREAU, J. P., and PURSER, B. H., 1973, Distribution and ultrastructure of Holocene ooids in the Persian Gulf, in Purser, B. H. (ed.), The Persian Gulf: Holocene carbonate sedimentation, and diagenesis in a shallow epicontinental sea: Springer-Verlag, Berlin, p. 279–328.

LOUCKS, R. G., and LONGMAN, M. W., 1982, Lower Cretaceous Ferry Lake Anhydrite, Fairway field, east Texas; product of shallow-subtidal deposition, in Handford, C. R., Loucks, R. G., Davies, C. R. (eds.), Depositional and diagenetic spectra of evaporites—a core workshop: Core Workshop 3, Society of Economic Paleontologists and Mineralogists, p. 130–173.

LOUTFI, G., and EL-BISHLAWI, S., 1986, Habitat of hydrocarbon in Abu Dhabi, UAE: in technical papers and case studies presented at the Symposium on the hydrocarbon potential of intense thrust zones. OAPEC/ADNOC Conference Abu Dhabi, United Arab Emirates, Dec. 14–18, 1986, v. 2, p. 65–124.

LOWE, D. R., 1983, Restricted shallow-water sedimentation of early Archean stromatolitic and evaporitic strata of the Strelley Pool Chert, Pilbara Block, Western Australia: Precambrian Research, v. 19, p. 239–283.

LOWELL, J. D., and GENIK, G. J., 1972, Sea-floor spreading and structural evolution of the southern Red Sea: American Association of Petroleum Geologists Bulletin, v. 56, p. 247–259.

LOWENSTEIN, T., 1982, Primary features in a potash evaporite deposit, the Permian Salado Formation of West Texas and New Mexico, in Handford, C. R., Loucks, R. G., Davies, C. R. (eds.), Depositional and diagenetic spectra of evaporites—a core workshop: Core Workshop 3, Society of Economic Paleontologists and Mineralogists, p. 276–304.

LUCIA, F. J., 1968, Recent sediments and diagenesis of South Bonaire, Netherlands Antilles: Journal of Sedimentary Petrology, v. 38, p. 845–858.

LYONS, W. B., HINES, M. E., and GAUDETTE, H. E., 1984, Major and minor pore water geochemistry of modern marine sabkhas: the influence of cyanobacterial mats, in Cohen, Y., Castenholz, R. W., and Halvorsor, H. O. (eds.), Microbial Mats: Stromatolites: A. R. Liss, New York, p. 411–424.

MACHEL, H. G. and MOUNTJOY, E. W., 1986, Chemistry and environments of dolomitization—a reappraisal: Earth Science Reviews, v. 23, p. 175–222.

MAGARITZ, M., GOLDENBURG, L., KAFRI, U. and ARAD, A., 1980, Dolomite formation in the seawater-freshwater interface: Nature, v. 287, p. 622–624.

MAIKLEM W. R., 1971, Evaporative drawdown—a mechanism for water level lowering and diagenesis in the Elk Point Basin: v. 19, p. 487–503.

MAIKLEM, W. R., BEBOUT, D. G., and GLAISTER, R.P., 1969, Classification of anhydrite—a practical approach: Canadian Petroleum Geology Bulletin, v. 17, p. 194–233.

MAISONNEUVE, J., 1982, The composition of the Precambrian ocean waters: Sedimentary Geology, v. 31, p. 1–11.

MANCINI, E. A., and BENSON, D. J., 1980, Regional stratigraphy of the Upper Jurassic Smackover carbonates of southwest Alabama: Transactions of Gulf Coast Association of Geological Societies, v. 30, p. 151–165.

MANSFIELD, C. F., 1980, A urolith of biogenic dolomite—another clue in the dolomite mystery: Geochimica et Cosmochimica Acta, v. 44, p. 829–839.

MARTIN, R. G., 1976, Geologic framework of northern and eastern continental margins, Gulf of Mexico, in Bouma, A. H., Moore, G. T., and Colement, J. M. (eds.), Beyond the shelf break: American Association of Petroleum Geologists Marine Geology Committee Short Course, v. 2, p. A21–A28.

MATTES, B. W. and MOUNTJOY, E. W., 1980, Burial dolomitization of the Upper Devonian Miette buildup, Jasper National Park, Alberta, in Zenger, D. H., Dunham, J. B., and Ethington, R. L. (eds.), Concepts and models of dolomitization, Society of Economic Paleontologists and Mineralogists, Special Publication 28, p. 259–320.

MATTHEWS, R. K., and POORE, R. Z., 1980, Tertiary $\partial^{18}O$ record and glacio-eustatic sea-level fluctuations: Geology, v. 8, p. 501–504.

MAWSON, SIR D., 1929, Some South Australian limestones in process of formation: Quarterly Journal of the Geological Society, v. 85, p. 613–623.

MAZZULLO, J. M., WILLIAMS, M, and MAZZULLO, S. J., 1984, The Queen Formation of Millard Field, Pecos County, Texas: its lithologic characteristics environment of deposition and reservoir petrophysics: Transactions of the southwest section of the American Association of Petroleum Geologists, Publication 84–78, p. 103–110.

MCCRARY, M. M., 1985, Depositional history and petrography of the Todilto Formation, New Mexico and Colorado: Unpublished thesis, University of Texas at Austin.

MCDONALD, G. T. F., 1953, Anhydrite-gypsum equilibrium reactions: American Journal of Science, v. 251, p. 884–898.

MCKENZIE, J. A., 1981, Holocene dolomitization of calcium carbonate sediments from the coastal sabkhas of Abu Dhabi: a stable isotope study: Journal of Geology, v. 89, p. 185–198.

MCKENZIE, J. A., HSU, K. J. and SCHNEIDER, J. F., 1980, Movement of subsurface waters under the sabkha, Abu Dhabi, U. A. E., and its relationship to evaporative dolomite genesis, in Zenger, D. H., Dunham, J. B. and Ethington, R. L. (eds.), Concepts and models of dolomitization, Society of Economic Paleontologists and Mineralogists, Special Publication 28, p. 11–30.

MCQUILLAN, H., 1985, Fracture-controlled production from the Oligocene-Miocene Asmari Formation in Gachsaran and Bibi Hakimeh fields, Southwest Iran, in Roehl, P. O., and Choquette, P. W., Carbonate petroleum reservoirs, Springer-Verlag, New York, p. 511–523.

MILLER, K. G., MOUNTAIN, G. S. and TUCHOLKE, B. E., 1985, Oligocene glacioeustacy and erosion on the margins of the North Atlantic: Geology, v. 13, p. 10–13.

MILNES, A. G. 1978, Structural zones and continental collision, Central Alps: Tectonophysics, v. 47, p. 369–392.

MISER, D. E., 1987, Microstructures in natural and synthetic dolomites: Ph.D. thesis, University of Texas at Austin.

MITCHELL, A. H. G. and READING, H. G., 1986, Sedimentation and Tectonics; Chapter 14, in Reading, H. G. (ed.) Sedimentary environments and facies, 2nd Edition Blackwell Scientific Publications, Boston, 615 p.

MOORE, C. H., 1984, The Upper Smackover of the Gulf Rim: depositional systems, diagenesis, porosity evolution and hydrocarbon production, in Jurassic of the Gulf Rim, 3rd annual Society of Economic Paleontologists and Mineralogists, Research Conference, p. 283–307.

MOORE, C. H. and DRUCKMAN, K., 1981, Burial diagenesis and porosity evolution, Upper Jurassic Smackover, Arkansas and Louisiana: American Association of Petroleum Geologists Bulletin, v. 65, p. 1297–1343.

MORROW, D. W., 1982, Diagenesis 2. Dolomite Part 2, Dolomitization models and ancient dolostones: Geoscience Canada, v. 9, p. 95–107.

MUIR, M. D., LOCK, D., and VON DER BORCH, C. C., 1980, The Coorong model for penecontemporaneous dolomite formation in the middle Proterozoic McArthur Group, Northern Territory, Australia, in Zenger, D. H., Dunham, J. B., and Ethington, R. L. (eds.), Concepts and models of dolomitization, Society of Economic Paleontologists and Mineralogists, Special Publication 28, p. 51–67.

MÜLLER, S. and TEITZ, G., 1971, Dolomite replacing "cement A" in biocalcarenites from Fuerteventura, Canary Islands, Spain, in Bricker, D. P. (ed), Carbonate cements: Johns Hopkins University Press, Baltimore, Md., 376 p.

MURRIS, R. J., 1980, Middle East: Stratigraphic Evolution and Oil Habitat: American Association of Petroleum Geologists Bulletin, v. 64, p. 597–618.

MYERS, D. M., and BONYTHON, C. W., 1958, The theory of recovering salt from sea water by solar evaporation: Journal of Applied Chemistry (Australia), v. 8, p. 207–219.

NEEV, D., and EMERY, K. O., 1967, The Dead Sea: Depositional processes and environments of evaporites: Israel Mineral Development Geological Survey Bulletin 41, 147 pp.

NEEV, D. and HALL, J. K., 1976, The Dead Sea geophysical survey, 19 July–1 August 1974. Geological Survey of Israel, Internal Report MGD 6/76, 21 pp.

NEEV, D. and HALL, J. K., 1979, Geophysical investigations in the Dead Sea: Sedimentary Geology, v. 25, p. 209–238.

NEWELL, N. D., RIGBY, J. K., FISCHER, A. G., WHITEMAN, A. J., HICKOX and BRADLEY, J. S., 1972, The Permian Reef Complex of the Guadalupe Mountain region, Texas and New Mexico; a study in paleoecology (2nd ed.): Hafner Publishing, New York, 236 p.

NURMI, R. D. and FRIEDMAN, G. M., 1977, Sedimentology and depositional environments of basin center evaporites, Lower Salina Group (Upper Silurian), Michigan Basin, in Fisher, J. H. (ed.), Reefs and Evaporites—concepts and depositional models, American Association of Petroleum Geologists, Studies in Geology 5, p. 23–52.

OGNIBEN, L., 1957, Petrografia della Serie Solifera Siciliana e considerzioni geolgische relative: Memorie descritt: Carta Geologica D'ltalia, v., 23, p. 1–275.

ORTI-CABO, F., and SHEARMAN, D. J., 1977. Estructuras y fabricas deposicionales en las evaporitas del Mioceno superior (Messiniense) de San Miguel de Salinas (Alicante, España): Instituto Investigaciones Geologicas Diputacion Provincial Universidad de Barcelone, v. 32, p. 5–54.

PARASCHIV, D. and OLTEANU, G., 1970, Oil Fields in the Mid-Pliocene zone of the eastern Carpathians (District of Ploiesti), in Halbouty, M. T. (ed.), Geology of giant petroleum fields, American Association of Petroleum Geologists Memoir 14, p. 399–427.

PARKER, J. M., 1967, Salt solution and subsidence structures, Wyoming, North Dakota, and Montana: American Association of Petroleum Geologists Bulletin, v. 51, p. 1929–1947.

PARNELL, J., 1983, Skeletal halites from the Jurassic of Massachusetts, and their significance: Sedimentology, v. 30, p. 711–715.

PARNELL, J., 1986, Devonian Magadi-type cherts in the Orcadian Basin, Scotland: Journal of Sedimentary Petrology, v. 56, p. 495–500.

PATTERSON, R. J., 1972, Hydrology and carbonate diagenesis of a coastal sabkha in the Persian Gulf: Ph.D. dissertation, Princeton University, 473 p.

PATTERSON, R. J., and KINSMAN, D. J. J., 1981, Hydrologic framework of a sabkha along the Arabian Gulf: American Association of Petroleum Geologists Bulletin, v. 65, p. 1457–1475.

PATTERSON, R. J. and KINSMAN, D. J. J., 1982, Formation of diagenetic dolomite in coastal sabkhas along the Arabian (Persian) Gulf: American Association of Petroleum Geologists Bulletin, v. 66, p. 28–43.

PAUL, J., 1980, Upper Permian algal stromatolite reefs, Harz Mountains (F. R. Germany), in Fuchtbauer, H. and Peryt, T. (eds.), The Zechstein Basin with emphasis on carbonate sequences. E. Schweizerbart'sche Verlagsbuchhandlung, Stuttgart, West Germany, p. 253–268.

PERTHUISOT, J. P., FLORIDIA, S., AND JAUZEIN A., 1972, Un modéle rècent de bassin côtier à sédimentation saline: la Sebkha el Melah (Zarzis, Tunisie). Revue de Geologie Dynamique et de Geographie Physique, v. 14, p. 67–83

PERYT, T. M., 1985, A Permian beach in the Zechstein dolomites of Western Poland: Influence on reservoirs: Journal of Petroleum Geology, v. 8, p. 463–474.

PHLEGER, F. B., 1969, A modern evaporite deposit in Mexico: American Association of Petroleum Geologists Bulletin, v. 53, p. 824–829.

PICARD, L., 1943, Structure and evolution of Palestine (with comparative notes on neighboring countries): Bulletin Geology Department Hebrew University Jerusalem, v. 4, no. 2-3-4, 187 pp.

PICHA, F., 1978, Depositional and diagenetic history of Pleistocene and Holocene oolitic sediments and sabkhas in Kuwait, Persian Gulf: Sedimentology, v. 25, p. 427–449.

PITT, W. D. and SCOTT, G. L., 1982, Porosity zones in the lower part of San Andres Formation, east-central New Mexico: New Mexico Bureau of Mines and Mineral Resources Circular 179, 19 p.

PITTMAN, J. G., 1985, Correlation of beds within the Ferry Lake Anhydrite of the Gulf Coastal Plain: Transactions of Gulf Coast Association of Geological Societies, v. 35, p. 251–260.

PRICE, P. E., KYLE, J. R., and WESSEL, G. R., 1983, Salt dome related zinc-lead deposits, in Kisvarsanyi, G. et al. (eds.), 1983 Proceedings of the International Conference on Mississippi Valley Type Lead Zinc Deposits, University of Missouri-Rolla, p. 558–571.

PURSER, B. H. (ed.), 1973a, The Persian Gulf: Holocene carbonate sedimentation and diagenesis in a shallow epicontinental sea: Springer-Verlag, New York, 471 p.

PURSER, B. H., 1973b, Sedimentation around bathymetric highs in the southern Persian Gulf, in Purser B. H. (ed.), The Persian Gulf: Holocene carbonate sedimentation and diagenesis in a shallow epicontinental sea, Springer-Verlag, New York, p. 157–177.

PURSER, B. H. and EVANS, G., 1973, Regional sedimentation along the Trucial Coast, S. E. Persian Gulf, in Purser, B. H. (ed.), The Persian Gulf: Holocene carbonate sedimentation and diagenesis in a shallow epicontinental sea, Springer-Verlag, New York, p. 211–232.

RADKE, B. M. and MATHIS, R. L., 1980, On the formation and occurrence of saddle dolomite: Journal of Sedimentary Petrology, v. 50, p. 1149–1168.

RAMONDETTA, P. J., 1982, Genesis and emplacement of oil in San Andres Formation, Northern Shelf of the Midland Basin, Texas: Bureau of Economic Geology University of Texas, Report of Investigation 116, 39 p.

RAUP, O. B., 1970. Brine mixing: An additional mechanism for formation of basin evaporites: American Association of Petroleum Geologists Bulletin, v. 54, p. 2246–2259.

READ, J. F., 1985, Carbonate platform facies models: American Association of Petroleum Geologists Bulletin, v. 69, p. 1–21.

REEDER, R. J., 1980, Phase transformations in dolomites: Ph.D. dissertation, University of California, Berkeley.

REEDER, R. J., 1981, Electron optical investigation of dolomites: Contributions to Mineralogy and Petrology, v. 76, p. 148–157.

REZAK, R., 1977, West Flower Garden Bank, Gulf of Mexico, in Frost, J. H., Weiss, M. P., and Saunders, J. B. (eds.), Reefs and related carbonates—ecology and sedimentology, American Association of Petroleum Geologists Studies in Geology 4, p. 27–35.

REZAK, R., 1985, Local carbonate production on a terrigenous shelf: Gulf Coast Association of Geological Societies Transactions, v. 35, p. 477–483.

REZAK, R. and BRIGHT, T. J., 1981, Seafloor instability at the East Flower Garden Bank, northwest Gulf of Mexico: Geo-marine Letters, v. 1, p. 97–103.

RICHARDSON M. and MICHAEL A. ARTHUR, 1986, Miocene synrift evaporites of the Red Sea—Their deposition and hydrocarbon source potential (abs): American Association of Petroleum Geologists Bulletin, v. 70, p.638–639.

RICHTER-BERNBURG, G., 1955, Stratigraphische Gliederung des deutschen Zechsteins: Zeitschrift der Deutschen Geologischen Gesellschaft, v. 105, p. 843–854.

RICHTER-BERNBURG, G., 1960, Zeitmessung geologischer Vorgänge nach Warvenkorrelierung im Zechstein: Geologische Rundschau, v. 49, p. 132–148.

RIDGLEY, J. L., 1984, Paleogeography and facies distribution of the Todilto limestone and Pony Express Limestone member of the Wanakah Formation, Colorado and New Mexico (abs.): Geological Society of America Abstracts with Program, v. 16, p. 252.

RISACHER, F., and EUGSTER, H. P., 1979, Holocene pisoliths and encrustations associated with spring-fed surface pools, Pasos Grandes, Bolivia: Sedimentology, v. 26, p. 253–270.

ROBERTSON, A. F. H. and WOODFORD, N. H., 1983, Genesis of the Batinah melange above the Semail ophiolite, Oman: Journal of Structural Geology, v. 5, p. 1–17.

ROEHL, P. O., 1967, Stony Mountain (Ordovician) and Interlake (Silurian) facies analogues of recent low-energy marine and subaerial carbonates, Bahamas: American Association of Petroleum Geologists Bulletin, v. 51, p. 1979–2032.

RONA, P. A., 1982, Evaporites at passive margins, in Scrutton, R. A. (ed.), Dynamics of Passive Margins, American Geophysical Union and Geological Society of America, Geodynamics Series, v. 6, p. 116–132.

RONOV, A, B., 1968, Probable changes in the composition of sea water in the course of geologic time: Sedimentology, v. 10, p. 25–43.

RONOV, A. B., KHAIN, K. E., BALUKHOVSKY, A. N. and SESLAVINSKY, K. B., 1980, Quantitative analysis of Phanerozoic sedimentation: Sedimentary Geology, v. 25, p. 311–325.

ROSEN, M. R., MISER, D. E., and WARREN, J. K., 1987, Sedimentology, mineralogy, and isotopic analysis of Pellet Lake, Coorong Region, South Australia: Sedimentology, v. 35, p. 105–123.

RUZYLA, K., and FRIEDMAN, G. M., 1981, Geological heterogeneities important to future enhanced recovery in carbonate reservoirs of Upper Ordovician Red River Formation at Cabin Creek field, Montana: Society of Petroleum Engineers/Department of Energy 2nd Joint Symposium on Enhanced Oil Recovery, p. 403–418.

RUZYLA, K., and FRIEDMAN, G. M., 1985, Factors controlling porosity in dolomite reservoirs of the Ordovician Red River Formation, Cabin Creek field, Montana, in Roehl, P. O., and Choquette, P. W. (eds.), Carbonate petroleum reservoirs: A case book, Springer-Verlag, New York, p. 39–58.

SALLER, A. H., 1984, Petrologic and geochemical constraints on the origin of dolomite: Subsurface, Enewetak Atoll: Geology, v. 12, p. 217–220.

SANDBERG, P. A., 1975, New interpretations of Great Salt Lake ooids and of ancient nonskeletal mineralogy: Sedimentology, v. 22, p. 497–537.

SARG, J. F., 1981, Petrology of the carbonate–evaporite transition of the Seven Rivers Formation (Guadalupian, Permian), southeast New Mexico: Journal of Sedimentary Petrology, v. 51, p. 73–95.

SASS and BEN-YAAKOV, 1977, The carbonate system in hypersaline solutions: Dead Sea brines: Marine Chemistry, v. 5, p. 83–199.

SCHENK, P. E., 1969, Carbonate-sulfate-red bed facies and cyclic sedimentation of the Windsorian stage (middle Carboniferous), Maritime Provinces: Canadian Journal of Earth Sciences, v. 6, p. 1037–1066.

SCHLAGER, W. and BOLZ, H., 1977, Clastic accumulation of sulfate evaporites in deep water: Journal of Sedimentary Petrology, v. 47, p. 600–609.

SCHREIBER, B. C., 1986, Arid shorelines and evaporites, in Reading, H. G. (ed.), Sedimentary Environments and Facies, 2nd Edition, Elsevier, New York.

SCHREIBER, B. C., 1977, Environments of subaqueous gypsum deposition, in Dean, W., and Schreiber, B. C. (eds.), Marine Evaporites, Society of Economic Paleontologists and Mineralogists short course Notes 4, p. 86–104.

SCHREIBER, B. C., and DECIMA, A., 1976, Sedimentary facies produced under evaporitic environments: A review. Memorie della Societa Geologica Italiana, v. 16, p. 111–126.

SCHREIBER, B. C., and DECIMA, A., 1978, Sedimentary facies produced under evaporitic environments: A review, in Catalano, R., Ruggieri, G., and Sprovieri, R. (eds.), Messinian Evaporites in the Mediterranean, Erice Seminar Oct. 1975: Societa Geologica Italiana, v. 16, p. 111–126.

SCHREIBER, B. C., and HSU, K. J., 1980, Evaporites, in Hobson, G. D. (ed.), Developments in petroleum geology—2., Applied Science Publishers, Barking, Essex, England, p. 87–138.

SCHREIBER, B. C. and KINSMAN, D. J. J., 1975, New observations on the Pleistocene evaporites of Montallegro, Sicily and a modern analog: Journal of Sedimentary Petrology, v. 45, p. 469–479.

SCHREIBER, B. C., and WARREN, J. K., 1984, Evaporite Short Course Notes, Sicily, June 1984.

SCHREIBER, B. C., FRIEDMAN, G. M., DECIMA, A., and SCHREIBER, E., 1976, Depositional environments of Upper Miocene (Messinian) evaporite deposits of the Sicilian Basin: Sedimentology, v. 23, p. 729–760.

SCLATER, J. G., and CHRISTIE, P. A. F., 1980, Continental stretching: an explanation of post Mid Cretaceous subsidence of the central North Sea Basin: Journal of Geophysical Research, v. 10, p. 3711–3739.

SEARS, S. O. and LUCIA, F. J., 1979, Reef growth model for Silurian pinnacle reefs, northern Michigan reef trend: Geology, v. 3, p. 299–302.

SEARS, S. O. and LUCIA, F. J., 1980, Dolomitization of Northern Michigan Niagara reefs by brine refluxion and fresh water/sea water mixing, in Zenger, D. H., Dunham, J. B., and Ethington, R. L. (eds.), Concepts and models of dolomitization, Society of Economic Paleontologists and Mineralogists, Special Publication 28, p. 215–235.

SELLEY, R. C., 1985, Elements of Petroleum Geology: W. H. Freeman and Co., New York, 449 p.

SELLI, R., 1973, An outline of the Italian Messinian, in Drooger, C. W. (ed.), Messinian events in the Mediterranean, North Holland Publishing, p. 150–171.

SENI, S. J., 1984, Role of diapirs in the loss of Louann salt, east Texas basin (abs): Geological Society of America Abstracts with Program, v. 16, p. 651.

SENI, S. J., and JACKSON, M. P. A., 1983a, Evolution of salt structures, east Texas diapir province, Part 1: Sedimentary record of halokinesis: American Association of Petroleum Geologists Bulletin, v. 67, p. 1219–1244.

SENI, S. J., and JACKSON, M. P. A, 1983b, Evolution of salt structures, east Texas diapir province, Part 2: Patterns and Rates of halokinesis: American Association of Petroleum Geologists Bulletin, v. 67, p. 1245–1274.

SENI, S. J., and JACKSON, M. P. A, 1983c, Sedimentary record of Cretaceous and Tertiary salt movement, east Texas Basin: Times, rates, and volumes of salt flow, implications to nuclear-waste isolation and petroleum exploration: Bureau of Economic Geology, the University of Texas at Austin, Report of Investigation 139.

SHAVER, R. H., 1977, Silurian reef geometry—new dimension to explore: Journal of Sedimentary Petrology, V. 47, p. 1409–1424.

SHAW, A. B., 1964, Time in stratigraphy, McGraw-Hill, New York, 365 p.

SHEARMAN, D. J., 1963, Recent anhydrite, gypsum, dolomite and halite from the coastal flats of the Arabian shore of the Persian Gulf: Geological Society of London, Proceedings, v. 1607, p. 63–65.

SHEARMAN, D. J., 1966, Origin of marine evaporites by diagenesis: Institute of Mining and Metallurgy Transactions Series B, v. 75, p. 208–215.

SHEARMAN, D. J., 1970, Recent halite rock, Baja California, Mexico: Institute of Mining and Metallurgy Transactions Series B, v. 79, p. 155–162.

SHERIFF, R. E., 1980, Seismic stratigraphy: IHRDC Publications, Boston, 227 p.

SHINN, E. A., 1973, Carbonate coastal accretion in an area of longshore transport, NE Qatar, Persian Gulf, in Purser B. H. (ed.), The Persian Gulf: Holocene carbonate sedimentation and diagenesis in a shallow epicontinental sea, Springer-Verlag, New York, p. 179–191.

SIBLEY, D. F., 1980, Climatic control of dolomitization, Seroe Domi Formation (Pliocene), Bonaire, N.A, in Zenger, D. H., Dunham, J. B., and Ethington, R. L. (eds.), Concepts and models of dolomitization, Society of Economic Paleontologists and Mineralogists, Special Publication 28, p. 247–258.

SKINNER, H. C. W., 1963, Precipitation of calcian dolomites and magnesian calcites in the southeast of South Australia: American Journal of Science, v. 261, p. 449–472.

SMITH, D. B., 1979, Rapid marine transgressions and regressions of the Upper Permian Zechstein Sea: Journal of the Geological Society London, v. 136, p. 155–156.

SMITH, D. C. and PULLEN, J. R., 1967, Hummingbird structure of Southeast Saskatchewan: Canadian Petroleum Geology Bulletin, v. 15, p. 468–482.

SMITH, D. L. and DEES, W. T., 1982, Heat flow in the Gulf coastal plain: Journal of Geophysical Research, v. 87, p. 7687–7693.

SMITH, G. I., 1979, Subsurface stratigraphy and geochemistry of Late Quaternary evaporites, Searles Lake, California: U. S. Geological Survey Professional Paper 1043.

SOLIMAN, S. M. and EL BADRY, O., 1970, Nature of Cretaceous sedimentation in Western Desert, Egypt, American Association of Petroleum Geologists Bulletin, v. 54, p. 2349–2370.

SOUTHGATE, P. N., 1982, Cambrian skeletal halite crystals and experimental analogues: Sedimentology, v. 29, p. 391–407.

SPENCER, A. W. and WARREN, J. K., 1986, Depositional styles in the Queen and Seven Rivers Formations, Yates Field, Pecos Co., Texas (extended abstract): Presented at Annual Meeting West Texas Geol. Soc., Midland Texas, Oct. 1986.

STANTON R. J., 1966, The solution brecciation process: Geol. Society of America Bulletin, v. 77, p. 843–848.

STEINHORN, I., 1981, Physical and hydrographic study of the Dead Sea during its transition period from a meromictic lake to a mixed lake: Ph.D. thesis, Weizmann Institute of Science, Rehovot Israel, 323 pp.

STEINHORN, I., 1983, In situ salt precipitation at the Dead Sea: Limnology and Oceanography, v. 28, p. 580–583.

STILLER, M. and CHUNG, Y. C., 1984, Radium in the Dead Sea: A possible tracer for the duration of meromixis: Limnology and Oceanography, v. 29, p. 574–586.

STOUDT, E. L., 1979, Smackover Formation of Gulf Coast region (abs.): American Association of Petroleum Geologists Bulletin, v. 63, p. 534–535.

SWENSEN, R. E., 1967, Trap Mechanisms in Nisku Formation of northern Montana, American Association of Petroleum Geologists Bulletin, v. 51, p. 1948–1958.

TALBOT, C. J., 1979, Fold trains in a glacier of salt in southern Iran: Journal of Structural Geology, v. 1, p. 5–18.

TALBOT, C. J. and JACKSON, M. P. A., 1987, Internal kinematics of salt diapirs: American Association of Petroleum Geologists Bulletin, v. 71, p. 1068–1093.

TANNER, W. F., 1970, Triassic-Jurassic lakes in New Mexico: Mountain Geologist, v. 7, p. 281–289.

TAYLOR, J. C. M., 1980, Origin of the Werraanhydrit in the U.K. Southern North Sea—a reappraisal, in Fuchtbauer, H. and Peryt, T. (eds.), The Zechstein Basin with emphasis on carbonate sequences. E. Schweizerbart'sche Verlagsbuchhandlung, Stuttgart, West Germany, p. 91–113.

TAYLOR, J. C. M., 1984, Late Permian - Zechstein, in Glennie, K. W. (ed.), Introduction to the petroleum geology of the North Sea, Blackwell Scientific Publishers, London, Chapter 4, p. 61–101.

TAZIEFF, 1970, The Afar Triangle: Scientific American, v. 22(2), p. 32–40.

TERRIERE, R. T., 1976, Geology of Fairway Field, east Texas: In North American oil and gas fields: American Association of Petroleum Geologists Memoir 24, p. 157–176.

TODD, R. G., 1976, Oolite bar progradation, San Andres Formation, Midland Basin, Texas: American Association of Petroleum Geologists Bulletin, v. 60, p. 907–925.

TRIPPET, A. R., 1981, Characteristics of diapirs on the outer continental shelf—upper continental slope boundary, northwest Gulf of Mexico: Transactions of Gulf Coast Association of Geological Societies, v. 31, p. 391–397.

TRUSHEIM, F., 1960, Mechanism of salt migration in northern Germany: American Association Petroleum Geologists Bulletin, v. 44, p. 1519–1540.

TUCKER, M. E., 1982, Precambrian dolomites: Petrographic and isotope evidence that they differ from Phanerozoic dolomites: Geology, v. 10, p. 7–12.

URAI, J. L., SPIERS, C. J., ZWART, H. J., and LISTER, G. S., 1986, Weakening of rock salt by water during long-term creep: Nature, v. 324, p. 554–557.

VAI, G. B., and RICCHI LUCCHI, F., 1977, Algal crusts, autochthonous and clastic gypsum in a cannibalistic evaporite basin: a case history from the Messinian of Northern Appennines: Sedimentology, v. 24, p. 211–244.

VON DER BORCH, C. C., 1965, The distribution and preliminary geochemistry of modern carbonate sediments of the Coorong area, South Australia: Geochimica et Cosmochimica Acta, v. 29, p. 781–799.

VON DER BORCH, C. C., 1976, Stratigraphy and formation of Holocene dolomitic carbonate deposits of the Coorong area, South Australia: Journal of Sedimentary Petrology, v. 46, p. 952–956.

VON DER BORCH, C. C., 1980, Evolution of late Proterozoic to early Palaeozoic foldbelt, Australia: Comparisons with post-Permian rifts and passive margins: Tectonophysics, v. 70, p. 115–134.

VON DER BORCH, C. C. and JONES, J. B., 1976, Spherular modern dolomite from the Coorong area, South Australia: Sedimentology, v. 23, p. 587–591.

VON DER BORCH, C. C. and LOCK, D., 1979, Geological significance of Coorong dolomites: Sedimentology, v. 26, p. 813–824.

VON DER BORCH, C. C., LOCK, D. and SCHWEBEL, D., 1975, Groundwater formation of dolomite in the Coorong region of South Australia: Geology, v. 3, p. 283–285.

Von der Borch, C. C., Bolton, B., and Warren, J. K., 1977, Environmental setting and microstructure of subfossil lithified stromatolites associated with evaporites, Marion Lake, South Australia: Sedimentology, v. 24, p. 693–708.

Walker, R. N., Muir, M. D., Diver, W. L., and Williams, N., 1977, Evidence of major sulfate evaporite deposits in the Proterozoic McArthur Group, N. T., Australia: Nature, v. 265, p. 526–529.

Wanless, H. R., 1979, Limestone response to stress; pressure solution and dolomitization: Journal of Sedimentary Petrology, v. 49, p. 437–462.

Ward, R. F., Kendall, C. G. St. C., and Harris, P. M., 1986, Late Permian (Guadalupian) facies and their association with hydrocarbons: American Association of Petroleum Geologists Bulletin, v. 70, p. 239–262.

Wardlaw, N. C., 1972, Unusual marine evaporites with salts of calcium and magnesium chloride in Cretaceous basins of Sergipe, Brazil: Economic Geology, v. 67, p. 156–168.

Wardlaw, N. C., and Schwerdtner, W. M., 1966, Halite-anhydrite seasonal layers in Middle Devonian Prairie evaporite Formation, Saskatchewan, Canada: Geological Society of America Bulletin, v. 77, p. 331–342.

Warren, J. K., 1982a, The hydrological setting, occurrence and significance of gypsum in late Quaternary salt lakes in South Australia: Sedimentology, v. 29, p. 609–637.

Warren, J. K., 1982b, The hydrological significance of Holocene tepees, stromatolites, and boxwork limestones in coastal salinas in South Australia: Journal of Sedimentary Petrology, v. 52, p. 1171–1201.

Warren, J. K., 1983a, Tepees, Modern (Southern Australia) and Ancient (Permian—Texas and New Mexico)—a comparison: Sedimentary Geology, v. 34, p. 1–19.

Warren, J. K., 1983b, On the significance of evaporite lamination: In Schreiber, B. C. (ed.), Proceedings of the 6th International Symposium on Salt, Toronto, 1983, v. 1, p. 161–170.

Warren, J. K., 1985, Coorong Dolomite, South Australia: the depositional setting (abs.): Society of Economic Paleontologists and Mineralogists Mid Year Meeting, Golden, Colorado, Aug., 1985.

Warren, J. K., 1986a, Source rock potential of shallow water evaporitic settings. Journal of Sedimentary Petrology, v. 56, p. 442–454.

Warren, J. K., 1986b, Tectonic setting of hydrocarbons in evaporite basins, in Technical papers and case studies presented at the Symposium on the hydrocarbon potential of intense thrust zones, Conference co-sponsored by Ministry of Petroleum and Mineral Resources UAE and OAPEC, in Abu Dhabi, United Arab Emirates, Dec. 14–18, 1986, Vol. 2, p. 1–62.

Warren, J. K., in press, Sea Marginal Evaporites: AAPG Memoir.

Warren, J. K., in press, Sedimentology of Coorong Dolomites: Sedimentology.

Warren J. K. and Hussain, M., 1984, Origin of dolomite in the Salt Flat Playa, West Texas (abs.): Annual Meeting, Southeast Section, Geological Society of America. Dallas, March 1984.

Warren, J. K. and Kendall, G. C. St. C., 1985, Comparison of marine sabkhas (subaerial) and salina (subaqueoûs) evaporites: Modern and ancient: American Association of Petroleum Geologists Bulletin, v. 69, p. 1013–1023.

WARREN, J. K., LOCK, D. and GRANGER, P., 1984, Coorong dolomites (South Australia) and models of ancient Coorong style dolomites (abs): American Association of Petroleum Geologists Bulletin, v. 68, p. 537–538.

WEILER, Y., SASS, E., and ZAK, I., 1974, Halite oolites and ripples in the Dead Sea, Israel: Sedimentology, V. 21, p. 623–632.

WENKERT, D. D., 1979, The flow of salt glaciers: Geophysical Research Letters, v. 6, p. 523–526.

WEST, I. M., BRANDON, A., and SMITH, M., 1968, A tidal evaporitic facies in the Visean of Ireland: Journal of Sedimentary Petrology, v. 38, p. 1079–1093.

WEST, I. M., ALI, Y. A. and HILMY, M. E., 1979, Primary gypsum nodules in a modern sabkha on the Mediterranean coast of Egypt: Geology, v. 7, p. 354–358.

WHITE, A. H., and YOUNGS, B. C., 1980, Cambrian alkali playa-lacustrine sequence in the northeastern Officer Basin, South Australia: Journal of Sedimentary Petrology, v. 50, p. 1279–1286.

WILLIAMSON, C. R., 1978, Depositional processes, diagenesis and reservoir properties of Permian Deep sea sandstones, Bell Canyon Formation, Texas—New Mexico: Ph.D. dissertation, University of Texas at Austin.

WILSON, A. O., 1985, Depositional and diagenetic Facies in the Jurassic Arab-C and -D Reservoirs, Qatif Field, Saudi Arabia, in Roehl, P. O., and Choquette, P. W. (eds.), Carbonate petroleum resevoirs, Springer-Verlag, New York, p. 319–340.

WILSON, J. L., 1967, Carbonate evaporite cycles in Lower Duperow Formation of the Williston Basin: Canadian Petroleum Geologists Bulletin, v. 15, p. 230–312.

WILSON, J. L., 1979, Limestone and dolomite reservoirs, in Hobson, G. D. (ed.), Developments in petroleum geology—2, Applied Scientific Publishing, Barking, Essex, England, p. 1–51.

WOLFBAUER, C. A. and SURDAM, R. C., 1974, Origin of non-marine dolomites Eocene Lake Gosiute, Green River Basin, Wyoming: Geological Society of America Bulletin, v. 85, p. 1733–1740.

WOOD, G. V. and WOLFE, M. J., 1969, Sabkha cycles in the Arab/Darb Formation off the Trucial Coast of Arabia: Sedimentology, v. 12, p. 165–191.

WOODBURY, H. O., MURRAY, JR., I. B., and OSBORNE, R. E., 1980, Diapirs and their relation to hydrocarbon accumulation, in Miall, A. D. (ed.), Facts and principles of world petroleum occurrence, Canadian Society of Petroleum Geologists Memoir 6, p. 119–142.

WORRALL, J. D. and WARREN, J. K., 1986, The occurrence of dolomite in Hatter's Pond field, Southwest Alabama, (abs): Presented at the Geological Society of America, Annual Meeting, San Antonio, Texas, Oct., 1986.

ZAAZA, M. W., 1978, The depositional facies, diagenesis, and reservoir heterogeneity of the upper San Andres formation in West Seminole field, Gaines County, Texas: University of Tulsa, Ph.D. dissertation, 210 p.

ZAK, I. and FREUND, R., 1981, Asymmetry and basin migration in the Dead Sea rift: Tectonophysics, v. 80, p. 27–38.

ZEIGLER, P. A., 1980, Hydrocarbon Provinces of the Northwest European Basin, in Miall, A. D. (ed.), Facts and principles of world petroleum occurrence; Canadian Society of Petroleum Geologists Memoir 6, p. 653–706.

ZEIGLER, P. A., 1982, Geological atlas of Western and Central Europe. Elsevier, Amsterdam, 130 p.

ZENGER D. H., 1972, Significance of supratidal dolomitization in the geologic record: Geological Society of America Bulletin, v. 83, p. 1–12.

ZENGER, D. H., 1983, Burial dolomitization in the Lost Burro Formation (Devonian), east-central California, and the significance of late diagenetic dolomitization: Geology, v. 11, p. 519–522.

ZHANG YI-GANG, 1981, Cool shallow origin of petroleum-microbial genesis and subsequent degradation: Journal of Petroleum Geology, v. 3, p. 427–444.

Index

A

B

C

D

E

F

G

H

M

N

O

P

Q

R

S

T

U

V

W

X

Y

Z